Antimicrobials in Animal Husbandry

Using antibiotics to promote growth, increase feed efficiency, and reduce mortality in indoor poultry farming is unsustainable and has been implicated in increased antibiotic resistance in humans. The use of antibiotics in food-producing animals is recognized as favoring the development of antibiotic resistance, and the ongoing emergence of antibiotic-resistant bacteria is a public health issue of increasing concern. The concern is that current life-saving antibiotics are becoming less effective. The requirement is for more toxic and costly antibiotics, reducing the available alternatives. Antibiotic use is significant in the emerging public health crisis of antibiotic resistance. Although most antibiotic use occurs in agricultural settings, relatively little attention has been paid to how antibiotic use in farm animals contributes to the overall problem of antibiotic resistance. The global spread of antimicrobial-resistant pathogenic bacteria is a continuing challenge to the healthcare of humans and domesticated animals.

This book discusses antimicrobial resistance mechanisms relevant to food-producing animals. Students, researchers, scientists, environmentalists, veterinary practitioners, academicians, stakeholders, and policymakers can benefit from using *Antimicrobials in Animal Husbandry* as a resource that addresses microbial biotechnology, microbiology, toxicology, and all disciplines related to antimicrobial research.

Features of the book:

- Covers antimicrobial resistance in animal husbandry with up-to-date research
- Describes the public health impact of the use of antibiotics in animal husbandry
- Includes recent references on each plausible antimicrobial resistance in human and animal health
- Describes the impact of improper use of antimicrobials on animal production
- Develops strategies for alternatives to antimicrobial application in food animals
- Aims to minimize the health-related problems of consumers resulting from antimicrobial residues in food-producing animals

Current Trends in Antimicrobial Research

Series Editor: Arti Gupta

Antimicrobials in Pharmaceutical and Medicinal Research
Edited by Arti Gupta and Ram Prasad

Antimicrobials in Food Science and Technology
Edited by Arti Gupta and Ram Prasad

Antimicrobial Resistance in Agriculture and its Consequences
Edited by Arti Gupta and Ram Prasad

Antimicrobials in Agriculture
Edited by Arti Gupta and Ram Prasad

Antimicrobials in Environment
Edited by Arti Gupta and Ram Prasad

Antimicrobials in Animal Husbandry
Edited by Arti Gupta and Ram Prasad

For more info: Current Trends in Antimicrobial Research - Book Series - Routledge & CRC Press

Antimicrobials in Animal Husbandry

Edited by Arti Gupta and Ram Prasad

CRC Press
Taylor & Francis Group
Boca Raton London New York

CRC Press is an imprint of the
Taylor & Francis Group, an **informa** business

Designed cover image: Shutterstock

First edition published 2026
by CRC Press
2385 NW Executive Center Drive, Suite 320, Boca Raton FL 33431

and by CRC Press
4 Park Square, Milton Park, Abingdon, Oxon, OX14 4RN

CRC Press is an imprint of Taylor & Francis Group, LLC

ISBN: 978-1-032-21747-5 (hbk)
ISBN: 978-1-032-21748-2 (pbk)
ISBN: 978-1-003-26984-7 (ebk)

DOI: 10.1201/9781003269847

Typeset in Times
by Apex CoVantage, LLC

Contents

About the Editors

Arti Gupta, PhD, obtained her PhD from Mahatma Jyotiba Phule Rohilkhand University, Bareilly, India, in 2010 in animal science. Gupta's current research interests are animal biotechnology, molecular plant biotechnology, molecular animal biotechnology, bioprocess technology, and microbiology. From July 2014 to date, she has been serving at Shri Avadh Raj Singh Smarak Degree College, Gonda, Uttar Pradesh, India, and has been engaged in some editorial work with Springer Nature, CRC Press, and De Gruyter. Gupta has 12 years of teaching and research experiences in plant and animal biotechnology and microbial biotechnology. She has worked as an external examiner for central evaluation in 2023–2024 at MJP Rohilkhand University, Bareilly.

Gupta has received the following awards: University Topper (Gold Medal), MSc (Biotechnology) from Ch.C.S. University, Meerut; Young Scientist Award (Gold Medal) from the Zoological Society of India, Lucknow; Best Poster Presenter by *Asian Journal of Experimental Science*, Jaipur; Best Poster Presenter from International Consortium of Contemporary Biologists and Madhawai-Shyam Educational Trust, Ranchi; Fellowship Award from International Consortium of Contemporary Biologist and Madhawi Shayam Educational Trust; and Dr. V.P. Agarwal (Gold Medal) from D.A.V. (P.G.) College, Muzaffarnagar. She has been involved with the following publications: 2 monographs and 15 edited books with Springer Nature, Switzerland, CRC Press/Taylor & Francis, United States, and De Gruyter, Germany; 7 book chapters; and around 50 national and international research papers. She has presented many abstracts in national and international symposia/seminars/conferences/workshops. Gupta has lifetime membership in the Indian Science Congress Association, Biotech Research Society of India, Zoological Society of India, and International Consortium of Contemporary Biologists. She has been serving as a member on several editorial boards. Previously, Gupta served as teaching personnel, G.B. Pantnagar University, Pantnagar, Uttarakhand, India; Visiting Academic Researcher, Biorefining and Advanced Material Research Center, Scotland's Rural College, University of Edinburgh, Scotland, United Kingdom.

Ram Prasad, PhD, is associated with the Department of Botany, Mahatma Gandhi Central University, Motihari, Bihar, India. His research interest includes applied and environmental microbiology, plant-microbe interactions, sustainable agriculture, and nanobiotechnology. Prasad has more than 300 publications to his credit, including research papers, review articles, and book chapters, and 7 patents issued or pending, and has edited or authored several books. He has guided one postdoc, two MPhil, nine PhD, and several master dissertation students. Prasad has 15 years of teaching/research experience and has been awarded the Sir CV Raman Scientist Award, Young Scientist Award, and Prof. J.S. Datta Munshi Gold Medal by the International Society for Ecological Communications; Fellow of Agricultural Technology Development Society; Fellow of Biotechnology Research Society of India; Fellow of the Society for Applied Biotechnology; the American Cancer Society UICC International Fellowship for Beginning Investigators, United States; Outstanding Scientist Award in the field of microbiology; BRICPL Science Investigator Award, Research Excellence Award, among others. He has been serving as a member of several editorial boards. Previously, Prasad served as Assistant Professor at Amity University Uttar Pradesh, India; Visiting Assistant Professor, Whiting School of Engineering, Department of Mechanical Engineering at Johns Hopkins University, Baltimore, United States; and Research Associate Professor at the School of Environmental Science and Engineering, Sun Yat-sen University, Guangzhou, China.

Contributors

Mohamed Morsi Mohamed Ahmed
Department of Nucleic Acids Research
Genetic Engineering and Biotechnology Research Institute [GEBRI]
City of Scientific Research and Technological Applications [SRTA-City]
Alexandria, Egypt

Al-Khayri JM
Department of Agricultural Biotechnology
College of Agriculture and Food Sciences
King Faisal University
Al-Ahsa 31982, Saudi Arabia

Amro Abd Al Fattah Amara
Department of Protein Research, Genetic Engineering and Biotechnology Research Institute [GEBRI]
City of Scientific Research and Technological Applications [SRTA-City]
Universities and Research Center District
New Borg El-Arab

V. Ambikapathy
Department of Botany and Microbiology
A.V.V.M Sri Puspham College (Autonomous)
(Affiliated to Bharathidasan University, Trichy-20)
Poondi-613503, Thanjavur, Tamil Nadu, India

S. Babu
Department of Botany
A.V.V.M Sri Puspham College (Autonomous), (Affiliated to Bharathidasan University, Trichy-20)
Poondi-613503, Thanjavur, Tamil Nadu, India

M. Chamundeeswari
Department of Biotechnology
St. Joseph's College of Engineering, OMR
Chennai-600119, India

N. Ganesh
Department of Botany
A.V.V.M Sri Puspham College (Autonomous), (Affiliated to Bharathidasan University, Trichy-20)
Poondi, Thanjavur, Tamil Nadu, India

Surendra Kumar Gautam
Department of Life Science
Faculty of Science
Shri Rawatpura Sarkar University
Raipur, Chhattisgarh, India

Arti Gupta
Department of Zoology
SARS Smarak Degree College
India

Charu Gupta
Amity Institute of Herbal Research & Studies
Amity University
Uttar Pradesh, India

Bhagyalakshmi N Hiremath
Department of Life Sciences
CHRIST (Deemed to be University)
Hosur Road, Bangalore-560029, India

Sunil H Koli
Department of Microbiology
Yashwantrao Chavan College of Science
Vidyanagar, Karad, 415124, Maharashtra, India

K. I. Leroy
Department of Biotechnology
St. Joseph's College of Engineering
OMR, Chennai-600119, India

Kiran R Marathe
School of Life Sciences
Kavayitri Bahinabai Chaudhari North Maharashtra University
Jalgaon, Maharashtra, India

Bhavana V Mohite
Department of Microbiology
Bajaj College of Science
Wardha-442001, Maharashtra, India

Saptadeepa Mukherjee
Department of Life Sciences
CHRIST (Deemed to be University)
Hosur Road, Bangalore-560029, India

Praveen Nagella
Department of Life Sciences
CHRIST (Deemed to be University)
Hosur Road, Bangalore, India

Chackrit Nuengjamnong
Department of Animal Husbandry
Faculty of Veterinary Science
Chulalongkorn University
Bangkok, Thailand

A. Panneerselvam
PG and Research Department of Botany and Microbiology
A.V.V.M Sri Puspham College (Autonomous), (Affiliated to Bharathidasan University, Trichy-20)
Poondi, Thanjavur-613503, Tamil Nadu, India

Swathi P
Institute for Stem Cell Science and Regenerative Medicine (inStem)
GKVK Campus
Bangalore-560065, India

Athrava S Patil
H R Patel Pharmacy College Shirpur
Maharashtra, India

Vikas S Patil
University Institute of Chemical Technology
Kavayitri Bahinabai Chaudhari North Maharashtra University
Jalgaon, 425001, Maharashtra, India

Satish V Patil
School of Life Sciences
Kavayitri Bahinabai Chaudhari North Maharashtra University
Jalgaon, 425001, Maharashtra, India

Natawadee Pinyosnit
Department of Animal Husbandry
Faculty of Veterinary Science
Chulalongkorn University
Bangkok, Thailand

Saranporn Poothong
Department of Animal Husbandry
Faculty of Veterinary Science
Chulalongkorn University
Bangkok 10330, Thailand

Bhoomika Prakash P
Department of Life Sciences
CHRIST (Deemed to be University)
Hosur Road, Bangalore-560029, India

P. Prakash
Department of Botany
A.V.V.M Sri Puspham College (Autonomous), (Affiliated to Bharathidasan University, Trichy-20)
Poondi, Thanjavur, Tamil Nadu, India

K.R. Preethy
Department of Biotechnology
Rajalakshmi Engineering College
Thandalam, Chennai, India

Rashmi Ramakrishnan
Department of Life Sciences
CHRIST (Deemed to be University)
Hosur Road, Bangalore, India

M. Ratha
Department of Microbiology
Sri Bharathi Arts and Science College for Women (Autonomous), (Affiliated to Bharathidasan University, Trichy-20)
Kaikurichi-622303, Pudukkottai, Tamil Nadu, India

Mounir Mohamed Salem-Bekhit
Department of Pharmaceutics
College of Pharmacy
King Saud University
Riyadh 11451, Saudi Arabia
Department of Microbiology and Immunology, Faculty of Pharmacy
Al-Azhar University
Cairo, Egypt

Jitendra D Salunkhe
School of Life Sciences
Kavayitri Bahinabai Chaudhari North Maharashtra University
Jalgaon, Maharashtra, India

S. Sharmila
Department of Biotechnology
St. Joseph's College of Engineering
OMR, Chennai-600119, India

Somboon Tanasupawat
Department of Biochemistry and Microbiology
Faculty of Pharmaceutical Sciences, Chulalongkorn University
Bangkok 10330, Thailand

Vasantha V L
Department of Life Sciences
CHRIST (Deemed to be University)
Hosur Road, Bangalore-560029, India

K. Vijayalakshmi
Assistant Professor
PG and Research Department of Microbiology
Jamal Mohamed College (Autonomous), (Affiliated to Bharathidasan University, Trichy-20)
Trichy-20, Thanjavur, Tamil Nadu, India

Bhavana C Yadav
Department of Life Sciences
CHRIST (Deemed to be University)
Hosur Road, Bangalore-560029, India

Khairy Mohamed Abdallah Zoheir
Department of Cell Biology
Biotechnology Research Institute
National Research Center
33 Bohouth St., Dokki, Cairo, Egypt

Preface

Antimicrobials are an indispensable part of veterinary medicine used for treating and controlling diseases, and they are a growth promoter in livestock production. Regular use of antimicrobials in veterinary practices may lead to residue in animal-originated products and create potential health problems. Antimicrobial residues in animal-originated foods may induce serious health problems such as allergic reactions and antimicrobial resistance, leading to carcinogenic and mutagenic effects in the human body. The misuse of antibiotics in human medicine is thought to be a principal cause of antimicrobial resistance. Still, some antimicrobial-resistant bacteria and their resistant genes originating from animals are also responsible for the development of antimicrobial resistance. Exposing current antimicrobial use in livestock production and its impacts on public health is imperative. Moreover, the safety levels of animal feeds and food products of animal origin must be strictly monitored, and public awareness should be developed against the indiscriminate use of antimicrobials in animal production. Therefore, this book summarizes antimicrobial use in livestock production and its hazardous residual impacts on the human body in emerging countries.

This book explores the various facets of antimicrobials in animal husbandry and their perceived impact on human health, including updated information about all genera of animal farming and pathogenic microbes. This valuable information may be exploited by policymakers and various groups of experts, including academic and research personnel engaged in clinical and medical microbiology, along with all veterinary and medical practitioners. Researchers, scientists, healthcare, and management practitioners will find this an essential resource for the legal use of antibiotics in animals and humans.

The first chapter by Poothong et al. describe antimicrobials in ruminant production, and the second chapter emphasizes the role of antibiotics in animal feed, prospects and future challenges. In Chapter 3, Fattah Amara et al. describe antimicrobial resistance in agriculture and its role in human and animal health; in Chapter 4, the role of β-lactams, β-lactamases inhibitors, and efflux pump inhibitors in the battle against antibiotic resistance is discussed by Marathe et al.; and in Chapter 5, Amro Abd Al Fattah Amara describes open traditional poultry production, all demonstrating an increasing interest toward healthier animals with less drug additives. In Chapter 6, Preethy et al. discuss perspectives on using antimicrobials in poultry and fiber production. In Chapter 7, Patil et al. describe bovine mastitis: types, diagnosis, treatments, antibiotic resistance, and consequences; and in Chapter 8, Charu Gupta highlights the natural defenses, exploring alternatives to antimicrobials in animal husbandry. In Chapter 9, Surendra Kumar Gautam provides a comprehensive assessment of recent advancements and future perspectives of antimicrobials in animal husbandry. Finally, in Chapter 10, Prakash et al. present novel approaches to combating microbial infections in animal husbandry.

Arti Gupta and Ram Prasad

Acknowledgments

First and foremost, we would like to thank God for his never-ending grace, mercy, and provision during what ended up being one of the toughest times of our lives.

We are indebted to the many people who helped bring this book to light. I wish to thank series editor Dr. Vijai Kumar Gupta, Dr. Marc Gutierrez and Dr. Bonita Glanville-Morris, Senior Production Editor Dr. Jyotsna Jangra, Commissioning Editor Dr. Renu Upadhyaya, and Taylor & Francis/CRC Press, for their generous assistance, constant support, and patience in initializing the series volume. The editors are very thankful to Dr. Pratibha Goyal, Vice Chancellor Faizabad University, Ayodhya and Dr. Rammanohar Lohia, Avadh University and Prof. Sanjay Srivastava, Vice Chancellor, Mahatma Gandhi Central University, Motihari, for their kind support and constant encouragement. Special thanks are due to well-wishers, family members, and friends.

Dr. Arti Gupta owes a debt of gratitude to her wonderful husband, Mr. Saket Agrawal, and their son, Reyansh. Their unrelenting support and motivation have been pivotal in inspiring her to undertake this endeavor, overcome all obstacles, and assemble all the pieces into a coherent whole. Dr. Ram Prasad gives his special thanks to his family for their constant invaluable support and motivation to publish. We both also give special thanks to all faculty colleagues and senior faculties of our team.

Arti Gupta
Shri Avadh Raj Singh Smarak Degree College, Gonda, UP, India

Ram Prasad
Mahatma Gandhi Central University, Motihari, Bihar, India

1 Antimicrobials in Ruminant Production

Saranporn Poothong, Natawadee Pinyosnit, Chackrit Nuengjamnong, and Somboon Tanasupawat

1.1 INTRODUCTION

1.1.1 Global Trend: Reducing and Substituting Antimicrobial Agents

Antimicrobial agents are widely used in livestock production to maintain health and productivity (Van Boeckel et al. 2015). Currently, the demand for animal food production is increasing globally. Nevertheless, animal food production still relies on the use of antimicrobial agents (Tiseo et al. 2020). The objective of using antibiotics is not only for disease treatment but also for growth promotion and disease prevention. The multidrug-resistant bacteria present in the ruminant industry include *Escherichia coli, Salmonella* spp., *Staphylococcus* spp., and *Enterococcus* spp. (Abdalhamed, Ghazy, and Zeedan 2021). Notable correlations were observed regarding the use of antimicrobials in food animals, alongside the relationship between human antimicrobial consumption and the emergence of antimicrobial resistance (Allel et al. 2023). Furthermore, the increasing number of animals raised for food production caused an increasing trend in antimicrobial consumption in 2010 and was projected to increase to 2030. Five countries with the greatest percentage increases in antimicrobial consumption by 2030 (i.e., Myanmar, Indonesia, Nigeria, Peru, and Vietnam) are predicted.

Owing to the potential risk to public health of the use of antimicrobial agents in animals, the Netherlands government and stakeholders of the livestock industry established the Taskforce Antibiotic Resistance in Animal Husbandry in 2008 to develop action plans for the production of each animal with the aim of controlling antimicrobial resistance in livestock. The results revealed that the Netherlands achieved a reduction of 68% in ten years when this action plan was used (Speksnijder et al. 2015). Similarly, the United States announced the Guidance for Industry in 2012, which led to a 36.99% decrease in antimicrobial use in animals. The United Kingdom published the UK five-year antimicrobial resistance strategy in 2013, which led to a decrease in antimicrobial use in food animals of 39.2% in the United Kingdom. China also released a national action plan to manipulate antimicrobial resistance in 2016 (Xiao and Li 2016). Denmark also banned all nontherapeutic antibiotics for a decade (Aarestrup et al. 2010). Currently, food-animal production companies in several countries rely on customer demands and raise animals without nontherapeutic antibiotics (Price, Koch, and Hungate 2015).

DOI: 10.1201/9781003269847-1

1.1.2 Mode of Action of Antimicrobial Agents in Ruminants

1.1.2.1 Growth Promoters

Antimicrobial agents have been widely used in the livestock industry, including in beef and dairy cattle, sheep, and goats. Antibiotics, which are known as antimicrobial growth promoters, can improve the growth rate when they are used at low levels (subtherapeutic dose) in feed (Cromwell 2002). Various classes of antibiotics have been used as antimicrobial growth promoters with different mechanisms of action. The trend of using antimicrobial agents as growth promoters has since become universal (Brown et al. 2017). Unfortunately, antimicrobial resistance (AMR) in pathogenic bacteria has emerged because of the administration of antimicrobial growth promoters (Aarestrup, Wegener, and Collignon 2008). The World Health Organization (WHO) has recommended that antimicrobial agents be discontinued as growth promoters (World Health Organization 2004). Recently, potential alternatives to antimicrobial growth promoters have been developed for animal administration, instead of the use of antimicrobial drugs.

1.1.2.2 Disease Prevention

Antimicrobial agents at intermediate doses have the potential to prevent disease (Cromwell 2002). Previous studies have shown that more than half of the antibiotics used in dairy cattle are aimed at preventing infectious disease during the dry period. Dry-cow therapy is related to udder health; therefore, the intramammary route was the most commonly administered route (63%). Antimicrobial agents used for dry-cow therapy are narrow-spectrum penicillins and procaine penicillin-aminoglycoside combinations (Kuipers, Koops, and Wemmenhove 2016). The most common route of antimicrobial agent administration for disease prevention is in feed with antimicrobials. However, the administration of antimicrobial agents at therapeutic levels to livestock for disease prevention purposes has been banned in many developed countries because of the awareness of AMR in both animals and humans (Landers et al. 2012).

1.1.2.3 Disease Treatment

Antimicrobial agents at high levels (therapeutic doses) are used for disease treatment in food animals. Alhaji and Isola (2018) reported that antimicrobial agents used for therapeutic purposes constituted the largest proportion (54.4%). Another study revealed that in-feed antimicrobial agents are frequently used to treat gastrointestinal disease, whereas injectable antimicrobial agents are frequently used for the treatment of respiratory and musculoskeletal disease (Badger et al. 2020).

1.1.3 List of Antibiotics Used in Ruminant Industries

Antimicrobial agents used in ruminant industries include several antibiotic classes with different mechanisms of action (Brown et al. 2017). There are three basic mechanisms of action of antibiotics: cell wall synthesis, nucleic acid synthesis, and protein synthesis. Antibiotics that target the cell wall include penicillins, cephalosporins, and carbapenems. Antibiotics that inhibit protein synthesis include aminoglycosides, macrolides, and tetracyclines. Antibiotics that inhibit nucleic acid synthesis include sulfonamides, trimethoprim, quinolones, and fluoroquinolones (Kapoor, Saigal, and Elongavan 2017). Examples of antibiotics used in ruminant production are shown in Table 1.1.

TABLE 1.1
List of Antibiotic Drugs Used in Ruminant Industries

Antibiotics	Classes	Animals	Applications	References
Amoxicillin	β-Lactam	Beef Dairy cattle	Disease treatment	(Badger et al. 2020) (Kuipers, Koops, and Wemmenhove 2016)
Ceftiofur	Cephalosporins	Beef	Disease prevention Disease treatment	(Badger et al. 2020)
Ceftriaxone	Cephalosporins	Sheep and goat	Disease treatment	(Scott and Menzies 2011)
Ciprofloxacin	Fluoroquinolones	Sheep and goat	Disease treatment	(Scott and Menzies 2011)
Erythromycin	Macrolide	Beef Sheep and goat	Disease treatment	(Badger et al. 2020) (Scott and Menzies 2011)
Flavophospholipol	Phosphoglycolipid	Beef	Growth promotion Disease prevention	(Badger et al. 2020)
Florfenicol	Amphenicols	Beef	Disease treatment	(Badger et al. 2020)
Lincomycin	Lincosamides	Dairy cattle	Disease treatment	(Saini et al. 2012)
Nalidixic acid	Quinolones	Sheep and goat	Disease treatment	(Scott and Menzies 2011)
Neomycin	Aminoglycosides	Sheep and goat	Growth promotion Disease treatment	(Alhaji and Isola 2018)
Oxytetracycline	Tetracyclines	Beef	Disease prevention Disease treatment	(Badger et al. 2020)
Penicillin	β-Lactam	Beef Dairy cattle Sheep and goat	Disease prevention Disease treatment Growth promotion Disease prevention Disease treatment Growth promotion Disease treatment	(Badger et al. 2020) (Kuipers, Koops, and Wemmenhove 2016) (Alhaji and Isola 2018)
Streptomycin	Aminoglycosides	Sheep and goat	Growth promotion Disease treatment	(Alhaji and Isola 2018)
Telithromycin	Ketolides	Sheep and goat	Disease treatment	(Scott and Menzies 2011)
Tetracycline	Tetracyclines	Dairy cattle Sheep and goat	Growth promotion Disease treatment	(Kuipers, Koops, and Wemmenhove 2016) (Scott and Menzies 2011)

(*Continued*)

TABLE 1.1 (*Continued*)
List of Antibiotic Drugs Used in Ruminant Industries

Antibiotics	Classes	Animals	Applications	References
Tilmicosin	Macrolide	Beef	Disease prevention Disease treatment	(Badger et al. 2020)
Trimethoprim sulfamethoxazole (co-trimoxazole)	Others	Beef Dairy cattle	Disease treatment Growth promotion	(Badger et al. 2020) (Kuipers, Koops, and Wemmenhove 2016)
Tulathromycin	Macrolide	Beef	Disease prevention Disease treatment	(Badger et al. 2020)
Tylosin	Macrolide	Beef	Disease prevention	(Badger et al. 2020)
Virginiamycin	Streptogramins	Beef	Growth promotion Disease prevention	(Badger et al. 2020)

1.2 PREVALENCE OF ANTIMICROBIAL RESISTANCE RESEARCH IN RUMINANTS

1.2.1 Beef

Antimicrobial resistance in food animals is not only a concern for human health but also a veterinary problem. In the past, ruminants were raised with antimicrobial use, but beef calves had only minimal exposure to antimicrobial agents; therefore, one of the studies hypothesized that there would be a low prevalence of antimicrobial-resistant bacteria in beef production (Call, Davis, and Sawant 2008). Antimicrobial resistance has developed in beef production due to subtherapeutic antimicrobial administration (Cameron and McAllister 2016). To address this issue, the Canadian beef industry has developed a surveillance program to detect antimicrobial resistance from respiratory and enteric bacteria (Hannon et al. 2020). Oxytetracycline resistance was most common in cattle, which is similar to reports that resistance to oxytetracycline was the highest across Pasteurellaceae isolated from beef and dairy cattle (Saini et al. 2012). Another study verified that the most frequently observed antimicrobial resistances in beef cattle were oxytetracycline (10%), tilmicosin (6.4%), and ampicillin (4.6%) (Andrés-Lasheras et al. 2021).

1.2.2 Dairy Cattle

Dairy production has been threatened by antibiotic-resistant pathogens. A previous study revealed that dairy-origin isolates were more likely to be resistant to many antimicrobials (Davis et al. 2007). Another study reported the prevalence of pathogenic bacteria and their resistance to different antibiotics. For example, the prevalence rates of *Streptococcus aureus*, *Escherichia coli*, and *Pseudomonas aeruginosa* were 30%, 17%, and 3.5%, respectively. Moreover, 90% of the *Streptococcus aureus* strains were resistant to penicillin, and 40% of the *E. coli* strains were resistant to streptomycin, whereas all *P. aeruginosa* strains were resistant to several antibiotics, including ceftriaxone, enrofloxacin, and levofloxacin (Ameen et al. 2019). Among antibiotics, oxytetracycline was the most frequently observed form of resistance in dairy cattle, followed by tilmicosin and tulathromycin. The resistance of *Pasteurella multocida* to tilmicosin and tulathromycin in dairy cattle is greater than that in beef cattle because macrolides are more frequently administered to dairy cattle (Andrés-Lasheras et al. 2021). Another study reported that *E. coli* isolated from sampled animals was resistant to six classes of antibiotics (Mainda et al. 2015). A previous study evaluated the prevalence of antimicrobial resistance in *E. coli* strains isolated from patients with clinical bovine mastitis. There were 28.31% resistant strains to at least one antibiotic. Moreover, 13 out of the 32 antibiotic-resistant isolates were identified as multidrug resistant, and 6 isolates were identified as extensively drug resistant. All the isolates were susceptible to ciprofloxacin and ofloxacin (Majumder et al. 2021).

1.2.3 Sheep and Goats

Antibiotic-resistant microorganisms from sheep have received less attention. Previously, 3% of isolates from sheep were resistant to one or more antimicrobials (Enne et al. 2008). In contrast, a more recent study revealed that 10% of isolates from sheep were resistant to one or more antimicrobials (Doidge, West, and Kaler 2021). Another study reported the isolation of cefotaxime-resistant strains from sheep from clinical diagnostic samples (Snow et al. 2011). The present study investigated the prevalence of antimicrobial *E. coli* isolated from the feces of dairy goats in Thailand. The results revealed that 141 of 180 isolates were resistant to at least one antimicrobial agent. The most frequent resistance of *E. coli* was to streptomycin, followed by tetracycline, kanamycin, and sulfamethoxazole-trimethoprim. Overall, the percentage of multidrug-resistant *E. coli* was 23.9% (Prapasawat and Intarapuk 2021).

1.3 ADVANCES AND ALTERNATIVE ANTIMICROBIALS USE IN RUMINANTS

1.3.1 Alternatives for Growth

1.3.1.1 Probiotics and Direct-Fed Microbes

Probiotics are defined as "live microorganisms which when administered in adequate amounts confer a health benefit on the host" (Hotel and Cordoba 2001). Similarly, direct-fed microbes (DFMs) are single or mixed live microorganisms that enhance beneficial health effects in the host (Elghandour et al. 2015). The difference between DFMs and probiotics is that probiotics are naturally inhabiting bacteria that are exclusively used for feed production (Brashears, Amezquita, and Jaroni 2005). The probiotics or DFMs used in ruminants include bacteria (*Bacillus, Bifidobacterium, Enterococcus, Lactobacillus, Propionibacterium, Megasphaera,* and *Prevotella*), yeast (*Saccharomyces cerevisiae*), and fungi (*Neocallimastix, Orpinomyces,* and *Piromyces*) (Seo et al. 2010; Song, Ibrahim, and Hayek 2012). Figure 1.1 shows the different mechanisms of probiotics or DFM in ruminants.

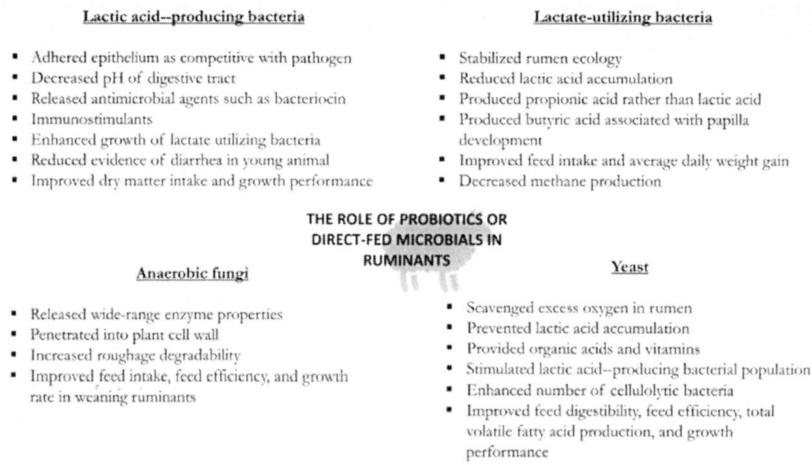

FIGURE 1.1 Probiotics or direct-feeding microbial mechanisms in ruminants as alternative antibiotics.

(*Sources*: Seo et al., 2010; Elghandour et al., 2015; Hess et al., 2020)

The bacterial DFMs usually used in ruminants include lactic acid bacteria, lactate-utilizing bacteria, and spore-forming bacteria. The most common DFM supplemented for growth performance is *Lactobacillus,* followed by bifidobacteria, *Enterococcus,* and *Bacillus*. Lactic acid bacteria target the small intestine to inhibit pathogens. Owing to their modes of action, such as antimicrobial effects, competitive attachment, and immune stimulation, this group of DFMs affects the balance of intestinal microbes. Therefore, lactic acid bacterial DFMs are considered immature ruminants (Doyle et al. 2019). Saleem, Zanouny, and Singer (2017) reported that a mixed culture of *P. acidilactici* and *P. pentosaceus* supplemented in postweaning lambs improved growth performance by increasing the final body weight (+3.16 kg) and average daily gain (ADG) (+ 25.2 g/lamb) and decreasing the feed conversion ratio (FCR) (−1.18) compared with those of the nonsupplemented group. Buffalo calves fed *L. acidophilus* at the first month of age similarly presented a positive response of ADG of 31.4% (Mudgal and Baghel 2010). Furthermore, lactate-utilizing bacteria, such as *Megasphaera elsdenii, Propionibacterium shermanii, Propionibacterium freudenreichii, Propionibacterium acidophilus,* and *Propionibacterium jensenii,* which produce propionic acid as the main product of lactate utilization, are also used as DFMs (Seo et al. 2010). A diet supplemented with *P. jensenii* 702 increased the weaning weight and total live weight of calves by 8.2 kg and 16.4 kg, respectively (Adams et al. 2008). *M. elsdenii* mainly produces butyric acid from lactate and glucose fermentation. It is an important volatile fatty acid for papillae development in preruminant animals. Additionally, *M. elsdenii* was supplemented with Angus calf feed. As a result, administration increased caloric density, leading to greater average daily weight gain and gain:feed efficacy. Moreover, the development of marbling was satisfactory in the group supplemented with *M. elsdenii,* without an adverse effect on the growth rate. In this study, the use of *M. elsdenii* as a feed additive was suggested to be a good precursor for enhancing the growth and quality of meat in calves and feedlots (DeClerck et al. 2020). When *M. elsdenii* was orally administered to 14-day-old dairy calves, the results indicated that *M. elsdenii* addition improved starter intake, weaning body weight, and feed efficiency (Muya et al. 2017). In addition, the ability of the Gram-positive spore-forming bacteria *Bacillus* spp. to resist hazardous environments and tolerate digestive enzymes has been studied. Several studies have defined the role of *Bacillus* as a probiotic by improving microbial balance, digestibility, and the immune response in animals. The introduction of *Bacillus subtilis* into the rumen stimulated the populations of *Thermoanaerobacteriaceae, Peptostreptococcaceae,* and *Alicyclobacillaceae,* which caused increases in fatty acid and ammonia emissions. Therefore, the promotion of the rumen product indicated a greater correlation with digestive efficiency (Ushakova et al. 2013). Supplementation with *B. subtilis* is not harmful to animals, and viable *B. subtilis* cells pass through the gastrointestinal tract (Choonkham et al. 2021). The addition of *B. licheniformis* to fattening lamb feed also markedly increased growth hormone, insulin-like growth factor, and insulin levels in the blood profile. There was no significant difference in ADG or dry matter intake (DMI) between probiotic-supplemented animals and antibiotic-treated animals (monensin treatment), resulting in a suitable alternative antibiotic candidate (Jia et al. 2018). Other antimicrobial substitutes include *B. amyloliquefaciens* and *B. pumilus,* which are well recognized for their ability to promote animal growth and immunologic responses. The growth performance of weaning goats supplemented with *Bacillus* was significantly different because of ruminal papilla and small intestinal villus growth (Zhang, Wang, and Wei 2020). For example, *Bacillus* species are suggested as potential spore-forming bacteria that increase the beneficial effects on hosts through their ability to alter rumen microbial communities.

Fungal DFMs have been chosen for adult ruminants and are aimed mostly at the rumen. Fungal feed additives modulate rumen microorganisms, which enhances beneficial activities to the host, such as increasing enzymatic activity. In addition, the anaerobic fungal DFM prevented excess lactic acid production in the rumen. Anaerobic rumen fungi have been widely investigated because of their potential enzyme functions. Owing to plant cell wall degradability by a broad range of enzymes from anaerobic fungi, this action increases the utilization of fibrous feed in terms of increasing the available energy supply for animals (Lee, Ha, and Cheng 2000). Anaerobic fungi penetrate into plant cells, increasing the space for bacterial attachment and digestion (Dagar et al. 2011). Consequently, coculture extracts of *Aspergillus oryzae* and *Aspergillus niger* improved the digestibility of roughage and total mixed ration diets in an in vitro experiment (Kong et al. 2021). Furthermore, supplementation with anaerobic fungi helps increase feed digestibility, feed efficiency, animal growth, milk yield, and volatile fatty acid production in the rumen (Saxena et al. 2010). The key to anaerobic fungus supplementation is stability because the microorganisms must be alive when they are administered to animals for the greatest benefit. Suitable product storage must be performed in dry and cool areas while avoiding heat, sunlight, and high humidity conditions

(Puniya et al. 2015). The demand for animal products has increased due to the increasing human population, whereas the available natural feed sources for animals have declined. Thus, the low quality of roughage treated with anaerobic fungi or their products can be an alternative way to improve feed quality and feed efficiency in ruminants (Hess et al. 2020).

In fact, yeast not only scavenges the remaining oxygen in the rumen, which results in an appropriate environment for rumen microbes, but also increases feed digestibility. Yeast cells are able to survive through the digestive tract and provide organic acids and vitamins (Khan et al. 2016). Numerous studies have indicated that yeast supplementation in ruminants enhances the growth performance and productivity of the animals by increasing the number of cellulolytic bacteria (Chaucheyras-Durand et al. 2012). Compared with the control, fattening calves supplemented with live *Saccharomyces cerevisiae* improved the average daily gain (+400 g/calf) and final body weight (Maamouri and Ben Salem 2021). Similarly, heat-stressed goats fed an *S. cerevisiae*–supplemented diet presented increases in total volatile fatty acids (TVFAs), DMI, digestibility, and ADG compared with nonsupplemented goats (Cai et al. 2021). Therefore, propionic acid plays an important role in the energy status of ruminants. Furthermore, beef cattle supplemented with *S. cerevisiae* gained weights ranging from 7.5% to 13%, depending on the feed formula (Puniya et al. 2015). Hence, yeast culture supplementation has been indicated for use in all ages of ruminants to improve growth and production, in contrast to the use of antibiotics.

1.3.1.2 Prebiotics

Prebiotics are nondigestible feed components that supply nutrients to promote beneficial bacteria in the digestive tract, providing health advantages to the host (Gibson et al. 2017). Indigestible saccharides are generally used as feed additives and are commonly found in fibrous plant materials.

The current application of prebiotics in immature ruminants has been studied. Several types of indigestible carbohydrates, i.e., mannan oligosaccharides (MOSs), fructo-oligosaccharides (FOSs), cello-oligosaccharides (CFs), and galactosyl-lactose (GL; galactose plus lactose), have been suggested for caloric supplementation (Uyeno, Shigemori, and Shimosato 2015). Influences of MOSs compared with those of antibiotics were evaluated in calves fed milk replacer. Starter intake was greater in calves fed MOSs, and supplementation also reduced diarrhea incidence (Kara et al. 2015). However, MOS and inulin supplemented in dairy claves did not improve growth performance or physiological changes (Tóth et al. 2020). Sheep fed MOSs presented increased nutrient efficiency due to increased nutrient digestibility, but MOSs had no effect on DMI, body weight, or ADG (Zheng et al. 2021). Previously, the growth-promoting effects of MOS supplementation did not have potential, as expected; however, prebiotics maintained good digestive health, similar to antibiotic supplementation. In addition, lambs supplemented with FOS presented significantly improved ADG and altered bacterial diversity. There are numerous lactic acid–producing bacteria, such as *Bifidobacterium, Enterococcus,* and *Lactobacillus,* in lambs fed FOS (Quijada et al. 2020). Moreover, the addition of FOS to milk replacer for calves had a positive effect on FCR and carcass weight, resulting in increased growth (Grand et al. 2013). Thus, prebiotics are effective candidates as alternative antimicrobial agents to encourage growth and gut health in immature ruminants.

1.3.1.3 Antimicrobial Peptides

Antimicrobial peptides (AMPs) are abundantly released by ruminant organs, as shown in Table 1.2. Antimicrobial peptides are the first-line defense offered by the natural innate immune response to pathogens. These peptides are categorized into two groups: anionic and cationic antimicrobial peptides. A small amount of anionic antimicrobial peptide was found in ruminants, whereas cationic antimicrobial peptides were detected in greater proportions. Furthermore, AMPs exhibit broad-range antimicrobial activities against Gram-positive bacteria, Gram-negative bacteria, fungi, and some viruses (Brogden et al. 2003). Mammalian AMPs include mainly defensins and cathelicidins. In addition, lysozyme, lactoferrin, and collectin are significant antimicrobial peptides. The rumen is a source of AMPs because of the presence of numerous microorganisms. Onime et al. (2021) reported novel antibacterial peptides called lubilisin. It demonstrated potential antimicrobial activity against methicillin-resistant *Staphylococcus aureus* without human or sheep erythrocyte cytotoxicity. In the era of antimicrobial-resistant bacteria worldwide, natural antimicrobial peptides are considered possible antibiotic substitutes (Rima et al. 2021; Kumar et al. 2020). The use of AMPs as a feed supplement has been less common in ruminant species but more common in poultry. Nevertheless, the effects of AMP feed supplementation on rumen fermentation and microorganism diversity were studied in goats by adding recombinant swine-specific PBD-mI and a fly antimicrobial peptide at a ratio of 1:1. As a result, supplementation with AMPs has positive effects on growth performance by increasing volatile fatty acid levels and enzyme activities (Ren et al. 2019). Furthermore, AMP (recombinant porcine β-defensin 2 and fly) treatment in juvenile goats also increased body weight, ADG, and enzymatic activity and altered rumen microorganism diversity (Liu et al. 2017). For example, recombinant swine β-defensin mixed with fly antimicrobial peptide is a promising feed addition for improving growth performance.

TABLE 1.2
Antimicrobial Peptides Found in Ruminants and Antimicrobial Activity

Animals	AMP Family	AMP Protein	Origin Source	Antimicrobial Activity	References
Cattle	Defensins	TAP	Mucosal epithelial cell, mammary gland	*Escherichia coli* *Klebsiella pneumoniae* *Pseudomonas aeruginosa* *Staphylococcus aureus*	(Isobe, Hosoda, and Yoshimura 2009; Yarus et al. 1996)
		LAP	Tongue, esophagus, rumen, reticulum, omasum, mammary gland	Broad-spectrum* antifungal activity	(Schonwetter, Stolzenberg, and Zasloff 1995; Tetens et al. 2010)

(*Continued*)

TABLE 1.2 (*Continued*)
Antimicrobial Peptides Found in Ruminants and Antimicrobial Activity

Animals	AMP Family	AMP Protein	Origin Source	Antimicrobial Activity	References
	Cathelicidins	1. BMAP-27 2. BMAP-28 3. BMAP-34	Myeloid cell	1. Broad spectrum* 2. MRSA, PDRAB 3. Broad spectrum*	(Scocchi, Wang, and Zanetti 1997) (Guo, Xun, and Han 2018)
Goat	Cathelicidins	ChBac3	Leukocyte	*Escherichia coli* *Listeria monocytogenes* *Pseudomonas aeruginosa*	(Shamova et al. 2009)
	Cathelicidins	miniChBac7.5 α, β	Neutrophils	*Acinetobacter baumannii* *Klebsiella* spp. *Listeria monocytogenes* *Micrococcus luteus* *Pseudomonas aeruginosa*	(Shamova et al. 2016)
	Cathelicidins	ChMAP-28	Mammary gland, leukocyte	*Acinetobacter baumannii* *Enterobacter cloacae* *Escherichia coli* *Klebsiella pneumoniae* *Pseudomonas aeruginosa*	(Emelianova et al. 2018; Zhang et al. 2014; Panteleev et al. 2018)
Sheep	Defensins	SBD-1	Tongue, trachea, gastrointestinal tract	*Escherichia coli* *Pseudomonas aeruginosa* *Proteus vulgaris* *Staphylococcus aureus* *Shigella flexneri*	(Zhao and Cao 2012)
	Cathelicidins	OaBac5 mini	Neutrophils	*Escherichia coli* *Pseudomonas aeruginosa* *Salmonella enterica* serovar Typhimurium	(Anderson, Hancock, and Yu 2004)
	Cathelicidins	SMAP29, 34	Bone marrow myeloid cell	*Enterococcus faecalis* *Escherichia coli* MRSA *Pseudomonas aeruginosa* *Salmonella enterica* serovar Typhimurium *Staphylococcus epidermidis*	(Skerlavaj et al. 1999; Anderson, Hancock, and Yu 2004)

TABLE 1.2 (*Continued*)
Antimicrobial Peptides Found in Ruminants and Antimicrobial Activity

Animals	AMP Family	AMP Protein	Origin Source	Antimicrobial Activity	References
Buffalo	Defensins	BNBD-1–4	Mastitis udder	*Candida albicans* *Escherichia coli*	(Das et al. 2010; Das et al. 2008)
	Defensins	LAP	Tongue, mastitis udder, teat canal, female reproductive tract	*Escherichia coli* *Listeria monocytogenes* *Salmonella enterica* serovar Typhimurium *Staphylococcus aureus*	(Kalita, Kumar, and Kumar 2009)
	Cathelicidins	BuMAP-28, 29,34	Myeloid cell	*Candida albicans* *Klebsiella pneumoniae* *Riemerella anatipestifer* *Salmonella enterica* serovar Typhimurium *Staphylococcus aureus*	(Panicker and George 2013; Panicker and George 2016)

Abbreviations: AMP, antimicrobial peptide; Bac, bactenecin; BD, β-defensin; BNBD, bovine neutrophil β-defensin; LAP, lingual antimicrobial peptide; MAP, myeloid antimicrobial peptide; MRSA, methicillin-resistant *Staphylococcus aureus*; PDRAB, pandrug-resistant *Acinetobacter baumannii*; TAP, tracheal antimicrobial peptide.

* AMPs against Gram-positive and Gram-negative bacteria.

1.3.1.4 Nonnutritive Phytochemicals

The alternative bactericidal activity and growth promotors of secondary metabolites from plants are widely known as phytochemical feed additives. The influences of phytochemicals in ruminants include rumen alteration and increased digestibility, resulting in increased feed efficiency and improved nutrient absorption. Phytochemicals play a significant role in animal feed intake and weight gain because of their feed antioxidant status, antimicrobial activities, and palatability stimulation by increasing flavor and smell (Valenzuela-Grijalva et al. 2017). Numerous studies of essential oils to improve growth performance in ruminants have been conducted in the finishing phases of production. The most popular essential oils include cinnamaldehyde, eugenol, ferulic acid, and hydroxycinnamic acid. Nevertheless, high-dose administration should involve the inhibition of rumen microorganism proliferation and function due to antimicrobial properties and antinutritive effects (Valenzuela-Grijalva et al. 2017; Kholif and Olafadehan 2021). One of the studies in finishing beef that used ferulic acid supplementation for the last 30 days demonstrated positive effects on meat quality and meat storage. Ferulic acid inhibited lipid oxidation in steer, suggesting extended storage and modification of fatty acid compositions (González-Ríos et al. 2016). Furthermore, treatment of lambs with ferulic acid improved their average daily gain and final weight due to skeletal muscle hypertrophy triggered by MHC-I upregulation (Peña-Torres et al. 2022). In small ruminants, a meta-analysis of numerous studies reported that essential oils were able to improve ADG without modifying DMI or FCR (Andri, Huda, and Marjuki 2020). However, essential oil supplementation increased DMI but did not affect ADG or FCR in lambs (Passetti, Passetti, and McAllister 2021). Moreover, thyme essential oil only enhances nitrogen metabolism and is not involved in growth performance in sheep (Ribeiro et al. 2019). On the other hand, finishing lambs supplemented with polyherbal additives (1 g PA/kg DM), including hydrolyzable tannins, flavonoids, and essential oils, presented greater average daily gains, greater backfat thicknesses, and lower feed conversion ratios than did the control group. Thus, a low dose of a polyherbal additive enhances growth performance and improves the feed conversion ratio without impairing the dry matter intake, carcass yield, meat composition, or health status of lambs (Dorantes-Iturbide et al. 2022).

Saponins, which are triterpene glycoside compounds, are found in plant materials. Saponins, especially propionate, increased VFA production. Owing to the Gram-positive antibacterial and defaunation activities of saponins, acetate and butyrate synthesis are reduced (Patra and Saxena 2009). Saponin derived from plants was added to steer feed to determine its effect on digestibility of low-quality roughage. The results showed that saponin supplementation decreased the feed passage rate, which encouraged DM digestibility and neutral detergent fiber digestibility (McMurphy 2014). Moreover, saponins suppress the protozoan population, which increases the amount of nutrients available for animals (Dai and Faciola 2019). In addition, the productive performance, carcass, and meat qualities of finishing lambs fed a polyherbal mixture (saponins, flavonoids, and polysaccharides) for 56 days were evaluated. As a result, there was no significant difference between groups in terms of DMI, ADG, FCR, or carcass characteristics; however, the addition of a polyherbal mixture resulted in better tenderness of the meat (Orzuna-Orzuna et al. 2021). Similarly, compared with nonsupplemented lambs, lambs supplemented with alfalfa saponin presented greater nutrient digestibility and final body weight without alterations in ADG or FCR (Liu et al. 2018). Therefore, saponins derived from plants also have positive effects on rumen fermentation and rumen microbial diversity by increasing nutrient digestibility and rumen defaunation.

Phenolic substances lead to improved gastrointestinal tract development, animal health, and growth performance. An alternative antibiotic, gallic acid, is a bioactive phenolic compound that was fed to Holstein calves during the preweaning period to determine its effect on growth performance. These findings demonstrated that administering gallic acid improved growth by increasing calf starter intake and average daily gain (Xu et al. 2022).

In conclusion, the previous results of phytochemical supplementation are inconsistent, possibly because of a lack of understanding of the appropriate dosage and the various concentrations of the active component in phytochemicals derived from different batches of plant materials. To maintain the efficacy of phytochemical extracts, their quality must be improved.

1.3.2 Alternatives for Disease Treatment and Prevention

1.3.2.1 Vaccines

Vaccination has been an integral part of ruminant preventive measures for quite a long time. The classical vaccines consisted of a live attenuated vaccine and an inactivated (killed) vaccine. However, both vaccine types have limitations. Inactivated vaccines are safe for both animals and users; however, they lack a cell-mediated immune response. On the other hand, attenuated vaccines developed to minimize the pathogenicity of pathogens effectively stimulate both humoral immune- and cell-mediated immune responses. Notably, live attenuated vaccines can cause pathogenic responses in animals after vaccination. Researchers have potentially developed next-generation vaccines, such as DNA vaccines, subunit vaccines, recombinant vaccines, and viral vector vaccines, which elicit both humoral and cellular immune responses (Meeusen et al. 2007; Hoelzer et al. 2018). Thus, vaccines have the opportunity to decrease antibiotic usage and have the potential to improve animal health, animal welfare, and public health.

Mastitis, an intramammary infection, occurs in dairy cattle and small ruminants. The gold standard methodology for the diagnosis of mastitis is bacterial culture, but the results depend on the sampling method and sample preservation. Currently, the popular tool for predicting mastitis in dairy species is the somatic cell count (SCC) method. When intramammary infection occurs, the SCC increases. For example, *Mycoplasma* spp. (contagious agalactia syndrome) is associated with elevated SCC but rarely shows clinical signs. Mastitis also affects human health through pathogen and leukocyte contamination in milk. To prevent or reduce the severity of clinical mastitis, several dairy farms are routinely vaccinated against typical udder infections. *S. agalactiae*, *S. aureus*, and *E. coli* are the most frequently targeted intramammary pathogens. Additionally, *S. agalactiae* and *S. aureus* vaccine types include cellular lysates and inactivated, attenuated, and subunit vaccines (i.e., toxins, proteins, polysaccharides), whereas the mutant antigen J5 has been extensively applied for *E. coli* vaccination (Ismail 2017). Experimental intramammary challenge with killed *S. aureus* was performed to determine the immunologic response of vaccinated cows. Vaccination reduced the severity of the inflammatory response, as shown by the decreased quarter milk SCC and polymorphonuclear neutrophilic leukocyte concentration in the vaccinated group compared with those in the unvaccinated group. However, intramammary infection with *S. aureus* still reduces milk production (Piepers et al. 2017). Similarly, a field trial in dairy herds with high somatic cell counts and *S. aureus* mastitis that were vaccinated with two polyvalent vaccines, namely, Startvac and Bestvac, was conducted to evaluate udder health sustainably. Both vaccinated cows subsequently presented a lower prevalence of *S. aureus,* although the vaccine compromised milk yield in both types of commercial vaccines. Thus, this study suggested that these two types of *S. aureus* vaccines are not appropriate measures for maintaining udder health in the field (Freick et al. 2016). Nevertheless, several studies revealed that this vaccine was ineffective at controlling mastitis in small ruminants.

Enterohemorrhagic *Escherichia coli* (EHEC, O157) is a zoonotic disease related to cattle and human transmission. Cattle are known as reservoirs of this hazardous microorganism because the pathogen is shed in their feces without causing any clinical symptoms. The vaccination successfully reduced EHEC shedding into the environment by 50%. Moreover, a predicted study of vaccinated cattle estimated that approximately 85% of EHEC infectious cases in humans were decreased because of blockage of the main pathogen source (Matthews et al. 2013). Additionally, two available vaccines were designed to minimize EHEC shedding. One vaccine targeted the type III secreted protein (TTSP) system, which is a bacterial component that supports bovine epithelial cell colonization. Another, the siderophore receptor and porin, is the target of vaccines against the bacterial cell membrane. A previous study compared immunized cattle with TTSP or the siderophore receptor and porin vaccine with unvaccinated cattle and reported odds ratios of 0.38 and 0.42 for TTSP and siderophore receptor and porin, respectively. The odds ratio explained the preventive factor for fecal shedding from both vaccines (Smith David, Sperandio, and Hovde Carolyn 2014). The development of EHEC-specific antigens in plant materials for the oral administration of vaccines is a candidate for injectable vaccines because they are cost effective. Initially, *Nicotiana benthamiana* expresses five recombinant proteins that can be designed as vaccines, after which the proteins can be transplanted into *Nicotiana tabacum*. Sheep fed transgenic transplastomic recombinant-EspA tobacco shed fewer EHEC strains than control animals did. The plant-made EHEC vaccine is recommended as an alternative for farmers to reduce EHEC shedding in the future (Miletic et al. 2017).

Studies on registration have shown the efficacy of the bacterial DNA immunostimulant Zelnate in the *Mannheimia haemolytica* challenge model. These investigations revealed a notable decrease in lung lesions and mortality rates. This study investigated the effects of Zelnate when it was administered simultaneously with a respiratory viral vaccine (Pyramid 5) in feedlots. These findings indicated that the immunostimulant was linked to a reduction in mortality due to bovine respiratory virus (Cusack 2024).

In small ruminants, caprine pleuropneumonia (CCPP) is a contagious disease and a serious illness caused by *Mycoplasma capricolum* subsp. *capripneumoniae*. The CCPP disturbs the respiratory system of animals. This disease has a morbidity rate of up to 100% and a mortality rate of approximately 60%. Infected animals have been treated with several antibiotic drugs. Tylosin is recommended, but it was unavailable in the CCPP spreading area. Although oxytetracycline has been applied to CCPP animals and shown to be an effective treatment, long-term oxytetracycline usage has increased the risk of antibiotic-resistant bacteria. Therefore, prevention and control of CCPP depend on vaccination. The use of the killed vaccine illustrated potential immune stimulation in goats, and the immune system remained stable for 12 months. Few studies on novel strain-based or recombinant specific–antigen vaccines to improve the immune response against CCPP have been conducted. However, this new generation of vaccine still needs more research in a large-scale population (Kabir et al. 2021).

One public health concern is brucellosis. The vaccination of female calves or female kids is the key to brucellosis control programs. Two effective vaccines for bovine brucellosis caused by *Brucella abortus* have been developed in cattle: live-modified S19 (Buck 1930; Koh and Morley 1981) and the RB51 vaccine (Olsen et al. 1999; Poester et al. 2006). Consequently, new technologies have led to the development of new candidate vaccines, such as subunit vaccines, DNA vaccines, and recombinant vector vaccines. Additionally, the Cu/Zn superoxide dismutase DNA vaccine elicited an immune response similar to that elicited by RB51 vaccination in an in vitro assay. The response result was impressive in laboratory-generated mice, but DNA vaccine application requires a better approach in cattle to obtain more information. The Omp28 subunit vaccine also induced immunity in mice against *B. abortus*; nevertheless, the response level was lower than that of the live attenuated vaccine. Moreover, recombinant vector vaccines have greater potential than DNA vaccines or subunit vaccines. The immunostimulant *B. abortus* gene was inserted into vectors containing viruses (e.g., forest virus) and bacteria (e.g., *E. coli, Lactococcus lactis,* and *Salmonella enterica*). Indeed, a vector vaccine expressing a Cu/Zn superoxide dismutase vaccine induced a cytokine response via helper T (Th1) cells, as did a Semliki Forest virus-based vector encoding *Brucella* translation initiation factor 3, which was highly resistant to *B. abortus* (Dorneles, Sriranganathan, and Lage 2015). In conclusion, next-generation vaccines that induce both humoral and cell-mediated immune responses are safe for humans and animals. In addition, the development of new vaccines can be an alternative strategy for the differentiation of infected animals and vaccinated animals (Kyriakis 2015).

Anaplasmosis is a tick-borne disease caused by *Anaplasma* species. Newly developed vaccines are designed to protect cattle from severe illness. The targeted mutagenesis *Anaplasma marginale* vaccine showed a protective effect against wild-type *A. marginale.* The cows vaccinated with targeted mutagenesis did not develop clinical signs of infection after challenge. Conversely, nonvaccinated and whole-cell-inactivated vaccinated cattle developed signs of anemia after 26 to 30 days of infection. Therefore, anaplasmosis vaccination via targeted mutagenesis successfully protects the host from infection with *A. marginale* (Hove et al. 2022).

1.3.2.2 Bacteriophages and Endolysins

Bacteriophages are viruses that infect specific bacterial species. Lytic phages infect the bacterial cell host and then lyse the bacterial cell wall, resulting in bacterial cell death. Additionally, bacteriophage endolysins, known as peptidoglycan hydrolases, can be categorized into three types: glycosidases, amidases, and endopeptidases. Bacteriophage endolysins, especially Gram-positive microorganisms, specifically break down the peptidoglycan of selective bacterial cell walls (Loessner 2005). Bacteriophages and phage endolysins are good competitive agents for treating and preventing disease in farm ruminants because of their high specificity, rapid target cell lysis, and reduced induction of resistant mutations (Gutiérrez et al. 2019).

Generally, ruminal acidosis can be divided into two categories on the basis of ruminal pH: the duration of ruminal pH under acidic conditions and clinical signs. First, subacute rumen acidosis is frequently caused by VFA accumulation followed by a decrease in the ruminal pH. Second, lactic acid accumulation is the main cause of acute rumen acidosis. *Streptococcus bovis*, a lactic acid–producing bacteria, is one of the microorganisms responsible for excessive lactic acid production in the rumen (Asanuma and Hino 2002). Endolysin has been used as an alternative tool against specific problematic bacteria. The in silico study of the endolysin LyJH307 supported the lytic effect of *S. bovis* on the ruminal pH and temperature conditions. Thus, a novel endolysin (LyJH307) may replace antimicrobial administration and reduce intensive labor protocols to prevent acute rumen acidosis (Kim et al. 2020).

AMR is notable in *Staphylococcus aureus,* which usually causes mastitis in ruminants. Therefore, phage therapy is an alternative biological control agent for antimicrobial activity. Previously, phage therapy for subclinical *S. aureus* mastitis was studied by intramammary infusion of phage K. However, the cure rate was similar to that of untreated udder because the phage was severely destroyed and rendered inactive by milk protein (Gill et al. 2006). The phage isolated from dairy farms still has the ability to infect *S. aureus* strains isolated from mastitis milk samples via the lysis plate method (Leite et al. 2019). The sequence analysis of the isolated bacteriophages subsequently revealed the highly conserved region of the endolysin gene. Interestingly, bacteriophage endolysins, such as *S. aureus* and coagulase-negative staphylococci (CNS), are potentially useful for the treatment of Gram-positive bacterial mastitis. Bacteriophage endolysins are effective protein antimicrobial agents because of their lytic activities. Additionally, the *S. aureus* bacteriophage phi11 endolysin exhibited potential lytic enzymes (i.e., amidases and endopeptidases), which killed *S. aureus* and the CNS in an in vitro assay (Donovan, Lardeo, and Foster-Frey 2006). The phage lytic protein endolysin LysRODI from the staphylophage philPHLA-RODI was suggested for use as a prophylactic treatment for mastitis. The LysRODI endolysin presented a good lysis rate with *S. aureus* and *S. epidermis* in mastitis mice (Gutiérrez et al. 2020). Indeed, phi11 and LysRODI endolysins were more effective than antibiotics were, and no resistant strains with phage endolysin were recognized. Overall, phage lytic endolysin represents a new replacement for antibiotics to treat and prevent staphylococcal mastitis.

Young ruminants with diarrhea are generally infected with *E. coli,* leading to an increased mortality rate and malabsorption in the gastrointestinal tract. Cattle and sheep are important reservoirs for *E. coli* O157:H7, which is a foodborne pathogen. Phage therapy was administered before the onset of diarrhea in calves and lambs. The results demonstrated that phage therapy effectively minimized mortality and greatly decreased the number of *E. coli* in feces. Enterohemorrhagic *E. coli* (O157:H7) was applied to Holstein cattle, which were then treated with phage therapy (KH1 and SH1) via the recto-anal junction mucosa. This phage suppressed the number of *E. coli* O157:H7 from feces shedding, although it did not eliminate this pathogen (Sheng et al. 2006). Hence, short-term phage administration in ruminants prior to slaughter might help reduce the number of *E. coli* and increase global food safety (Callaway et al. 2008).

1.3.2.3 Predator Bacteria

Predator bacteria were discovered accidentally when scientists discovered bacteriophages in the 1960s (Stolp and Petzold 1962). *Bdellovibrio* strains are found in different ecosystems worldwide. All predator strains, called *Bdellovibrio*-like organisms (BALOs), are Gram-negative bacteria that consume other Gram-negative strains. BALOs can be divided into two groups according to their ability to attack Gram-negative bacterial prey.

First, epibiotic predators (e.g., *Bdellovibrio exovorus* and *Micavibrio aeruginosavorus*) adhere to the outer membrane of prey and then ingest all the nutrients in the prey cell. After that, epibiotic predators are divided by binary fission and remain in the extracellular habitat. Another predator, known as a periplasmic predator (e.g., *B. bacteriovorus*), attaches to the outer membrane and invades the periplasmic space, subsequently consuming prey nutrients and reproducing themselves. For example, prey cells are lysed, and their progeny are subsequently released into the extracellular environment (Jurkevitch 2014; Madhusoodanan 2019). An in vitro study of predator bacteria revealed thc potential ability of these bacteria to act against multidrug-resistant Gram-negative bacteria such as *E. coli, K. pneumoniae,* and *P. aeruginosa* (Kadouri et al. 2013). Thus, the theory of biological control using predator bacteria was proposed. Many in vitro and in vivo studies have been performed to investigate the pathogen elimination ability and host safety of predator bacteria. Previous results revealed that *Bdellovibrio* was nontoxic to all animal models (Bonfiglio et al. 2020; Atterbury and Tyson 2021). In an in vitro study in cattle, *B. bacteriovorus* significantly reduced the number of antimicrobial-resistant *Salmonella* strains in fecal samples; however, there was no remarkable reduction in *E. coli* O157:H7 in either the rumen fluid or the fecal samples (Page et al. 2015). In addition, *B. bacteriovorus* 109J was used to kill *Moraxella bovis,* which causes infectious bovine keratoconjunctivitis, in an in vitro model. Treatment with *B. bacteriovorus* decreased *M. bovis* adherence to Madin-Darby bovine kidney cells by sixfold (Boileau, Clinkenbeard, and Iandolo 2011). This study provides potential ideas for manipulating *M. bovis* corneal infection via effective biological control. Consequently, the efficacy of *B. bacteriovorus* 109J was evaluated in calves to examine corneal lesions and maintain living cells of predator bacteria in tears and ocular discharge after challenge with *M. bovis*. Nonetheless, the results demonstrated that *B. bacteriovorus* 109J was not successful as an alternative antibiotic for keratoconjunctivitis treatment (Boileau et al. 2016). Given the inconsistent response in animal models, further research is needed to support the efficacy of *Bdellovibrio* application in animals.

1.3.3 Alternatives for Disease Resistance Control

Genome engineering technologies have been developed for several decades. Transgenic farm animals are purposed to defend against serious illness. Furthermore, the general methods of genome editing–based genomic alterations in ruminants include zinc-finger nucleases, translation activator-like effector nucleases, and clustered regulatory interspaced short palindromic repeats (CRISPR). The genes of transgenic ruminants, such as those of tuberculosis, Johne's disease, brucellosis, and mastitis, were edited for disease resistance.

Tuberculosis is a zoonotic disease found worldwide in cattle and buffalo. Tuberculosis can be transmitted from ruminant to human and from human to human. Indeed, *Mycobacterium bovis* is a common pathogen of bovine tuberculosis and infects humans via airborne transmission, direct contact with infected animals, and consumption of contaminated animal products. Tuberculosis has remained widespread without efficient control or protocols to eliminate the disease. In support of reports of bovine tuberculosis prevalence in southern Italy, positive identification of infection in dairy heads represents a threat to the economy and human health (Abbate et al. 2020). The new era of genetic modulation is focused on bacterial resistance genes, which regulate the immunologic response. The translation activator-like effector nucleases technique was applied to Holstein-Friesian cattle by inserting the mouse SP110 gene into their genome. This strategy aimed to prevent *M. bovis* infection. Transgenic cattle in which the SP110 gene was edited demonstrated that this genetic modification suppressed the proliferation of *M. bovis* via a macrophage mechanism. The infected macrophages of transgenic cattle respond to apoptosis rather than necrotic activity. Moreover, transgenic cattle offspring presented heterogeneous SP110 genes from their mothers (Wu et al. 2015). In addition, a specific gene-targeting method, the CRISPR system, was developed for genetic engineering. This system reduced off-target effects in transgenic animals. For single CRISPR-associated protein 9 (CAS9), the natural resistance-associated macrophage protein-1 (NRAMP1) gene was inserted into bovine fetal fibroblasts, and these transgenic cattle could increase resistance to bovine tuberculosis (Gao et al. 2017). Moreover, the CRISPR-CAS9 insertion technique and homology-mediated end-joining based approach for DNA integration improved the capacity of NRAMP1 transgenic cattle to resist *M. bovis* infection without off-target modifications. Additionally, gene-edited cattle not only exhibit tuberculosis resistance but also sustainably present higher rates of pregnancy and birth (Yuan et al. 2021). Therefore, the transgenic method in cattle is an optional procedure to diminish and control global tubterculosis.

One disease on dairy farms is Johne's disease, which may cause economic loss. It is a chronic enteritis in cattle. The infection of Johne's disease is caused by *Mycobacterium avium* subsp. *paratuberculosis,* which is a highly contagious pathogen. The candidate gene related to the disease phenotype of Johne's disease has been revealed in several studies (Mucha et al. 2009; Verschoor et al. 2010; Sharma et al. 2015). The CRISPR-CAS9 technique was applied to knock out single-nucleotide polymorphisms of the interleukin-10 receptor alpha gene (IL10RA) to confirm the immunologic function of IL10RA. IL10RA plays an important role in anti-inflammatory activity and responsiveness to *M. avium* subsp. *paratuberculosis* lysate (Mallikarjunappa et al. 2020). This knowledge of genetic approaches is a future opportunity for developing effective control methods in the dairy industry.

Brucellosis, caused by *Brucella* spp., is distributed worldwide. The common zoonotic *Brucella* is *Brucella melitensis,* which infects mainly ruminants, especially sheep and goats (von Bargen, Gorvel, and Salcedo 2012). Brucellosis affects the reproductive system of animals, inducing infertility in males and abortion in pregnant animals. Brucellosis may be transmitted via both direct contact with infected animals and indirect exposure to contaminated foods. Currently, vaccination is the only procedure used to protect livestock from infection. However, brucellosis has not been eliminated from parts of Asia, Africa, America, or Mediterranean countries. This disease has subsequently attracted both global public health and economic interest for livestock holders because of its high abortion rate (Gomez et al. 2013). Additionally, *Brucella* has the ability to maintain virulence for several months outside the host's cell and in the environment. In the immunologic approach, this microorganism can escape phagocytosis by macrophages, in which bacteria multiply. An alternative therapeutic focused on gene therapy involving the modification of ovine macrophage cells via CRISPR-CAS9 technology to inactivate genes associated with *Brucella* cell replication. Gene therapy has been shown to have a bactericidal effect on *Brucella* number in ovine macrophages (Karponi et al. 2019). Nevertheless, gene therapy has only been investigated in vitro, and further in vivo experiments against zoonotic diseases are needed.

Mastitis is an inflammatory response of the mammary gland caused by intramammary infection. Bacteria are common causes of disease, and the main treatment protocol relies on antibiotic usage. Consequently, the substantial use of antibiotics on farms is considered to increase the prevalence of antibiotic-resistant microbes (Cheng and Han 2020). The antimicrobial properties of human lysozyme, which is found in human milk, include the lysis of bacterial cell walls. The transgenic goats were generated by adding DNA constructed from human lysozyme (HLZ) via microinjection (Maga, Shoemaker, et al. 2006). Transgenic goat milk decreased bacterial growth in both in vitro and in vivo studies. The transgenic-HZL milk suppressed the growth of *E. coli* and *S. aureus*. In addition, the lysozyme in transgenic milk acts as a preservative agent in cold milk storage because of the slow proliferation of *Pseudomonas fragi* (Maga, Cullor, et al. 2006). Knockin-HZL was applied to transgenic cattle to produce a biopharmaceutical within the mammary gland (Yang et al. 2011). The transgenic-HZL cows with zinc-finger nucleases inserted at the bovine β-casein locus subsequently presented with human lysozyme in their milk and exhibited bactericidal effects on Gram-positive bacteria such as *S. aureus* (Liu et al. 2014). However, the antimicrobial ability of transgenic-HZL milk to kill Gram-negative bacteria (e.g., *E. coli*) was lower than that of transgenic cows harboring recombinant human lactoferrin (Hyvönen et al. 2006). Hence, genetic engineering can provide a valuable procedure for enhancing mastitis resistance and animal welfare on dairy farms without altering milk composition.

Since the introduction of genetic engineering to the scientific sector, the number of studies has increased progressively. The ethical problem of animal welfare has become a matter of public concern. There are three principles guiding the use of animals in scientific research (three *R*s): animal number reduction, refinement of husbandry practices, and animal replacement. In addition, the survival rate of the transgenic embryos ranged between 1% and 30%, indicating that many animals were needed. Furthermore, duplicate experiments are the consequence of unshared information in the scientific community regarding new species generation through genetic engineering. Therefore, the demand for a large number of animals opposes the concept of reduction. To avoid the entry of transgenic carcasses into the ecological system, Canada has restrictions regarding the disposal and euthanasia of genetically modified animals (Ormandy, Dale, and Griffin 2011). Another ethical challenge associated with transgenic animals is refinement, which is described as the pain and suffering of the animals. To develop new genetically engineered animals, invasive procedures such as inducing the superovulation of donors via intraperitoneal or subcutaneous administration, surgically implanting transgenic embryos into recipients, performing vasectomy on male animals, and genotyping genetically engineered offspring are necessary. In addition, the insertion or deletion of a gene may be detrimental to an animal and result in animal mutation. In addition, genetic knockouts may impair the metabolism, growth, and behavior of animals. Accordingly, researchers should minimize the adverse effects of genetically modified animals (Resnik 2011). Alternative considerations to reduce the number of animals in an experiment were introduced as part of the 4*R*s. The procedure included reduction, refinement, replacement, and responsibility. As society develops agreement on the acceptability of genetically engineered animals, there must also be efficient and appropriate communication among the scientific community, industry, and government (Mahatthanadull and Thanaboripat 2021).

Overall, genome engineering technology is an alternative method for controlling and eradicating detrimental zoonotic diseases. Moreover, the gene editing method is also a possible way to improve animal health and animal welfare in livestock. Nevertheless, these applications potentially accelerate the genetic engineering of animal products and address ethical issues.

1.4 CONCLUSION AND FUTURE PROSPECTS

People have become increasingly aware of food safety and disease outbreaks. Consequently, there was a considerable correlation between human antimicrobial resistance and livestock productivity. The development of antimicrobial alternatives for both humans and animals has risen. Furthermore, alternatives as feed additives that promote animal growth effectively include probiotics, prebiotics, antimicrobial peptides, and nonnutritive phytochemicals. Due to their enzymatic activity, spore-forming bacteria, such as *Bacillus* spp., seem beneficial for growth performance and nutrient digestibility in ruminants. In addition, a combination of lactic acid bacteria and prebiotics is intended to be administered to immature animals to improve rumen health and stimulate villi growth. Currently, antimicrobial peptides and nonnutritive phytochemicals have been hot topics of discussion. Additionally, both of them show efficacy in bactericidal activity against pathogens. However, the antimicrobial peptide characterization, manufacturing, and animal application require further investigation. Correspondingly, research on nonnutritive phytochemicals is ongoing to determine the active component, mechanism, and maintenance of their potency. Vaccines may replace prophylactic antibiotics for preventive purposes. Mastitis vaccines may be used in this approach to prevent intramammary infection and minimize the use of intramammary antibiotics in dry cows. Even though the mastitis vaccine lessens the inflammatory response, it cannot inhibit the milk production decline that causes economic loss. Moreover, vaccination against caprine pleuropneumonia is promising. New-generation vaccines, such as subunit and recombinant vaccines, efficiently stimulate the immune response. However, these vaccines need more research in large populations. The alternatives to infectious treatment consisted of bacteriophage, endolysin, and predator bacteria. Bacteriophage and phage endolysin are capable of lysing bacterial cells rapidly with a high degree of target specificity. The most interesting use is in the reduction of *E. coli* shedding, which has a positive influence on food safety and public health. Conversely, predator bacteria show inconsistent effects when applied to animal models. Some infections are difficult to cure with medicine, and the disease persists in ruminant herds. Subsequently, genetically engineered animals are designed to resist crucial diseases and control zoonotic diseases. Even though transgenic ruminants might be resistant to tuberculosis, there are ethical concerns about the welfare of livestock animals.

REFERENCES

Aarestrup, FM, VF Jensen, HD Emborg, E Jacobsen, and HC Wegener. 2010. "Changes in the use of antimicrobials and the effects on productivity of swine farms in Denmark." *Am J Vet Res* 71 (7):726–733. https://doi.org/10.2460/ajvr.71.7.726.

Aarestrup, FM, HC Wegener, and P Collignon. 2008. "Resistance in bacteria of the food chain: Epidemiology and control strategies." *Expert Rev Anti Infect Ther* 6 (5):733–750. https://doi.org/10.1586/14787210.6.5.733.

Abbate, JM, F Arfuso, C Iaria, G Arestia, and G Lanteri. 2020. "Prevalence of bovine tuberculosis in slaughtered cattle in Sicily, southern Italy." *Animals* 10 (9). https://doi.org/10.3390/ani10091473.

Abdalhamed, AM, AA Ghazy, and GSG Zeedan. 2021. "Studies on multidrug-resistance bacteria in ruminants with special interest on antimicrobial resistances genes." *Adv Anim Vet Sci* 9 (6):835–844.

Adams, MC, J Luo, D Rayward, S King, R Gibson, and GH Moghaddam. 2008. "Selection of a novel direct-fed microbial to enhance weight gain in intensively reared calves." *Anim Feed Sci Technol* 145 (1–4):41–52.

Alhaji, NB, and TO Isola. 2018. "Antimicrobial usage by pastoralists in food animals in North-central Nigeria: The associated socio-cultural drivers for antimicrobials misuse and public health implications." *One Health* 6:41–47. https://doi.org/10.1016/j.onehlt.2018.11.001.

Allel, K, L Day, A Hamilton, L Lin, L Furuya-Kanamori, CE Moore, T Van Boeckel, R Laxminarayan, and L Yakob. 2023. "Global antimicrobial-resistance drivers: An ecological country-level study at the human & animal interface." *Lancet Planet Health* 7 (4):e291-e303. https://doi.org/10.1016/S2542–5196(23)00026–8.

Ameen, F, SA Reda, SA El-Shatoury, EM Riad, ME Enany, and AA Alarfaj. 2019. "Prevalence of antibiotic resistant mastitis pathogens in dairy cows in Egypt and potential biological control agents produced from plant endophytic actinobacteria." *Saudi J Biol Sci* 26 (7):1492–1498. https://doi.org/10.1016/j.sjbs.2019.09.008.

Anderson, RC, REW Hancock, and P-L Yu. 2004. "Antimicrobial activity and bacterial-membrane interaction of ovine-derived cathelicidins." *Antimicrob Agents Chemother* 48 (2):673–676. https://doi.org/10.1128/AAC.48.2.673–676.2004.

Andrés-Lasheras, S, R Ha, R Zaheer, C Lee, CW Booker, C Dorin, JV Donkersgoed, R Deardon, S Gow, SJ Hannon, S Hendrick, M Anholt, and TA. McAllister. 2021. "Prevalence and risk factors associated with antimicrobial resistance in bacteria related to bovine respiratory disease—A broad cross-sectional study of beef cattle at entry into Canadian feedlots." *Front Vet Sci* 8 (710). https://doi.org/10.3389/fvets.2021.692646.

Andri, F, AN Huda, and M Marjuki. 2020. "The use of essential oils as a growth promoter for small ruminants: A systematic review and meta-analysis." *F1000Res* 9 (486). https://doi.org/10.12688/f1000research.24123.2.

Asanuma, N, and T Hino. 2002. "Regulation of fermentation in a ruminal bacterium, Streptococcus bovis, with special reference to rumen acidosis." *Anim Sci J* 73 (5):313–325. https://doi.org/10.1046/j.1344–3941.2002.00044.x.

Atterbury, RJ, and J Tyson. 2021. "Predatory bacteria as living antibiotics – Where are we now?" *Microbiology* 167 (1). https://doi.org/10.1099/mic.0.001025.

Badger, SM, KF Sullivan, D Jordan, C Caraguel, SW Page, P Cusack, D Frith, and DJ Trott. 2020. "Antimicrobial use and stewardship practices on Australian beef feedlots." *Aust Vet J* 98 (1–2):37–47. https://doi.org/10.1111/avj.12889.

Boileau, MJ, KD Clinkenbeard, and JJ Iandolo. 2011. "Assessment of Bdellovibrio bacteriovorus 109J killing of Moraxella bovis in an in vitro model of infectious bovine keratoconjunctivitis." *Can J Vet Res* 75 (4):285–291.

Boileau, MJ, R Mani, MA Breshears, M Gilmour, JD Taylor, and KD Clinkenbeard. 2016. "Efficacy of Bdellovibrio bacteriovorus 109J for the treatment of dairy calves with experimentally induced infectious bovine keratoconjunctivitis." *Am J Anim Vet Sci* 77 (9):1017–1028. https://doi.org/10.2460/ajvr.77.9.1017.

Bonfiglio, G, B Neroni, G Radocchia, M Marazzato, F Pantanella, and S Schippa. 2020. "Insight into the possible use of the predator Bdellovibrio bacteriovorus as a probiotic." *Nutrients* 12 (8):2252. https://doi.org/10.3390/nu12082252.

Brashears, MM, A Amezquita, and D Jaroni. 2005. "Lactic acid bacteria and their uses in animal feeding to improve food safety." In *Advances in Food and Nutrition Research*, 1–31. Academic Press.

Brogden, KA, M Ackermann, PB McCray, and BF Tack. 2003. "Antimicrobial peptides in animals and their role in host defences." *Int J Antimicrob Agents* 22 (5):465–478. https://doi.org/10.1016/s0924–8579(03)00180–8.

Brown, K, RRE Uwiera, ML Kalmokoff, SPJ Brooks, and GD Inglis. 2017. "Antimicrobial growth promoter use in livestock: A requirement to understand their modes of action to develop effective alternatives." *Int J Antimicrob Agents* 49 (1):12–24. https://doi.org/10.1016/j.ijantimicag.2016.08.006.

Buck, JM. 1930. "Studies of vaccination during calfhood to prevent bovine infectious abortion." *J Agric Res* 41 (9):667–689.

Cai, L, J Yu, R Hartanto, and D Qi. 2021. "Dietary supplementation with Saccharomyces cerevisiae, Clostridium butyricum and their combination ameliorate rumen fermentation and growth performance of heat-stressed goats." *Animals* 11 (7):2116.

Call, DR, MA Davis, and AA Sawant. 2008. "Antimicrobial resistance in beef and dairy cattle production." *Anim Health Res Rev* 9 (2):159–167. https://doi.org/10.1017/s1466252308001515.

Callaway, TR, TS Edrington, AD Brabban, RC Anderson, ML Rossman, MJ Engler, MA Carr, KJ Genovese, JE Keen, ML Looper, EM Kutter, and DJ Nisbet. 2008. "Bacteriophage isolated from feedlot cattle can reduce *Escherichia coli* O157:H7 populations in ruminant gastrointestinal tracts." *Foodborne Pathog Dis* 5 (2):183–191. https://doi.org/10.1089/fpd.2007.0057.

Cameron, A, and TA McAllister. 2016. "Antimicrobial usage and resistance in beef production." *J Anim Sci Biotechnol* 7:68–68. https://doi.org/10.1186/s40104-016-0127-3.

Chaucheyras-Durand, F, E Chevaux, C Martin, and E Forano. 2012. "Use of Yeast Probiotics in Ruminants: Effects and Mechanisms of Action on Rumen pH, Fibre Degradation, and Microbiota According to the Diet." In *Probiotic in Animals*, 119–152. IntechOpen. https://doi.org/10.5772/50192.

Cheng, WN, and SG Han. 2020. "Bovine mastitis: Risk factors, therapeutic strategies, and alternative treatments - A review." *Asian-Australas J Anim Sci* 33 (11):1699–1713. https://doi.org/10.5713/ajas.20.0156.

Choonkham, W, AAH van Ruitenbeek, JT Schonewille, and W Suriyasathaporn. 2021. "Preliminary assessment of probiotic Bacillus subtilis C-3102 in feces: Evaluation of their survival after oral supplementation in goats." *Vet Integr Sci* 19 (2):153–159.

Cromwell, GL. 2002. "Why and how antibiotics are used in swine production." *Anim Biotechnol* 13 (1):7–27. https://doi.org/10.1081/ABIO–120005767.

Cusack, PMV. 2024. "Alternatives to conventional antibiotics for the prevention and treatment of commonly occurring diseases in feedlot cattle." *Aust Vet J* 102 (5):229–241. https://doi.org/10.1111/avj.13314.

Dagar, SS, S Kumar, P Mudgil, R Singh, and AK Puniya. 2011. "D1/D2 domain of large-subunit ribosomal DNA for differentiation of Orpinomyces spp." *Appl Environ Microbiol* 77 (18):6722–6725. https://doi.org/10.1128/AEM.05441–11.

Dai, X, and AP Faciola. 2019. "Evaluating strategies to reduce ruminal protozoa and their impacts on nutrient utilization and animal performance in ruminants – A meta-analysis." *Front Microbiol* 10 (2648). https://doi.org/10.3389/fmicb.2019.02648.

Das, H, SU Ahmed, SK Shukla, S Shukla, A Latif, and D Sharma. 2010. "Two β defensin cationic peptides from mastitic milk of *Bubalus bubalis*." *Asian J Anim Sci* 4 (1):1–12.

Das, H, N Swamy, G Sahoo, SU Ahmed, and T More. 2008. "β-defensin antibiotic peptides in the innate immunity of the buffalo: In vivo and in vitro studies." *Altern Lab Anim* 36 (4):429–440. https://doi.org/10.1177/026119290803600404.

Davis, MA, DD Hancock, TE Besser, JB Daniels, KN Baker, and DR Call. 2007. "Antimicrobial resistance in *Salmonella enterica* serovar Dublin isolates from beef and dairy sources." *Vet Microbiol* 119 (2–4):221–230. https://doi.org/10.1016/j.vetmic.2006.08.028.

DeClerck, JC, ZE Wade, NR Reeves, MF Miller, BJ Johnson, GA Ducharme, and RJ Rathmann. 2020. "Influence of *Megasphaera elsdenii* and feeding strategies on feedlot performance, compositional growth, and carcass parameters of early weaned, beef calves." *Transl Anim Sci* 4 (2):863–875. https://doi.org/10.1093/tas/txaa031.

Doidge, C, H West, and J Kaler. 2021. "Antimicrobial resistance patterns of *Escherichia coli* isolated from sheep and beef farms in England and Wales: A comparison of disk diffusion interpretation methods." *Antibiotics (Basel)* 10 (4). https://doi.org/10.3390/antibiotics10040453.

Donovan, DM, M Lardeo, and J Foster-Frey. 2006. "Lysis of staphylococcal mastitis pathogens by bacteriophage phi11 endolysin." *FEMS Microbiol Lett* 265 (1):133–139. https://doi.org/10.1111/j.1574–6968.2006.00483.x.

Dorantes-Iturbide, G, JF Orzuna-Orzuna, A Lara-Bueno, LA Miranda-Romero, GD Mendoza-Martínez, and PA Hernández-García. 2022. "Effects of a polyherbal dietary additive on performance, dietary energetics, carcass traits, and blood metabolites of finishing lambs." *Metabolites* 12 (5):413. https://doi.org/10.3390/metabo12050413.

Dorneles, EMS, N Sriranganathan, and AP Lage. 2015. "Recent advances in *Brucella abortus* vaccines." *Vet Res* 46 (1):76. https://doi.org/10.1186/s13567-015-0199-7.

Doyle, N, P Mbandlwa, WJ Kelly, G Attwood, Y Li, RP Ross, C Stanton, and S Leahy. 2019. "Use of lactic acid bacteria to reduce methane production in ruminants, a critical review." *Front Microbiol* 10 (2207). https://doi.org/10.3389/fmicb.2019.02207.

Elghandour, MMY, AZM Salem, JSM Castañeda, LM Camacho, AE Kholif, and JC Vázquez Chagoyán. 2015. "Direct-fed microbes: A tool for improving the utilization of low quality roughages in ruminants." *J Integr Agric* 14 (3):526–533. https://doi.org/10.1016/S2095–3119(14)60834–0.

Emelianova, AA, DV Kuzmin, PV Panteleev, M Sorokin, AA Buzdin, and TV Ovchinnikova. 2018. "Anticancer activity of the goat antimicrobial peptide ChMAP–28." *Front Pharmacol* 9 (1501). https://doi.org/10.3389/fphar.2018.01501.

Enne, VI, C Cassar, K Sprigings, MJ Woodward, and PM Bennett. 2008. "A high prevalence of antimicrobial resistant *Escherichia coli* isolated from pigs and a low prevalence of antimicrobial resistant *E. coli* from cattle and sheep in Great Britain at slaughter." *FEMS Microbiol Lett* 278 (2):193–199. https://doi.org/10.1111/j.1574–6968.2007.00991.x.

Freick, M, Y Frank, K Steinert, A Hamedy, O Passarge, and A Sobiraj. 2016. "Mastitis vaccination using a commercial polyvalent vaccine or a herd-specific *Staphylococcus aureus* vaccine." *Tierarztl Prax Ausg G Grosstiere Nutztiere* 44 (04):219–229.

Gao, Y, H Wu, Y Wang, X Liu, L Chen, Q Li, C Cui, X Liu, J Zhang, and Y Zhang. 2017. "Single Cas9 nickase induced generation of NRAMP1 knockin cattle with reduced off-target effects." *Genome Biol* 18 (1):13. https://doi.org/10.1186/s13059-016-1144-4.

Gibson, GR, R Hutkins, ME Sanders, SL Prescott, RA Reimer, SJ Salminen, K Scott, C Stanton, KS Swanson, PD Cani, K Verbeke, and G Reid. 2017. "Expert consensus document: The International Scientific Association for Probiotics and Prebiotics (ISAPP) consensus statement on the definition and scope of prebiotics." *Nat Rev Gastroenterol Hepatol* 14 (8):491–502. https://doi.org/10.1038/nrgastro.2017.75.

Gill, JJ, JC Pacan, ME Carson, KE Leslie, MW Griffiths, and PM Sabour. 2006. "Efficacy and pharmacokinetics of bacteriophage therapy in treatment of subclinical *Staphylococcus aureus* mastitis in lactating dairy cattle." *Antimicrob Agents Chemother* 50 (9):2912–2918. https://doi.org/10.1128/aac.01630–05.

Gomez, G, J Pei, W Mwangi, LG Adams, A Rice-Ficht, and TA Ficht. 2013. "Immunogenic and invasive properties of *Brucella melitensis* 16M outer membrane protein vaccine candidates identified via a reverse vaccinology approach." *PLoS One* 8 (3):e59751. https://doi.org/10.1371/journal.pone.0059751.

González-Ríos, H, JL Dávila-Ramírez, EA Peña-Ramos, M Valenzuela-Melendres, L Zamorano-García, TY Islava-Lagarda, and NV Valenzuela-Grijalva. 2016. "Dietary supplementation of ferulic acid to steers under commercial feedlot feeding conditions improves meat quality and shelf life." *Anim Feed Sci Technol* 222:111–121. https://doi.org/10.1016/j.anifeedsci.2016.10.011.

Grand, E, F Respondek, C Martineau, J Detilleux, and G Bertrand. 2013. "Effects of short-chain fructooligosaccharides on growth performance of preruminant veal calves." *J Dairy Sci* 96 (2):1094–1101. https://doi.org/10.3168/jds.2011–4949.

Guo, Y, M Xun, and J Han. 2018. "A bovine myeloid antimicrobial peptide (BMAP-28) and its analogs kill pan-drug-resistant *Acinetobacter baumannii* by interacting with outer membrane protein A (OmpA)." *Medicine (Baltimore)* 97 (42):e12832. https://doi.org/10.1097/md.0000000000012832.

Gutiérrez, D, L Fernández, A Rodríguez, and P García. 2019. "Role of bacteriophages in the implementation of a sustainable dairy chain." *Front Microbiol* 10 (12). https://doi.org/10.3389/fmicb.2019.00012.

Gutiérrez, D, V Garrido, L Fernández, S Portilla, A Rodríguez, MJ Grilló, and P García. 2020. "Phage lytic protein LysRODI prevents staphylococcal Mastitis in Mice." *Front Microbiol* 11 (7). https://doi.org/10.3389/fmicb.2020.00007.

Hannon, SJ, SA Brault, SJG Otto, PS Morley, TA McAllister, Calvin W. Booker, and Sheryl P. Gow. 2020. "Feedlot cattle antimicrobial use surveillance network: A Canadian journey." *Front Vet Sci* 7 (968). https://doi.org/10.3389/fvets.2020.596042.

Hess, M, SS Paul, AK Puniya, M van der Giezen, C Shaw, JE Edwards, and K Fliegerová. 2020. "Anaerobic fungi: Past, present, and future." *Front Microbiol* 11 (2621). https://doi.org/10.3389/fmicb.2020.584893.

Hoelzer, K, L Bielke, DP Blake, E Cox, SM Cutting, B Devriendt, E Erlacher-Vindel, E Goossens, K Karaca, S Lemiere, M Metzner, M Raicek, MC Suriñach, NM Wong, C Gay, and F Van Immerseel. 2018. "Vaccines as alternatives to antibiotics for food producing animals. Part 1: Challenges and needs." *Vet Res* 49 (1):64. https://doi.org/10.1186/s13567-018-0560-8.

Hotel, ACP, and A Cordoba. 2001. "Health and nutritional properties of probiotics in food including powder milk with live lactic acid bacteria." *Prevention* 5 (1):1–10.

Hove, P, S Madesh, A Nair, D Jaworski, H Liu, J Ferm, MD Kleinhenz, MA Highland, AK Curtis, JF Coetzee, SM Noh, Y Wang, D Genda, and RR Ganta. 2022. "Targeted mutagenesis in *Anaplasma marginale* to define virulence and vaccine development against bovine anaplasmosis." *PLoS Pathog* 18 (5):e1010540. https://doi.org/10.1371/journal.ppat.1010540.

Hyvönen, P, L Suojala, T Orro, J Haaranen, O Simola, C Røntved, and S Pyörälä. 2006. "Transgenic cows that produce recombinant human lactoferrin in milk are not protected from experimental *Escherichia coli* intramammary infection." *Infect Immun* 74 (11):6206–6212. https://doi.org/10.1128/iai.00238–06.

Ismail, ZB. 2017. "Mastitis vaccines in dairy cows: Recent developments and recommendations of application." *Vet World* 10 (9):1057–1062. https://doi.org/10.14202/vetworld.2017.1057–1062.

Isobe, N, K Hosoda, and Y Yoshimura. 2009. "Immunolocalization of lingual antimicrobial peptide (LAP) in the bovine mammary gland." *Anim Sci J* 80 (4):446–450. https://doi.org/10.1111/j.1740–0929.2009.00652.x.

Jia, P, K Cui, T Ma, F Wan, W Wang, D Yang, Y Wang, B Guo, L Zhao, and Q Diao. 2018. "Influence of dietary supplementation with *Bacillus licheniformis* and *Saccharomyces cerevisiae* as alternatives to monensin on growth performance, antioxidant, immunity, ruminal fermentation and microbial diversity of fattening lambs." *Sci Rep* 8 (1):16712. https://doi.org/10.1038/s41598-018-35081-4.

Kabir, A, AH Mirani, M Rasheed, DK Bhuptani, HA Kaleri, R Kumar, MB Abro, M Khatri, ZH Bozdar, and AW Solangi. 2021. A review on contagious caprine Pleuro-pneumonia disease in small ruminants; Sheep and goats. *Pure Appl Biol* 11 (2):531–542.

Kadouri, DE, K To, RM Shanks, and Y Doi. 2013. "Predatory bacteria: A potential ally against multidrug-resistant Gram-negative pathogens." *PLoS One* 8 (5):e63397. https://doi.org/10.1371/journal.pone.0063397.

Kalita, DJ, A Kumar, and S Kumar. 2009. "Structure-function studies of Bubalus bubalis lingual antimicrobial peptide analogs." *Vet Res Commun* 33 (2):149–161. https://doi.org/10.1007/s11259-008-9081-7.

Kapoor, G, S Saigal, and A Elongavan. 2017. "Action and resistance mechanisms of antibiotics: A guide for clinicians." *J Anaesthesiol Clin Pharmacol* 33 (3):300–305. https://doi.org/10.4103/joacp.JOACP_349_15.

Kara, C, H Cihan, M Temizel, S Catik, Y Meral, A Orman, A Yibar, and H Gencoglu. 2015. "Effects of supplemental Mannanoligosaccharides on growth performance, faecal characteristics and health in dairy calves." *Asian-Australas J Anim Sci* 28 (11):1599–1605. https://doi.org/10.5713/ajas.15.0124.

Karponi, G, SK Kritas, G Papadopoulou, E-K Akrioti, E Papanikolaou, and E Petridou. 2019. "Development of a CRISPR/Cas9 system against ruminant animal brucellosis." *BMC Vet Res* 15 (1):422–422. https://doi.org/10.1186/s12917-019-2179-z.

Khan, RU, N Shabana, D Kuldeep, K Karthik, T Ruchi, MM Abdelrahman, IA Alhidary, and Z Arshad. 2016. "Direct-fed microbial: Beneficial applications, modes of action and prospects as a safe tool for enhancing ruminant production and safeguarding health." *Int J Pharm* 12 (3):220–231.

Kholif, AE, and OA Olafadehan. 2021. "Essential oils and phytogenic feed additives in ruminant diet: Chemistry, ruminal microbiota and fermentation, feed utilization and productive performance." *Phytochem Rev* 20 (6):1087–1108. https://doi.org/10.1007/s11101-021-09739-3.

Kim, H, HG Lee, I Kwon, and J Seo. 2020. "Characterization of Endolysin LyJH307 with Antimicrobial activity against *Streptococcus bovis*." *Animals: An Open Access Journal from MDPI* 10 (6):963. https://doi.org/10.3390/ani10060963.

Koh, SH, and FH Morley. 1981. "The effect of calfhood vaccination with strain 19 on the serological diagnosis and eradication of bovine brucellosis." *Aust Vet J* 57 (12):551–553. https://doi.org/10.1111/j.1751–0813.1981.tb00432.x.

Kong, F, N Lu, Y Liu, S Zhang, H Jiang, H Wang, W Wang, and S Li. 2021. "*Aspergillus oryzae* and *Aspergillus niger* co-cultivation extract affects In Vitro degradation, fermentation characteristics, and bacterial composition in a diet-specific manner." *Animals: An Open Access Journal from MDPI* 11 (5):1248. https://doi.org/10.3390/ani11051248.

Kuipers, A, WJ Koops, and H Wemmenhove. 2016. "Antibiotic use in dairy herds in the Netherlands from 2005 to 2012." *J Dairy Sci* 99 (2):1632–1648. https://doi.org/10.3168/jds.2014–8428.

Kumar, R, SA Ali, SK Singh, V Bhushan, M Mathur, S Jamwal, AK Mohanty, JK Kaushik, and S Kumar. 2020. "Antimicrobial peptides in farm animals: An updated review on its diversity, function, modes of action and therapeutic prospects." *Vet Sci* 7 (4):206.

Kyriakis, CS. 2015. "Tomorrow's vector vaccines for small ruminants." *Vet Microbiol* 181 (1–2):47–52. https://doi.org/10.1016/j.vetmic.2015.08.003.

Landers, TF, B Cohen, TE Wittum, and EL Larson. 2012. "A review of antibiotic use in food animals: Perspective, policy, and potential." *Public Health Rep* 127 (1):4–22. https://doi.org/10.1177/003335491212700103.

Lee, SS, JK Ha, and K Cheng. 2000. "Relative contributions of bacteria, protozoa, and fungi to in vitro degradation of orchard grass cell walls and their interactions." *Appl Environ Microbiol* 66 (9):3807–3813. https://doi.org/10.1128/aem.66.9.3807-3813.2000.

Leite, JA, HP Pereira, CAV Borges, BRC Alves, AIAP Ramos, MF Martins, and EF Arcuri. 2019. "Lytic bacteriophages as a potential alternative to control *Staphylococcus aureus*." *Pesqui Agropecu Bras* 54.

Liu, C, Y Qu, P Guo, C Xu, Y Ma, and H Luo. 2018. "Effects of dietary supplementation with alfalfa (*Medicago sativa* L.) saponins on lamb growth performance, nutrient digestibility, and plasma parameters." *Anim Feed Sci Technol* 236:98–106.

Liu, Q, S Yao, Y Chen, S Gao, Y Yang, J Deng, Z Ren, L Shen, H Cui, Y Hu, X Ma, and S Yu. 2017. "Use of antimicrobial peptides as a feed additive for juvenile goats." *Sci Rep* 7 (1):12254. https://doi.org/10.1038/s41598-017-12394-4.

Liu, X, Y Wang, Y Tian, Y Yu, M Gao, G Hu, F Su, S Pan, Y Luo, Z Guo, F Quan, and Y Zhang. 2014. "Generation of mastitis resistance in cows by targeting human lysozyme gene to β-casein locus using zinc-finger nucleases." *Proc R Soc Lond B Biol Sci* 281 (1780):20133368. https://doi.org/10.1098/rspb.2013.3368.

Loessner, MJ. 2005. "Bacteriophage endolysins–current state of research and applications." *Curr Opin Microbiol* 8 (4):480–487. https://doi.org/10.1016/j.mib.2005.06.002.

Maamouri, O, and M Ben Salem. 2021. "The effect of live yeast *Saccharomyces cerevisiae* as probiotic supply on growth performance, feed intake, ruminal pH, and fermentation in fattening calves." *Vet Med and Sci* 8:631. https://doi.org/10.1002/vms3.631.

Madhusoodanan, J. 2019. "Inner workings: Probing predatory bacteria as an antibacterial remedy." *Proc Natl Acad Sci* 116 (46):22887–22890. https://doi.org/10.1073/pnas.1917513116.

Maga, EA, JS Cullor, W Smith, GB Anderson, and JD Murray. 2006. "Human lysozyme expressed in the mammary gland of transgenic dairy goats can inhibit the growth of bacteria that cause mastitis and the cold-spoilage of milk." *Foodborne Pathog Dis* 3 (4):384–392. https://doi.org/10.1089/fpd.2006.3.384.

Maga, EA, CF Shoemaker, JD Rowe, RH BonDurant, GB Anderson, and JD Murray. 2006. "Production and processing of milk from transgenic goats expressing human lysozyme in the mammary gland." *J Dairy Sci* 89 (2):518–524. https://doi.org/10.3168/jds.S0022-0302(06)72114-2.

Mahatthanadull, S, and D Thanaboripat. 2021. "Animal biotechnology and ethical issues." *Curr Appl Sci Technol* 408–417.

Mainda, G, PR Bessell, JB Muma, SP McAteer, ME Chase-Topping, J Gibbons, MP Stevens, DL Gally, and BM deC. Bronsvoort. 2015. "Prevalence and patterns of antimicrobial resistance among *Escherichia coli* isolated from Zambian dairy cattle across different production systems." *Sci Rep* 5 (1):12439. https://doi.org/10.1038/srep12439.

Majumder, S, D Jung, J Ronholm, and S George. 2021. "Prevalence and mechanisms of antibiotic resistance in *Escherichia coli* isolated from mastitic dairy cattle in Canada." *BMC Microbiol* 21 (1):222. https://doi.org/10.1186/s12866-021-02280-5.

Mallikarjunappa, S, UK Shandilya, A Sharma, K Lamers, N Bissonnette, NA Karrow, and KG Meade. 2020. "Functional analysis of bovine interleukin-10 receptor alpha in

response to *Mycobacterium avium* subsp. *Paratuberculosis* lysate using CRISPR/Cas9." *BMC Genetics* 21 (1):121. https://doi.org/10.1186/s12863-020-00925-4.
Matthews, L, R Reeve, DL Gally, JC Low, MEJ Woolhouse, SP McAteer, ME Locking, ME Chase-Topping, DT Haydon, LJ Allison, MF Hanson, GJ Gunn, and SWJ Reid. 2013. "Predicting the public health benefit of vaccinating cattle against *Escherichia coli* O157." *Proc Natl Acad Sci* 110 (40):16265–16270. https://doi.org/10.1073/pnas.1304978110.
McMurphy, CP. 2014. "Effects of including saponins (Micro-Aid®) on intake, rumen fermentation and digestibility in steers fed low-quality prairie hay." *Anim Feed Sci Technol* 190:47–58. https://doi.org/10.1016/j.anifeedsci.2014.01.007.
Meeusen, ENT, J Walker, A Peters, P-P Pastoret, and G Jungersen. 2007. "Current status of veterinary vaccines." *Clin Microbiol Rev* 20 (3):489–510. https://doi.org/10.1128/CMR.00005-07.
Miletic, S, M Hünerberg, A Kaldis, J MacDonald, A Leuthreau, T McAllister, and R Menassa. 2017. "A plant-produced candidate subunit vaccine reduces shedding of Enterohemorrhagic *Escherichia coli* in ruminants." *Biotechnol J* 12 (10). https://doi.org/10.1002/biot.201700405.
Mucha, R, MR Bhide, EB Chakurkar, M Novak, and I Mikula. 2009. "Toll-like receptors TLR1, TLR2 and TLR4 gene mutations and natural resistance to *Mycobacterium avium* subsp. *paratuberculosis* infection in cattle." *Vet Immunol Immunopathol* 128 (4):381–388. https://doi.org/10.1016/j.vetimm.2008.12.007.
Mudgal, V, and RPS Baghel. 2010. "Effect of probiotic supplementation on growth performance of pre-ruminant buffalo (*Bubalus bubalis*) calves." *Buffalo Bull* 29 (3):225–228.
Muya, MC, LJ Erasmus, K Miller, C Aperce, FV Nherera, and PM Moshidi. 2017. "Performance of Holstein calves having free access to milk and dosed with *Megasphaera elsdenii*." *Sci Agric* 74:189–194.
Olsen, SC, B Bricker, MV Palmer, AE Jensen, and NF Cheville. 1999. "Responses of cattle to two dosages of *Brucella abortus* strain RB51: Serology, clearance and efficacy." *Res Vet Sci* 66 (2):101–105.
Onime, LA, LB Oyama, BJ Thomas, J Gani, P Alexander, KE Waddams, A Cookson, N Fernandez-Fuentes, CJ Creevey, and SA Huws. 2021. "The rumen eukaryotome is a source of novel antimicrobial peptides with therapeutic potential." *BMC Microbiol* 21 (1):1–13.
Ormandy, EH, J Dale, and G Griffin. 2011. "Genetic engineering of animals: Ethical issues, including welfare concerns." *Can Vet J* 52 (5):544–550.
Orzuna-Orzuna, JF, G Dorantes-Iturbide, A Lara-Bueno, GD Mendoza-Martínez, LA Miranda-Romero, R López-Ordaz, and PA Hernández-García. 2021. "Productive performance, carcass traits, and meat quality in finishing lambs supplemented with a polyherbal mixture." *Agriculture* 11 (10):942.
Page, JA, B Lubbers, J Maher, L Ritsch, and SE Gragg. 2015. "Investigation into the efficacy of *Bdellovibrio bacteriovorus* as a novel preharvest intervention to control *Escherichia coli* O157:H7 and *Salmonella* in cattle using an in vitro model." *J Food Prot* 78 (9):1745–1749. https://doi.org/10.4315/0362–028x.jfp-15–016.
Panicker, VP, and S George. 2013. "Identification of three novel myeloid cathelicidin cDNAs and their predicted peptides in buffalo (*Bubalus bubalis*)." *Indian J Biochem Biophys* 50 (4):273–277.
Panicker, VP, and S George. 2016. "Chemistry and antimicrobial potential of the buffalo myeloid cathelicidin, BuMAP–34." *Int J Pept Res Ther* 22 (3):413–420.
Panteleev, PV, IA Bolosov, AÀ Kalashnikov, VN Kokryakov, OV Shamova, AA Emelianova, SV Balandin, and TV Ovchinnikova. 2018. "Combined antibacterial effects of goat

cathelicidins with different mechanisms of action." *Front Microbiol* 9:2983–2983. https://doi.org/10.3389/fmicb.2018.02983.

Passetti, LCG, RAC Passetti, and TA McAllister. 2021. "Effect of essential oil blends and a nonionic surfactant on rumen fermentation, anti-oxidative status, and growth performance of lambs." *Transl Anim Sci* 5 (3). https://doi.org/10.1093/tas/txab118.

Patra, AK, and J Saxena. 2009. "The effect and mode of action of saponins on the microbial populations and fermentation in the rumen and ruminant production." *Nutr Res Rev* 22 (2):204–219. https://doi.org/10.1017/s0954422409990163.

Peña-Torres, EF, C Castillo-Salas, I Jiménez-Estrada, A Muhlia-Almazán, EA Peña-Ramos, A Pinelli-Saavedra, L Avendaño-Reyes, C Hinojosa-Rodríguez, M Valenzuela-Melendres, U Macías-Cruz, and H González-Ríos. 2022. "Growth performance, carcass traits, muscle fiber characteristics, and skeletal muscle mRNA abundance in hair lambs supplemented with ferulic acid." *J Anim Sci Technol* 64 (1):52–69. https://doi.org/10.5187/jast.2022.e3.

Piepers, S, A Prenafeta, J Verbeke, A De Visscher, R March, and S De Vliegher. 2017. "Immune response after an experimental intramammary challenge with killed *Staphylococcus aureus* in cows and heifers vaccinated and not vaccinated with Startvac, a polyvalent mastitis vaccine." *J Dairy Sci* 100 (1):769–782. https://doi.org/10.3168/jds.2016–11269.

Poester, FP, VSP Gonçalves, TA Paixão, RL Santos, SC Olsen, GG Schurig, and AP Lage. 2006. "Efficacy of strain RB51 vaccine in heifers against experimental brucellosis." *Vaccine* 24 (25):5327–5334. https://doi.org/10.1016/j.vaccine.2006.04.020.

Prapasawat, W, and A Intarapuk. 2021. "Prevalence of antimicrobial resistance and integrons in *Escherichia coli* isolated from feces of dairy goats in nong chok, bangkok, thailand." *Vet Integr Sci* 19:223–236.

Price, LB, BJ Koch, and BA Hungate. 2015. "Ominous projections for global antibiotic use in food-animal production." *Proc Natl Acad Sci U S A* 112 (18):5554–5555. https://doi.org/10.1073/pnas.1505312112.

Puniya, AK, AZM Salem, S Kumar, SS Dagar, GW Griffith, M Puniya, SR Ravella, N Kumar, T Dhewa, and R Kumar. 2015. "Role of live microbial feed supplements with reference to anaerobic fungi in ruminant productivity: A review." *J Integr Agric* 14 (3):550–560. https://doi.org/10.1016/S2095–3119(14)60837–6.

Quijada, NM, R Bodas, JM Lorenzo, S Schmitz-Esser, D Rodríguez-Lázaro, and M Hernández. 2020. "Dietary supplementation with sugar beet fructooligosaccharides and garlic residues promotes growth of beneficial bacteria and increases weight gain in neonatal lambs." *Biomolecules* 10 (8). https://doi.org/10.3390/biom10081179.

Ren, Z, R Yao, Q Liu, Y Deng, L Shen, H Deng, Z Zuo, Y Wang, J Deng, H Cui, Y Hu, X Ma, and J Fang. 2019. "Effects of antibacterial peptides on rumen fermentation function and rumen microorganisms in goats." *PloS One* 14 (8):e0221815–e0221815. https://doi.org/10.1371/journal.pone.0221815.

Resnik, DB. 2011. "Ethical issues concerning transgenic animals in biomedical research." In *The Ethics of Animal Research: Exploring the Controversy*, 169–179. Cambridge, MA: The MIT Press.

Ribeiro, ADB, MVC Ferraz, DM Polizel, AA Miszura, LGM Gobato, JPR Barroso, I Susin, and AV Pires. 2019. "Thyme essential oil for sheep: Effect on rumen fermentation, nutrient digestibility, nitrogen metabolism, and growth." *Arq Bras Med Vet Zootec* 71:2065–2074.

Rima, M, M Rima, Z Fajloun, J-M Sabatier, B Bechinger, and T Naas. 2021. "Antimicrobial peptides: A potent alternative to antibiotics." *Antibiotics* 10 (9):1095.

Saini, V, JT McClure, D Léger, S Dufour, AG Sheldon, DT Scholl, and HW Barkema. 2012. "Antimicrobial use on Canadian dairy farms." *J Dairy Sci* 95 (3):1209–1221. https://doi.org/10.3168/jds.2011–4527.

Saleem, AM, AI Zanouny, and AM Singer. 2017. "Growth performance, nutrients digestibility, and blood metabolites of lambs fed diets supplemented with probiotics during pre- and post-weaning period." *Asian-Australas J Anim Sci* 30 (4):523–530. https://doi.org/10.5713/ajas.16.0691.

Saxena, S, JP Sehgal, AK Puniya, and K Singh. 2010. "Effect of administration of rumen fungi on production performance of lactating buffaloes." *Benef Microbes* 1 (2):183–188. https://doi.org/10.3920/bm2009.0018.

Schonwetter, BS, ED Stolzenberg, and MA Zasloff. 1995. "Epithelial antibiotics induced at sites of inflammation." *Science* 267 (5204):1645–1648. https://doi.org/10.1126/science.7886453.

Scocchi, M, S Wang, and M Zanetti. 1997. "Structural organization of the bovine cathelicidin gene family and identification of a novel member." *FEBS Lett* 417 (3):311–315. https://doi.org/10.1016/s0014–5793(97)01310–0.

Scott, L, and P Menzies. 2011. "Antimicrobial resistance and small ruminant veterinary practice." *Vet Clin North Am Food Anim Pract* 27:23–32. https://doi.org/10.1016/j.cvfa.2010.10.015.

Seo, JK, S-W Kim, MH Kim, SD Upadhaya, DK Kam, and JK Ha. 2010. "Direct-fed microbials for ruminant animals." *Asian-Australas J Anim Sci* 23 (12):1657–1667. https://doi.org/10.5713/ajas.2010.r.08.

Shamova, O, D Orlov, C Stegemann, P Czihal, R Hoffmann, K Brogden, N Kolodkin, G Sakuta, A Tossi, H-G Sahl, V Kokryakov, and RI Lehrer. 2009. "ChBac3.4: A novel proline-rich antimicrobial peptide from goat leukocytes." *Int J Pept Res Ther* 15 (2):107–119. https://doi.org/10.1007/s10989-009-9170-7.

Shamova, OV, DS Orlov, MS Zharkova, SV Balandin, EV Yamschikova, D Knappe, R Hoffmann, VN Kokryakov, and TV Ovchinnikova. 2016. "Minibactenecins ChBac7.Nα and ChBac7. Nβ - Antimicrobial peptides from leukocytes of the goat *Capra hircus*." *Acta Naturae* 8 (3):136–146. https://doi.org/10.32607/20758251-2016-8-3-136-146.

Sharma, BS, MK Abo-Ismail, FS Schenkel, Q You, CP Verschoor, SD Pant, and NA Karrow. 2015. "Association of TLR4 polymorphisms with *Mycobacterium avium* subspecies *paratuberculosis* infection status in Canadian Holsteins." *Anim Genet* 46 (5):560–565. https://doi.org/10.1111/age.12333.

Sheng, H, HJ Knecht, IT Kudva, and CJ Hovde. 2006. "Application of bacteriophages to control intestinal *Escherichia coli* O157:H7 levels in ruminants." *Appl Environ Microbiol* 72 (8):5359–5366. https://doi.org/10.1128/AEM.00099–06.

Skerlavaj, B, M Benincasa, A Risso, M Zanetti, and R Gennaro. 1999. "SMAP-29: A potent antibacterial and antifungal peptide from sheep leukocytes." *FEBS Lett* 463 (1–2):58–62. https://doi.org/10.1016/s0014–5793(99)01600–2.

Smith David, R, V Sperandio, and JH Carolyn. 2014. "Vaccination of Cattle against *Escherichia coli* O157:H7." *Microbiol Spectr* 2 (6):2.6.09. https://doi.org/10.1128/microbiolspec.EHEC-0006–2013.

Snow, LC, H Wearing, B Stephenson, CJ Teale, and NG Coldham. 2011. "Investigation of the presence of ESBL-producing *Escherichia coli* in the North Wales and West Midlands areas of the UK in 2007 to 2008 using scanning surveillance." *Vet Rec* 169 (25):656. https://doi.org/10.1136/vr.100037.

Song, D, S Ibrahim, and S Hayek. 2012. "Recent application of probiotics in food and agricultural science." *Probiotics* 10:1–34.

Speksnijder, DC, DJ Mevius, CJ Bruschke, and JA Wagenaar. 2015. "Reduction of veterinary antimicrobial use in the Netherlands. The Dutch success model." *Zoonoses Public Health* 62 (Suppl 1):79–87. https://doi.org/10.1111/zph.12167.

Stolp, H, and H Petzold. 1962. "Untersuchungen über einen obligat parasitischen Mikroorganismus mit lytischer Aktivität für Pseudomonas-Bakterien." *J Phytopathol* 45 (4):364–390. https://doi.org/10.1111/j.1439–0434.1962.tb02050.x.

Tetens, J, JJ Friedrich, A Hartmann, M Schwerin, E Kalm, and G Thaller. 2010. "The spatial expression pattern of antimicrobial peptides across the healthy bovine udder." *J Dairy Sci* 93 (2):775–783. https://doi.org/10.3168/jds.2009–2729.

Tiseo, K, L Huber, M Gilbert, TP Robinson, and TP Van Boeckel. 2020. "Global trends in antimicrobial use in food animals from 2017 to 2030." *Antibiotics (Basel)* 9 (12). https://doi.org/10.3390/antibiotics9120918.

Tóth, S, M Kovács, B Bóta, J Szabó-Fodor, G Bakos, and H Fébel. 2020. "Effect of mannan oligosaccharide (MOS) and inulin supplementation on the performance and certain physiological parameters of calves reared on milk replacer." *J Appl Anim Res* 48 (1):228–234. https://doi.org/10.1080/09712119.2020.1770096.

Ushakova, NA, RV Nekrasov, NA Meleshko, GY Laptev, LA Il'ina, AA Kozlova, and AV Nifatov. 2013. "Effect of *Bacillus subtilis* on the rumen microbial community and its components exhibiting high correlation coefficients with the host nutrition, growth, and development." *Microbiology* 82 (4):475–481. https://doi.org/10.1134/S0026261713040127.

Uyeno, Y, S Shigemori, and T Shimosato. 2015. "Effect of probiotics/prebiotics on cattle health and productivity." *Microbes Environ* 30 (2):126–132. https://doi.org/10.1264/jsme2.ME14176.

Valenzuela-Grijalva, NV, A Pinelli-Saavedra, A Muhlia-Almazan, D Domínguez-Díaz, and H González-Ríos. 2017. "Dietary inclusion effects of phytochemicals as growth promoters in animal production." *J Anim Sci Technol* 59:8–8. https://doi.org/10.1186/s40781-017-0133-9.

Van Boeckel, TP, C Brower, M Gilbert, BT Grenfell, SA Levin, TP Robinson, A Teillant, and R Laxminarayan. 2015. "Global trends in antimicrobial use in food animals." *Proc Natl Acad Sci* 112 (18):5649. https://doi.org/10.1073/pnas.1503141112.

Verschoor, CP, SD Pant, Q You, FS Schenkel, DF Kelton, and NA Karrow. 2010. "Polymorphisms in the gene encoding bovine interleukin-10 receptor alpha are associated with *Mycobacterium avium* ssp. *paratuberculosis* infection status." *BMC Genet* 11:23. https://doi.org/10.1186/1471-2156-11-23.

von Bargen, K, JP Gorvel, and SP Salcedo. 2012. "Internal affairs: Investigating the *Brucella* intracellular lifestyle." *FEMS Microbiol Rev* 36 (3):533–562. https://doi.org/10.1111/j.1574–6976.2012.00334.x.

World Health Organization (WHO). 2004. "Second Joint FAO/OIE/WHO expert workshop on non-human antimicrobial usage and antimicrobial resistance: Management options." World Health Organization. https://catalog.nlm.nih.gov/permalink/01NLM_INST/1o1phhn/alma9912277213406676

Wu, H, Y Wang, Y Zhang, M Yang, J Lv, J Liu, and Y Zhang. 2015. "TALE nickase-mediated SP110 knockin endows cattle with increased resistance to tuberculosis." *Proc Natl Acad Sci* 112 (13):E1530–E1539.

Xiao, Y, and L Li. 2016. "China's national plan to combat antimicrobial resistance." *Lancet Infect Dis* 16 (11):1216–1218. https://doi.org/10.1016/s1473–3099(16)30388–7.

Xu, HJ, QY Zhang, LH Wang, CR Zhang, Y Li, and YG Zhang. 2022. "Growth performance, digestibility, blood metabolites, ruminal fermentation, and bacterial communities in response to the inclusion of gallic acid in the starter feed of preweaning dairy calves." *J Dairy Sci* 105 (4):3078–3089. https://doi.org/10.3168/jds.2021-20838.

Yang, B, J Wang, B Tang, Y Liu, C Guo, P Yang, T Yu, R Li, J Zhao, L Zhang, Y Dai, and N Li. 2011. "Characterization of bioactive recombinant human lysozyme expressed in milk of cloned transgenic cattle." *PLoS One* 6 (3):e17593. https://doi.org/10.1371/journal.pone.0017593.

Yarus, S, JM Rosen, AM Cole, and G Diamond. 1996. "Production of active bovine tracheal antimicrobial peptide in milk of transgenic mice." *Proc Natl Acad Sci* 93 (24):14118–14121. https://doi.org/10.1073/pnas.93.24.14118.

Yuan, M, J Zhang, Y Gao, Z Yuan, Z Zhu, Y Wei, T Wu, J Han, and Y Zhang. 2021. "HMEJ-based safe-harbor genome editing enables efficient generation of cattle with increased resistance to tuberculosis." *J Biol Chem* 296. https://doi.org/10.1016/j.jbc.2021.100497.

Zhang, G-W, S-J Lai, Y Yoshimura, and N Isobe. 2014. "Expression of cathelicidins mRNA in the goat mammary gland and effect of the intramammary infusion of lipopolysaccharide on milk cathelicidin-2 concentration." *Vet Microbiol* 170 (1–2):125–134.

Zhang, N, L Wang, and Y Wei. 2020. "Effects of *Bacillus amyloliquefaciens* and *Bacillus pumilus* on rumen and intestine morphology and microbiota in weanling Jintang Black goat." *Animals: An Open Access Journal from MDPI* 10 (9):1604. https://doi.org/10.3390/ani10091604.

Zhao, P, and G Cao. 2012. "Production of bioactive sheep β-defensin-1 in Pichia pastoris." *J Ind Microbiol Biotechnol* 39 (1):11–17. https://doi.org/10.1007/s10295-011-0992-x.

Zheng, C, J Zhou, Y Zeng, and T Liu. 2021. "Effects of mannan oligosaccharides on growth performance, nutrient digestibility, ruminal fermentation and hematological parameters in sheep." *PeerJ* 9:e11631.

2 Role of Antibiotics in Animal Feed

Prospects and Future Challenges

Vasantha V L, Bhoomika Prakash P, Bhavana C Yadav, Bhagyalakshmi N Hiremath, Saptadeepa Mukherjee, Swathi P, Rashmi Ramakrishnan, Praveen Nagella, and Al-Khayri JM

2.1 INTRODUCTION

Antibiotics work by either killing or stopping the spread of bacteria, and they are frequently used to treat a wide range of bacterial infections. Antibiotic resistance has developed over time as a result of the widespread use of antibiotics in a variety of settings, including hospitals, livestock, and aquaculture farms, as well as the careless disposal of waste into the environment from municipal waste, animal farms, and pharmaceutical companies (Sivagami et al. 2020; Mithuna et al. 2024). Animal feed is produced using antimicrobial medications as growth promoters, prophylactics, metaphylactics, and medicines. The amount that is administered to animals for growth enhancement significantly surpasses the amount used to treat diseases. It has been verified that livestock and poultry fed with antibiotic-containing feed achieve desired growth faster than those fed with regular feed. Furthermore, administering antibiotics lowers costs of animal healthcare, which is an important aspect of animal husbandry, and has increased their demand in the market (Mithuna et al. 2024).

Asia's per capita consumption of animal protein has grown from 7 to 25 g over the last 50 years ("Antibiotic Use and Resistance in Food Animals Current Policy and Recommendations 2006"). The global demand for animal protein in the human diet is at a high demand. Furthermore, studies have estimated that antibiotic use in animal foods in India is expected to rise by 82% by 2030, which is a significant increase (van Boeckel et al. 2015). According to a 2010 statistical analysis, China accounted for 23% of the world's antibiotic intake in the production of animal feed, followed by the United States (13%), Brazil (9%), India (3%), and Germany (3%). According to research, this rating is expected to be closer to Mexico (2%), Brazil (8%), India (4%), China (30%), and the United States (10%). By 2030, a few

DOI: 10.1201/9781003269847-2

countries supposedly will have an increase in the consumption of antibiotics: Asia (205%), Republic of Indonesia (202%), Africa (163%), Republic of Peru (160%), and Vietnam (157%), which were among the most antibiotic-intense countries in 2010. According to van Boeckel et al. (2015), in Brazil, Russia, India, China, and South Africa, animal consumption of antibiotics is expected to rise by nearly 99% by 2030, while the human population is only expected to grow by 13% during the same time period. The current state of affairs is a result of the growing demand for animal-based foods brought on by modernization, rapid population growth, and rising affluence. This has pressured farmers to produce and supply more in confined spaces, less time, and with fewer resources (Mutua et al. 2020). Antibiotics are therefore incorporated in animal feeds to improve animal health and promote their growth.

However, antibiotics are not metabolized efficiently by animals. Complete or partial metabolism occurs depending on the complexity of the antibiotic, which is then excreted into the environment. The antibiotics would be present in the excreta in their indigenous state or as active/inactive metabolites (Ho et al. 2014; Awad et al. 2014; Zhang et al. 2015). Accordingly, the discharge into soil and water bodies alters the natural habitats of soil microbes, leading to the rise of "superbugs" that have detrimental effects on the environment and human health (Washer and Joffe 2006). Because they can spread from animal food to people through contaminated food, soil, and water, antibiotic-resistant bacterial strains will become a serious threat to human health (Argudín et al. 2017; Muloi et al. 2018).

The European Union (EU) gradually phased out all antimicrobial growth promoters in the livestock industry due to widespread public health concerns (Butaye et al. 2003; Brown et al. 2017). According to the World Health Organization's (WHO) Library Cataloguing-in-Publication Data Global Action Plan on Antimicrobial Resistance (WHO 2015), "raising awareness and knowledge of antimicrobial resistance (AMR) through effective communication, education, and training" is one of the strategies that nations can use to reduce the risk. Designing effective preventive measures is challenging due to the lack of data on antibiotic use, even though several studies have documented patterns of bacterial isolate resistance ("Antibiotic Use and Resistance in Food Animals Current Policy and Recommendations 2006"). Therefore, little is known about the extent of the environmental threat that AMR organisms pose to public health and the effects that antibiotic residues have on soil ecosystem processes like biogeochemical cycles.

Various novel strategies have shown potential in replacing and/or supplementing antibiotics with vaccines, bacteriocins, phage therapy, probiotics, prebiotics, and symbiotics (Yang et al. 2024). They improve feed efficiency and encourage animal growth, are nontoxic, are easy to remove from the body, are stable during gastrointestinal transit, easily decompose, are environmentally friendly, are selectively active against pathogens with little to no effect on resident gut flora, and most importantly, provide resistance to infection (Mutua et al. 2020). Furthermore, the long-term sustainability of livestock farming depends on the prudent use of antibiotics. Simultaneously, there is a need to improve the oversight and severe enforcement of laws as well as regulations governing their application.

2.2 NEED FOR ANTIBIOTICS IN ANIMAL FEED

Antibiotics are used in animal feeds considering the major benefits they provide, such as improved animal health, meat quality and quantity, growth efficiency, economic productivity, and public health (van Boeckel et al. 2015). They are generally used to treat infectious diseases in animals, protecting humans from serious illnesses that can be spread to humans through contact with diseased animals, contaminated animal-based food, or environmental persistence. It enhances the performance of the animal by lessening the physiological growth restriction that occurs during the illness. Antibiotics are efficient against *Escherichia coli, Leptospira, Campylobacter fetus,* and *Salmonella* infections to mention a few (Bacanlı and Başaran 2019).

The studies suggest that animals with antibiotic-containing feed require less feed to attain the same level of growth as animals fed on non-antibiotic-containing feed. Feed costs account for a major portion of animal husbandry revenue. Thus, adding antibiotics lowers cost expenditures significantly. In addition to improving the health of the animal, antibiotics in the feed allow for a more efficient conversion of feed into animal products. The daily growth rate of animals fed antibiotic-supplemented food is known to be 1% to 10% higher than that of animals fed antibiotic-free feed (Chattopadhyay 2014). According to Hughes and Heritage (2013), meat from animals fed antibiotics is also of superior quality, having more protein and less fat than meat from animals not fed antibiotics. Furthermore, tetracycline and penicillin added to chicken feed increased hatchability and egg production (Gustafson and Bowen 1997). The rate of respiratory morbidity in cows, as well as the rate of relapses and mortality of animals with severe respiratory disease, was also reduced when sulfamethazine and chlortetracycline were added to feed (Gallo and Berg 1995). Overall, it was discovered that antibiotics decreased morbidity and mortality while also improving body weight gain. Antibiotic-supplemented feed also significantly decreased the prevalence of preclinical illnesses at all pig growth stages (Cromwell 2002). The anti-inflammatory properties of antibiotics, particularly macrolides, validate their substantial advantages (Buret 2010). By thinning the mucous membranes in the gut, changing gut mobility, removing harmful bacteria to create favorable conditions for beneficial intestinal microorganisms, and separating proteins for muscle growth, it also helps to improve an animal's gut microbiota (Darwish et al. 2013). For a variety of purposes, β-lactams, polymyxins, tetracyclines, quinolones, and chloramphenicols are used in veterinary medicine (Cháfer-Pericás et al. 2010). Therefore, if properly managed, antibiotics play a crucial role in the efficient and profitable production of livestock.

Administration of antibiotics to animals must be safe and efficient. According to Landers et al. (2012), antibiotics are difficult to approve for use in humans or animals because they can cause resistant strains or encourage drug resistance. Antibiotics are typically administered to livestock through feed or drinking water when treatment is necessary (Bacanlı and Başaran 2019).

2.3 CLASSIFICATION OF ANTIBIOTICS

Antibiotics are drugs that are used to safeguard the health and welfare of both humans and animals. This process inhibits and kills the majority of microbial pathogens. More than 250 chemical entities are currently authorized for use in veterinary and medical care. Humans and animals almost exclusively use a small number of antibiotic classes. For instance, instead of receiving treatment, cattle with bovine tuberculosis are usually destroyed. Some, like ionophores and flavophospholipols, are used exclusively in veterinary medicine (McEwen and Collignon 2018). These are utilized in animal feed to address clinical issues, prevent and control disease outbreaks, and support animal development. Animal feed contains antibiotics for a number of purposes, such as therapeutic, preventive, and medical purposes (Landers et al. 2012). The use of antibiotics is strictly regulated in a number of nations with advanced animal health systems. In addition to treating sick animals, medical use can also involve treating a group of animals with antibiotic-fused feed or water (Han et al. 2024; Barton 2000).

Antibiotics usually contain complex chemicals with multiple functionalities within a single molecule. Antibiotics may be cationic, anionic, zwitter ionic, or neutral in nature with respect to pH. These are also categorized according to their chemical structure and mode of action. Major classification groups are lactams, tetracyclines, aminoglycosides, amphenicols, macrolides, and other subclasses (Menkem et al. 2019).

2.3.1 β-Lactam

By covalently attaching to important penicillin-binding proteins (PBPs), which are enzymes involved in the last stages of peptidoglycan cross-linking in both Gram-positive and Gram-negative bacteria, β-lactam antibiotics, including penicillin, ampicillin, cloxacillin, and amoxicillin, are bactericidal medications that stop bacteria from forming cell walls (Bush and Bradford 2016). This class of antibiotics contains a β-lactam ring atom (a heteroatomic ring configuration with a nucleus that contains three carbon atoms and one nitrogen atom).

2.3.2 Amphenicols

Antibiotics with a phenylpropanoid structure are known as amphenicols. Chloramphenicol, thiamphenicol, azidamfenicol, and florfenicol block the peptidyl transferase enzyme on the bacteria's 50S ribosome (Menkem et al. 2019).

2.3.3 Tetracyclines

Tetracyclines have been found in *Streptomyces* species. Duggar discovered the first tetracycline, aureomycin (chlortetracycline) (Duggar 1948). Subsequently, Finlay et al. (1950) isolated Terramycin (oxytetracycline) from fermentation broths of *Streptomyces aureofaciens* and *S. rimosus* (Mohr 2016). They are antibiotics with four hydrocarbon rings and an octahydrotetracene-2-carboxamide skeleton, and they are classified as a polyketide with an octahydrotetracene-2-carboxamide skeleton. Oxytetracycline and tetracycline cure blood-related infections (Menkem et al. 2019). A number of functional groups are attached to a fused tetracyclic nucleus that is present in tetracycline. In the protein manufacturing process, this makes them bacteriostatic inhibitors (Mohr 2016).

2.3.4 Macrolides

The significant macrolide class was discovered by James M. McGuire when he isolated and reported the first lead compound, erythromycin (Ilotycin), from *Streptomyces erythreus*. A 1952 soil sample from a small island in the Philippines contained the bacterial strain (McGuire et al. 1952). According to Mohr (2016), macrolides are macrocyclic polyketides that have a 12- to 16-chained macrocyclic lactone ring and an amino sugar that is glycosidically linked. These are used to treat infectious diseases in sheep and cattle and are effective against both Gram-positive and Gram-negative bacteria, including cell wall–lacking mycoplasmas, spirochaetes, *Haemophilus influenzae, Bordetella pertussis, Chlamydia,* and *Legionella pneumophila* (Menkem et al. 2019). Macrolides attach to the bacterial ribosome's 50S subunit, blocking the translocase, and as a result, protein synthesis is arrested. They destroy metabolically active microorganisms (bacteriostatic) (Mohr 2016).

2.3.5 Aminoglycosides

An aminocyclitol ring is connected to amino sugars through glycosidic bonds. For instance, streptomycin, gentamicin, neomycin, and spectinomycin inhibit the production of bacterial proteins by attaching to the 50s ribosomal subunits (Menkem et al. 2019). Streptomycin, like many other aminoglycoside antibiotics, is a highly effective broad-spectrum antibiotic that works by inhibiting bacterial protein synthesis.

2.3.6 Fluoroquinolones

Fluoroquinolones are growth promoters that have a fluorine atom joined to the core ring structure, usually at the 6-position. Ciprofloxacin, enrofloxacin, and norfloxacin work by preventing the synthesis of nucleic acid (Menkem et al. 2019).

2.4 CLASSIFICATION OF FEED FOR ANIMALS

Feed can be the substances derived from plants, animals, and any different sources that are given to animals in order to provide nutritional support. Nutrition is necessary for the body. It is especially important in food-producing animals considering their vital role in providing essential nutrients through their products, like milk, eggs, and meat ("Introduction to Nutrition" 2019).

An animal feed needs components from a variety of plants and animals in order to be complete. The formulations of ingredients in feed differ depending on the animal. Feed may be made of one material or a combination of substances. These different substances can be feed processed, raw, or semi-processed (Malomo and Ihegwuagu 2017; Sapkota et al. 2007; Wang et al. 2024). The two primary categories of feeds are concentrates and forages.

2.4.1 Concentrates

Feeds found to provide high levels of energy and protein content but lack fiber content (soya, corn, oats, wheat, and molasses) are known as concentrate feeds (Saha

TABLE 2.1
Classification of Antibiotics and Their Use in Animals

Category	Class of Antibiotic	Structure	Antibiotics	Animals Used	References
Critically important	Aminoglycosides	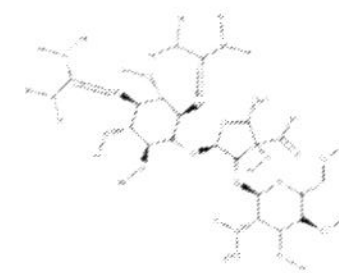	Hygromycin B Neomycin Streptomycin	Sheep, pigs, cattle, goats, and poultry	(McEwen and Collignon 2018; Salvatore and Katz 1993; Call et al. 2008; Mingeot-Leclercq et al. 1999)
	Aminoglycoside-like antibiotics	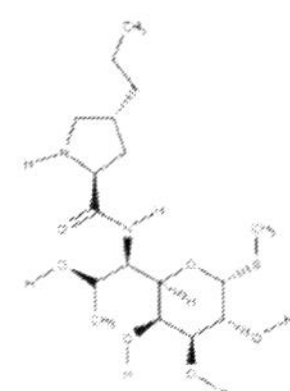	Lincomycin Spectinomycin	Dogs, cats, swine, and ferrets	(McEwen and Collignon 2018; Salvatore and Katz 1993;
	Amphenicols	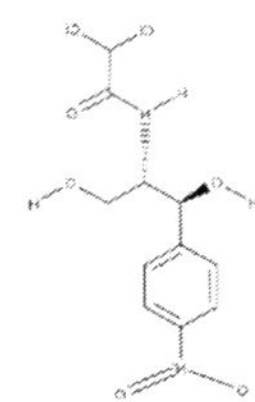	Chloramphenicol Thiamphenicol	Pork, beef, lamb, and chicken	(McEwen and Collignon 2018; Wu et al. 2021)
	Cephalosporins (third and fourth generations)	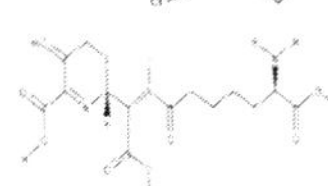	Cefazolin	Cattle, swine, chickens, and turkeys	(McEwen and Collignon 2018; Schnirring and Roos 2012)

(Continued)

TABLE 2.1 (*Continued*)
Classification of Antibiotics and Their Use in Animals

Category	Class of Antibiotic	Structure	Antibiotics	Animals Used	References
	Macrolides	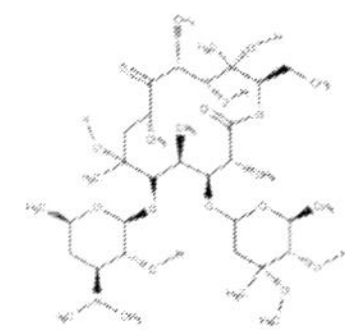	Erythromycin Oleandomycin Tylosin Erythromycin Kitasamycin	Cattle and pigs	(McEwen and Collignon 2018; Barton 2000; Salvatore and Katz 1993; Pyörälä et al. 2014)
	Penicillins (natural, aminopenicillins, aminopenicillins with β-lactamase inhibitor, antistaphylococcal)	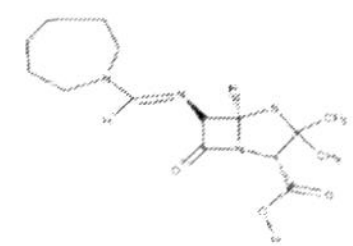	Mecillinam Flucloxacillin Benzathine-benzylpenicillin Phenoxymethyl-penicillin	Egg production, feed efficiency, and hatchability	(Effect of Antibiotics on Animal Feed)
	Fluoroquinolones	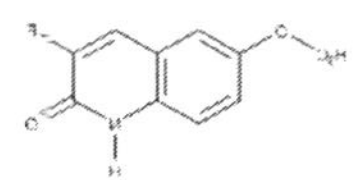	Ciprofloxacin Gemifloxacin Levofloxacin Moxifloxacin Ofloxacin	Poultry	(McEwen and Collignon 2018; Collignon 2005)
	Tetracyclines	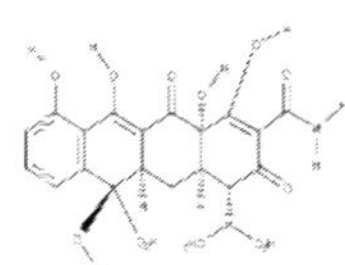	Chlortetracycline Oxytetracycline	Turkeys and broilers	(McEwen and Collignon 2018; Salvatore and Katz 1993; "National Research Council [US] Committee to Study the Human Health Effects of Subtherapeutic Antibiotic Use." 1980)

	Sulfonamides	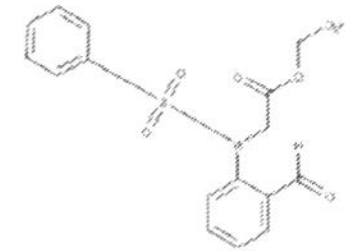	Sulfadiazine Sulfamethizole Sulfamethoxazole Trimethoprim	Beef cattle, swine, dogs, quail, chicken, carp, horses, nonlactating dairy cattle, and turkeys	(McEwen and Collignon 2018; Floron and Fajt 2018)
	β-Lactam	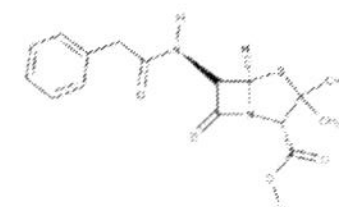	Penicillin G	Cattle	(McEwen and Collignon 2018; Salvatore and Katz 1993; Fejzic et al. 2014)
Highly important	Ionophores	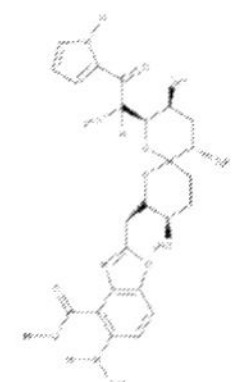	Lasalocid Monensin Narasin Salinomycin	Beef cattle diets	(McEwen and Collignon 2018)
	Lincosamides	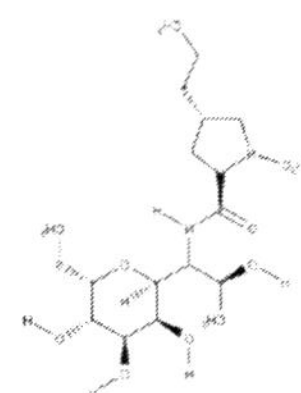	Lincomycin Clindamycin Pirlimycin	Cattle and pigs	(McEwen and Collignon 2018; Pyörälä et al. 2014)
	Phosphonic acid		Fosfomycin	Pigs and chickens	(McEwen and Collignon 2018)
	Pleuromutilins	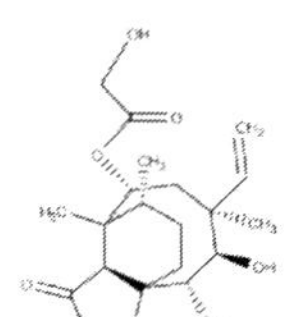	Retapamulin	Swine (majorly); poultry and rabbits (to some extent)	(McEwen and Collignon 2018)

(Continued)

TABLE 2.1 (*Continued*)
Classification of Antibiotics and Their Use in Animals

Category	Class of Antibiotic	Structure	Antibiotics	Animals Used	References
	Polymyxins (including bacitracin and other polypeptides)	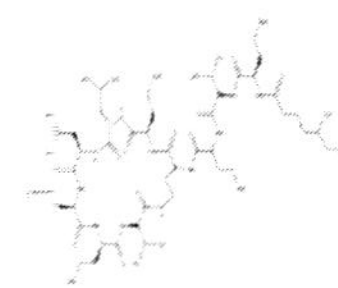	Colistin Polymyxin B	Cattle and swine	(McEwen and Collignon 2018; Varadaraj 2010)
	Pleuromutilins (first generation)	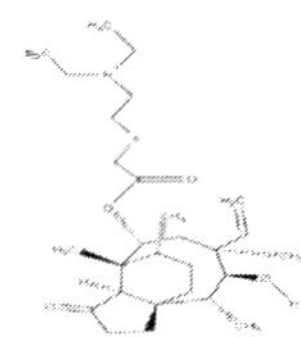	Tiamulin	Swine (majorly), poultry and rabbit (to a lesser extent)	(McEwen and Collignon 2018; Barton 2000; Van Duijkeren et al. 2014)
Important	Aminocoumarin	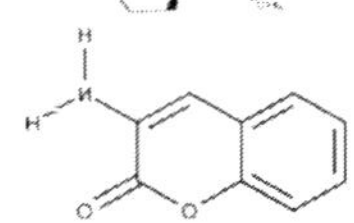	Novobiocin	Cattle farm and sheep farm	(McEwen and Collignon 2018)
	Thiostrepton	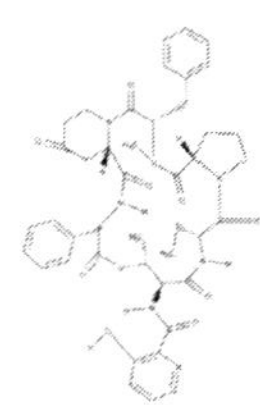		Dogs and cats	(McEwen and Collignon 2018)
	Streptogramins	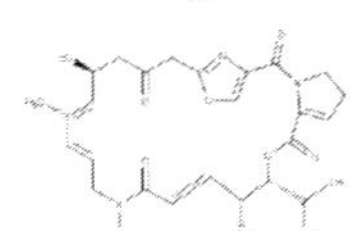	Virginiamycin	Swine, turkey, and broiler chicken	(McEwen and Collignon 2018)

Arsenicals	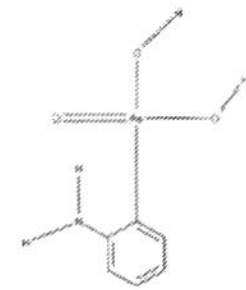	Roxarsone Nitarsone 3-Nitro-arsonic acid	Chicken and swine	(McEwen and Collignon 2018; Barton 2000; "Center for Veterinary Medicine" 2021)
Bicyclomycin	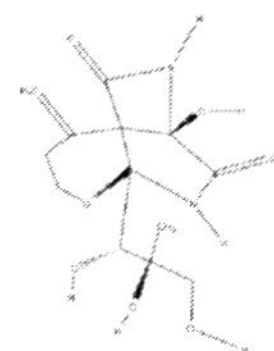	Bicozamycin	Calves and pigs	(McEwen and Collignon 2018: Kohn and Widger 2005)
Fusidane	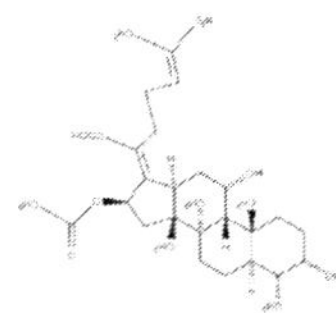	Fusidic acid	Poultry chicken	(McEwen and Collignon 2018; Butaye et al. 2003)
Orthosomycins	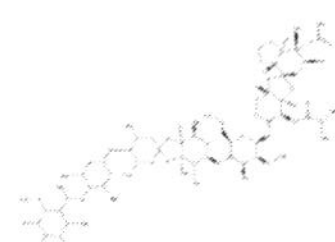	Avilamycin2	Pigs and calves	(McEwen and Collignon 2018; Butaye et al. 2003)
Quinoxalines	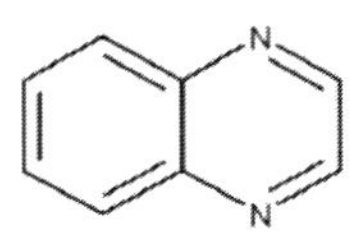	Carbadox Olaquindox	Pigs, dogs, and rats	(McEwen and Collignon 2018; Barton 2000)

(Continued)

TABLE 2.1 (*Continued*)
Classification of Antibiotics and Their Use in Animals

Category	Class of Antibiotic	Structure	Antibiotics	Animals Used	References
	Polypeptides	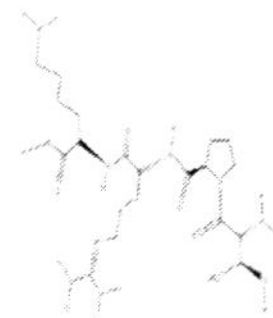	Bacitracin	Pigs, poultry, and fish	(McEwen and Collignon 2018)
	Moenocinol		Bambermycin	Cattle	(McEwen and Collignon 2018; Salvatore and Katz 1993)
	Polyether		Monensin Lasolocid Monensin Narasin Salinomycin	Dogs and cats	(McEwen and Collignon 2018; Barton 2000; Salvatore and Katz 1993)
	Oligosaccharides	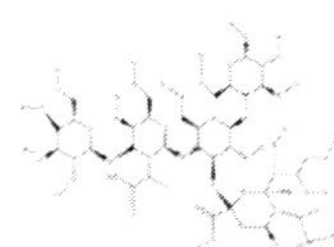	Avilamycin	Broilers, pullets, piglets, and calves	(McEwen and Collignon 2018; Barton 2000; Ioannis Mavromichalis 2017)
	Glycopeptides	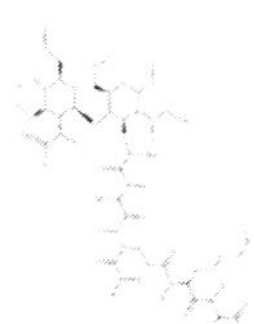	Avoparcin	Sheep, cow, pig	(McEwen and Collignon 2018; Barton 2000)

et al. 2010). It is observed that concentrate feeds contain less than 18% fiber and has about 60% of total digestible nutrients. Compared to other types of feeds, this feed is less heavy and has a high level of digestion capacity. Sincc thcse feeds are sources of concentrated nutrients, they have high levels of nutrients compared to roughage. Protein content in this type of feed can vary from 2% to 88%. Concentrate feeds can be further divided into subcategories as well (Matsushima 1979).

2.4.1.1 Energy Concentrate Feed

Energy concentrate feeds are rich in energy but low in fiber and usually have 20% protein content. This generally consists of cereal grains, sorghums, bran, barley, and molasses. Barley, corn, oats, rye, and sorghums are particularly cultivated for animal feed. These grains can be fed whole or can be ground, either alone or in combination with high-protein oil meals, minerals, and vitamins, which will form a wholesome feed for pigs and poultry and provide the necessary diet for the animals. Due to the high energy content in the cereal grains, these are used in a large proportion to make the feed. Commonly used cereal grains in feeding cattle are triticale, corn, rye, milo, barley, wheat, barley, and oats, among which corn is the most common (Matsushima 1979). These grains have a high amount of starch with 8% to 12% of crude protein, 2% to 5% fat, and calcium in a minute amount of about 0 to 15%. Corn is the most suitable feed for cows, due to its taste and its contribution to reducing overeating, which can cause digestive system problems. Feeding cows with only wheat grain can result in acidosis (Matsushima 1979). Poultry also use barley, maize, wheat, sorghum as their energy sources (Mallick et al. 2020). Molasses can be added in fish feed to make it soft and palatable, which makes it a major source of energy (Yadav et al. 2021). Barley is rich in fiber, with a fiber content of about 6% to 14%, and has less lysine and oil content. This is considered the important concentrate feed in the process of making pigs fatter.

2.4.1.2 Protein-Concentrated Feed

Protein-concentrated feed is mainly classified based on the source as plant and animal. Protein supplements derived from animal protein are usually rich in calcium and phosphorus.

2.4.1.2.1 Plant-Based Proteins

Plant-based proteins consist of crude protein 47%, 1% calcium, 1.5% phosphorous, and 2.5% fiber. According to Bist et al. (2024) and Matsushima (1979), the most widely used protein sources in animal feed include soybean, peanut, flaxseed, cottonseed, corn gluten meal, peanut oil meal, sunflower seeds, canola, and many more. Poultry needs certain amino acids, like methionine and lysine. When poultry is raised in conventional methods, amino acids are usually supplemented as synthetic amino acids. But when raised in organically, synthetic amino acids cannot be used. An alternative for synthetic amino acids is the compounds obtained from plants and vegetables (Tufarelli et al. 2018). Fish feeds are also designed from plant-based protein sources, such as soybean meal, but these are low in amino acids that have to be supplemented by providing them with lipids and omega-3-fatty acids from oils (Steve

Craaig 2017). Among plant proteins, soybean has a high amount and good quality of protein (41% to 48%).

2.4.1.2.2 Animal-Based Proteins

Animal-based proteins are the products or by-products obtained from animals, including fish meal, meat scraps, poultry waste, feather meal, milk, tissues, and blood meal. The highest protein source provider is blood meal. It provides about 84% to 86% of protein and has low calcium and phosphorus content (Bist et al. 2024). Animal-based proteins should be evaluated before using to avoid any kind of infection. To obtain protein from feather meal, it has to be hydrolyzed in order to make it available for animals (Matsushima 1979). Poultry excreta can be dried and used as feed for ruminants. Hatchery waste is a combination of unhatched, infertile eggs and eggshells. This waste plays a role in chicken feed to help weight gain.

2.4.1.3 Vitamins and Minerals

Mineral supplements can be calcium carbonate and limestone, while vitamins can be from fish oil. Some vitamins must be supplemented through feed, and some vitamins are synthesized by the microbes present in the rumen and hindgut. Vitamin A is a necessary component in animal feed. It can be given through plant sources containing carotenoids (Cherian 2020). Minerals are also important for animals' physiological functioning. Minerals constitute about 4% of body weight. Animals need trace amounts of cobalt, copper, iron, iodine, manganese, molybdenum, and zinc (López-Alonso 2012).

2.4.2 Forage

Forage (or roughage) can be defined as herbaceous plant species or a part of the plant that is given to animals (Putnam and Orloff 2014). This is also the feed that consists of heavy parts of the plants with good fiber content and less total digestible nutrients (Matsushima 1979). Forage usually consists of plants belonging to the family

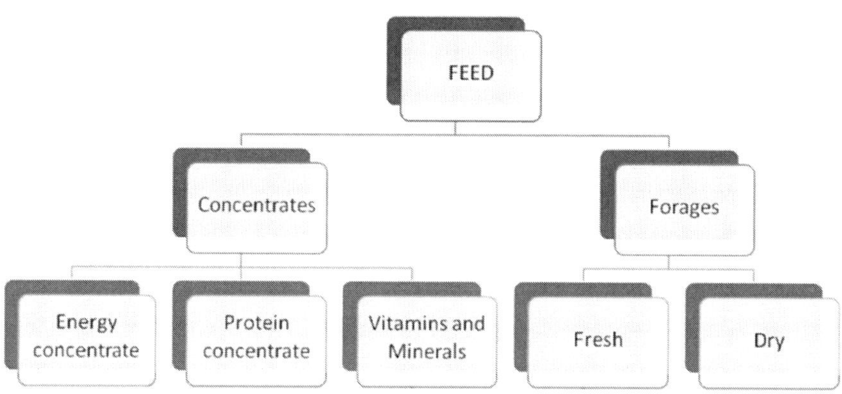

FIGURE 2.1 Types of animal feeds.

Poaceae or Fabaceae (legumes). These plants can be either fed directly or partially digested. When forages are being used in a feed formulation, it is necessary to look into the nutritional status of the forage crop. Depending on the amount of proteins, lipids, and carbohydrates, the digestibility value is determined. This calculation is important when feeding nonruminants or ruminants (Capstaff and Miller 2018).

2.4.2.1 Fresh Forages

Proteinaceous forages include legumes. Legumes are a rich source of proteins and other mineral contents, like calcium, phosphorous, magnesium, copper, and cobalt. A good example of a legume can be alfalfa. This is commonly used as feed for cattle, mainly calves that require high amounts of protein. It has a small amount of calcium and also phosphorous that are required for the cattle. Pasture is another type of fresh forage; it is the grass grown on land for grazing purposes. Cattle, sheep, and goats mainly feed on pasture. The protein content can vary from 3% to 30%. Grass is the most suitable and easily available feed for all livestock animals. Grass contains crude protein of about 5% to 10%, 0.3% to 0.5% calcium, and 0.2% to 0.3% phosphorous (Matsushima 1979). Silage, including silages in feed, can stimulate digestion and will prevent bloat in animals. Commonly used silage includes green sorghum fodder and green corn fodder. They provide the necessary nutrients to cows and other ruminants (Matsushima 1979). Roots and tubers are sources of carbohydrates. Tubers and the starchy roots of plants store edible starch in their rhizomes, corms, and tubers. Along with carbohydrates, they also have a high amount of dietary fiber (Chandrasekara and Josheph Kumar 2016; Siregar et al. 2024).

2.4.2.2 Dry Forage

Dry forage includes hay and straw. The moisture content in plants is reduced for inhibition of microbes and other pathogens. This is the main purpose of dry forages. A combination of cereal crops and legumes constitutes straws – it contains high fiber but less nutritive value. This is not used to feed pigs and poultry (Matsushima 1979).

2.5 ANTIBIOTIC DRUGS FOR LIVESTOCK

Antibiotics are used primarily as (1) therapeutics to treat established infections, (2) prophylactics/metaphylactics to prevent the spread and development of bacterial diseases in healthy animals, and (3) growth promoters to increase body weight and feed conversion. Even so, a number of nations have passed laws that forbid the use of antibiotics to prevent disease and promote growth (Ben Lagha et al. 2017). Antimicrobial use helps to increase profitability and productivity by reducing loss associated with illness and death due to bacterial infections. Antimicrobials also aid in feed conversion efficiency; hence, they are commonly employed as growth boosters by livestock producers. This has made a significant contribution to increasing animal protein supply for food security as a countermeasure against hunger and malnutrition, particularly in developing countries (Ojo et al. 2016).

2.5.1 Antibiotics in Dairy Cattle Feed

2.5.1.1 Antibiotics for Cows

India is the nation that produces the most milk, according to the Food and Agriculture Organization (FAO). India is responsible for about 22% of global milk production. Next in line are China, Pakistan, Brazil, and the United States. According to "Gateway to Dairy Production and Products" (FAO 2021), cattle alone account for 81% of the world's milk production, followed by buffaloes (15%), goats (2%), sheep (1%), and camels (0.5%). For a nursing cow to meet her needs for milk and milk components produced by the mammary glands, she needs to consume carbohydrates, water, amino acids, minerals, fatty acids, and vitamins (Erickson and Kalscheur 2019). Using antibiotics has led to healthier, more productive animals as well as a decrease in animal diseases, illness, and mortality, in addition to producing large quantities of nutrient-dense, high-quality, and reasonably priced food for human consumption (Oliver et al. 2011). A triangular model of epidemiology indicates that diseases that are infectious can be caused by the interaction of disease-causing agents, diseased animals, and the environment. Some diseases that affect cows include mastitis, bovine tuberculosis, anthrax, nocardiosis, and contagious bovine pleuropneumonia (Abdelhay Kaoud 2019). Some of the antibiotics incorporated in animal feed are penicillin, macrolides, and aminoglycosides; oxytetracycline, chloramphenicol, sulfonamides and trimethoprim, and tetracyclines (Macdiarmid 1978). Dairy cows are susceptible to three major diseases affecting production: reproductive disorders, mastitis, and lameness. Antibiotics are a potential weapon to fight against these diseases. A small amount of antibiotics is added to animal feed each year at a rate of 2 to 50 g/ton to improve animal performance. The reasons include improving feed conversion into products of animals, a greater growth rate, and a lower mortality rate ("Use of Antibiotics in Dairy Cows" 2014).

2.5.1.2 Antibiotics for Goat and Sheep

Main products such as milk, wool, and meat are obtained from sheep and goats. According to studies in 2019, China ranks first in the production of goat meat, i.e., 2,357,205 tons. India ranks second by producing 553,380 tons. If we look into goat milk production, India ranks first by producing 5,400,400 tons of milk as of the 2019 survey ("World's Best Meat Producing Countries Based on FAO" 2021). Like poultry, beef, and swine, the goat industry has not been more focused on using antibiotics. The withdrawal time of antibiotics if not maintained properly will result in high levels of antibiotics in the meat or milk (Landfried et al. 2018). Different microorganisms that are in high concentrations in animal feed will consume most of the nutrients present in the feed and also have an effect on the intestinal absorption of food materials. It is also noted that the animals provided antibiotics tend to show a 1% to 10% improvement in growth rate and feed efficiency over the ones without antibiotics. Animal meat will have less fat and a higher protein content when the animal is fed with antibiotics (Chattopadhyay 2014). Goat milk was found to be the first milk product used by humans (in around 8000 BC). Goat milk is found to have many advantages over cow's milk in its properties, like varied alkalinity, high digestive capacity, and medicinal use for humans (Sepe and Argüello 2019).

Due to many factors, such as large herd size, less aeration, and poor maintenance, infectious diseases can easily spread. Some of the bacterial diseases that affect goats are tetanus, caseous lymphadenitis, tuberculosis, anthrax, foot rot, and enterotoxaemia. All these different pathogens can affect different organs, which can also lead to death (Kumaragurubaran Karthik and Manimuthu Prabhu 2021). The term *maximum residue limit* (MRL) refers to a measure of the maximum quantity of antibiotics residue present in milk, meat, cheese, etc. Antibiotics residues present in raw milk are seen to be heat resistant, which cannot be removed easily. Some studies have shown that antibiotics were present in cheese and yogurt even after processing at high-temperature treatments and pasteurization (Quintanilla et al. 2019). Goats usually feed on leaves and grasses, and they also feed on roughage. Antibiotics used are chlortetracycline, danofloxacin, benzyl-penicillin, sulfonamides, and many others (Zhang et al. 2021a).

2.5.2 Antibiotics in Fish Feed

Aquaculture is fast advancing in terms of protein source. A survey showed that globally, 17% of protein consumed is from aquatic animals (Schar et al. 2020). A statistical study performed in 2015–2016 showed that India is producing about 10.79 million tonnes of fish per year. As of now, India is responsible for 6% of global fisheries production, ranking second globally, after China (Sivaraman 2018). This increasing demand for fish products led to extensive production, leading to crowding. When fisheries are crowded, the spread of disease is likely to increase, and that is when antibiotics come into the picture. Diseases in fish can be due to any kind of stress. In their natural conditions, they are less affected by any diseases because they do not experience any stress. In rearing places, due to overcrowding, there will be an inadequate amount of dissolved oxygen and competition for feed, which leads to stress. This stress can make fish weak, and it can cause immunosuppression, leading to increased chance of infections (Kelly 2013). Some of the common diseases affecting them are ich, dropsy, columnaris, gill disease, white-spot disease, cloudy eye, lymphocytis, and many more. Some of the bacteria affecting them are *Enterococcus faecalis*, *Lactococcus garvieae*, and *Aeromonas hydrophila* (Sharma et al. 2012). These antibiotics aid in preventing or treating any kind of bacterial infection. A study has shown that giving fish antibiotics will reduce the population of gut animals, which will in turn maximize intestinal absorption of nutrients. This will lead to less co-consumption of nutrients. Nutrients in fish feed are less effectively used in their overall body growth. Antibiotics also decrease the population of nonpathogenic gut microbes, leading to a more efficient digestive system and improved nutrient intake from feed (FAO 2015, "Responsible Use of Antibiotics in Aquaculture").

Different antibiotics approved to use in fish feed are Terramycin, Romet, and florfenicol (Kelly 2013). When a fish is infected by any bacteria, first it should be diagnosed in order to determine suitable antibiotics for treatment. When providing fish with antibiotics, care should be taken and the duration of feeding should be recorded. It is observed that fish will stop consuming food when they are infected, and treatment should be provided as soon as possible. Later, depending on the degree of infection, the dose can be decided on and given accordingly. While talking about the dose and the duration, we should also look into withdrawal time. It is a very important step so as to ensure that no antibiotics reach the consumer's plate. Continuous usage of these antibiotics will lead to resistance to infections or diseases (Kelly 2013; Yuan et al. 2023).

2.5.3 Antibiotics in Poultry Feed

Over the past 50 years, antibiotics have gained popularity along with biosecurity and hygiene practices, which have helped the poultry industry grow by halting the spread of various avian diseases. In poultry production, antibiotics are used to boost feed conversion, encourage growth, and fend off illness (Bist et al. 2024). By making some modifications to the immune system of broiler chickens, antibiotics can be successfully used at subtherapeutic levels in poultry, protecting their health. In addition to controlling gastrointestinal infections, the administration of antibiotics alters the intestinal microbiota (Mehdi et al. 2018). In poultry, giving antibiotics to the entire flock for treatment, disease prevention (metaphylaxis), and growth promotion is usually acceptable.

Antibiotics are typically used to treat intestinal infections brought on by *Salmonella, Escherichia coli,* or *Clostridium*, including colibacillosis and necrotic enteritis. According to Roth et al. (2019), poultry infections are a serious problem that result in significant financial losses. Some of the diseases caused are chronic respiratory disease, coccidiosis, infectious coryza, fowl cholera, infectious synovitis, botulism, and pullorum disease (Cruz-Rincon 2021). Some studies showed that chickens fed with tylosin had less population of bacteria *Lactobacillus* spp. in the ileum (Mehdi et al. 2018). By using antibiotics in poultry production, an increase in bacteria resistant to antibiotics was shown. During 2016, chicken herds were usually treated with polymyxins, tetracyclines, and penicillin, and later were treated with sulfonamides and trimethoprim (Roth et al. 2019). Chickens nurture a population of *Enterobacteriaceae* in the digestive tract, which is resistant to aminosides and meat has resistance to tetracyclines (Mehdi et al. 2018).

2.6 EFFECT OF ANTIBIOTICS ON OTHER LIVESTOCK

The human population is predicted to reach 9.15 billion by 2050, and it is still growing. This could enhance consumer demand for organic livestock products like milk and meat. In developing countries, livestock plays a significant role as an intrinsic aspect of agricultural and rural life, supplying food and income for many farmers. One of the biggest issues facing animal husbandry is the scarcity of feed to satisfy the dietary requirements of the existing animal population (Balami et al. 2018). The major duty of the agricultural industry is to provide high-quality animal-based nourishment to the population (Miroshnik et al. 2021).

2.6.1 Antibiotics in Pork Feed

Pigs are recognized as the most common livestock farmed because the costs of maintenance are comparatively lower. According to Rizaldi et al. (2019), the body grows fairly quickly and does not need extensive maintenance cages. The production of meat has rapidly increased in emerging nations, and this trend is anticipated to continue, especially for swine (Chen et al. 2019). Pork meat is in high demand all around the world. Pork is frequently farmed under harsh conditions. The production of meat as a source of protein is one of several agricultural sectors that are constantly under pressure to increase productivity due to the growing human population (Monger et al. 2021). Pigs commonly feed on agricultural waste and also litter in large sizes. Their manure is commonly applied to the soil to increase its fertility. In terms of nutrition, culture, and economy, pork is one of the most important commodities (Rizaldi et al. 2019). Pig breeding is one of the world's fastest growing animal husbandry fields (Wang and Li 2024). Antibiotics such as tetracyclines, macrolides, β-lactams, aminoglycosides, and quinolones are frequently utilized in pig farming (Miroshnik et al. 2021).

Infection prevention and treatment are essential for productivity and the welfare of farmed animals raised for meat. Depending on the animal density and the biosecurity measures in place, infectious diseases can spread swiftly on farms, infecting one or more herds and resulting in high mortality. If left untreated, this could lead to a decrease in the producers' productivity and profit (Monger et al. 2021). The amount and quality of animal products produced and consumed have increased due to the high demand for animal protein. Developed countries like China, Brazil, and the United States estimate that cattle alone consume 50% to 80% of all antibiotics generated. The establishment of intensive farming systems in South and Southeast Asian developing nations has led to an increase in the number of livestock and the consumption of antibiotics by animals that produce food (Hosain et al. 2021).

2.6.2 Antibiotics in Lamb Feed

Ruminants raised by rural farmers, as well as those sold in markets across the country, have been found to be contaminated with one or more diseases/pests (Balami et al. 2018). The relationship between ruminants and rumen bacteria is symbiotic; the ruminant gives the microorganisms nutrients and a favorable rumen environment, and the rumen microorganisms aid in fermentation and digestion. Ruminants get their protein and energy from fermentation by-products like volatile fatty acids and microbial protein (Gouws 2019).

Appropriate dietary components in the lamb diet may have a direct impact on meat quality (Garcia-Galicia et al. 2020). Feed additives have long been used as rumen enhancers to lower nitrogen excretion or methane emissions, modify rumen pH, increase feed intake and efficiency, improve fiber digestibility, and lower the risk of milk fat depression. Enhancing output and efficiency is the aim of feed additive supplementation in ruminant production systems (Gouws 2019). Polyether ionophore, like monensin antibiotic, is commonly used in animal feed to improve energy efficiency and control rumen fermentation. Increased milk supply, greater feed digestibility, higher live weight gain and improved energy metabolism are all advantages of feeding monensin to ruminant animals (Lin et al. 2020). The majority of cattle diseases detected in the Northeastern Nigeria were infections caused by bacteria, viruses, fungi, and parasites (Balami et al. 2018). Metabolic diseases are also found, the most common of which are rumen acidosis and bloat. In the rumen, Gram-positive and Gram-negative bacteria can be found. Gram-positive bacteria ferment to produce hydrogen ions, lactate, ammonia, acetate, and formate. The majority of these by-products contribute to the production of methane, which is undesirable due to energy waste. Rumen acidosis is largely caused by lactic acid–producing bacteria because of their acidic pH. Bloat is divided into two categories. The first is gas buildup caused by a physical impediment that prevents gas eructation. The other type of bloat is called foamy bloat, which happens when a layer of foam forms on top of the digesta in the rumen, preventing gas eructation. The animal suffocates in both cases due to pressure on the diaphragm and lungs (Gouws 2019).

Long-term livestock productivity and the control of animal diseases that could be transferred to humans through the food chain have been demonstrated to depend heavily on antimicrobials (Ojo et al. 2016). Animals are commonly administered lower concentrations of antibiotics used as growth promoters than those used for therapeutic purposes. Increased concentrations are more likely to result in AMR development (Hosain et al. 2021). Antibiotics used as in-feed additives have an impact on every stage of pork production, from gestation, to nursing, to growth, and to slaughter, yet the effects vary by stage. The use of antibiotics in animal feed has become a major global issue due to the significant risks to human, animal, and environmental health (Lin et al. 2020). Antibiotics, probiotics, prebiotics, and other growth-stimulating medications are utilized in big pig breeding operations to boost production, natural resistance, and farm animal genetic potential (Miroshnik et al. 2021).

2.7 EMERGENCE AND PERSISTENCE OF ANTIBIOTIC RESISTANCE

Globally, AMR is a significant public health concern. One important resistance selector is the use of antibiotics (Ma et al. 2021). Antimicrobial growth promoters is the term used to describe antibiotics used as growth supplements (Han et al. 2024; Mathew et al. 2007). According to current estimates, the amount of antibiotics consumed per kilogram of animal produce is 148 mg/kg for poultry, 45 mg/kg for cattle, and 172 mg/kg for pigs. In 2018, China, the world's biggest producer and consumer of antibiotics, used about 29,774.09 tonnes of antibiotics for animal husbandry alone, of which 53.2% were used as growth promoters for antibiotics (Ma et al. 2021). In dairy operations, a variety of drugs are used to treat illnesses that impact dairy cows. According to González Pereyra et al. (2015), 41% of all medications used in livestock are antimicrobials.

AMR occurs as a consequence of reckless use of antibiotics, endangering the health of both animals and people. The use of antibiotics in veterinary medicine, particularly as antimicrobial growth promoters, has a significant impact on the prevalence of antibiotic resistance in animal pathogens and is a risk factor for the advent of antibiotic resistance in human pathogenic microbes (Han et al. 2024; Marshall and Levy 2011). The World Health Organization (WHO 2019) defines AMR as "an increase in the minimum inhibitory concentration of a drug for a previously susceptible strain." Antibiotic resistance is caused by four mechanisms: target modification, drug inactivation, decreased permeability, and enhanced efflux (Yichao Yang et al. 2019). AMR is expected to cause 10 million deaths per year by 2050, costing the world economy $100 trillion, according to research. It is becoming increasingly prevalent in the world. Antibiotics used in feed animals are closely related to use of antibiotics in human treatment, and the use in animals can lead to resistance. Cross-species transmission from animals to humans can occur (Han et al. 2024; Boerlin 2004).

The use of antibiotics in animals worldwide has promoted development of antibiotic resistance. The genes of animal populations as transposons and on plasmids show resistance. This is exacerbated by the ease with which resistant strains can be introduced between animals by fecal contact. In 1951, after streptomycin was administered to turkeys, the first incidence of antibiotic resistance in food animals was documented. Since then, antibiotic resistance to tetracyclines, sulfonamides, β-lactams, and penicillin has become more common. Antibiotic usage in animal food for long periods generates optimal circumstances for the establishment and transmission of resistant strains (Ma et al. 2021). Small dairy producers in Kenya produce milk with β-lactam residues exceeding the extreme levels. In response to widespread antibiotic use, bacteria have developed antibiotic resistance. An organism with a novel mutation or newly acquired characteristic will survive and propagate under selective pressure, which refers to the factors influencing the environment that allow this to happen. Farm products can retain antibiotic residues due to the use of antibiotics. Through farm effluents, antibiotics can enter aquatic and terrestrial ecosystems. Antimicrobial-resistant bacteria and bioactive veterinary drug residues present in farm wastes can seep into water and soil. This can be a source of antibiotic-resistant bacteria in the environment after they are applied to the land (Han et al. 2024; Sawant et al. 2005). A study comprising of 374 *E. coli* isolates from bovine mastitic milk were subjected to antimicrobial susceptibility tests, and tetracycline (23.3%) showed the most frequently observed drug resistance, followed by streptomycin (17.1%), ampicillin (16.6%), neomycin (11.8%), and trimethoprim/sulfamethoxazole (11.2%). Agriculture is using antimicrobials extensively, raising concerns about the potential for the appearance of antibiotic-resistant bacteria on farms and in human populations (Tark et al. 2017).

There are two primary ways that AMR can spread. First, mediate an AMR phenotype by acquiring resistance genes, causing de novo mutations, or both, using plasmid mobile genetic elements. A number of vectors, including humans, insects, birds, agriculture, and water, can also spread AMR. A number of other factors are probably involved, but antibiotic abuse is a major contributor to the emergence and prevalence of AMR. Poor management, poor hygiene, poverty, literacy, and community and health infrastructure are all known to affect health outcomes, and they are thought to have an equal or greater influence on AMR. According to Tang et al. (2017), Pormohammad et al. (2019), and McEwen and Collignon (2018), this environment also contributes to the development and dissemination of resistance in insects, animals, and birds via water. Since livestock waste is used as fertilizer, it may be a means for antibiotic resistance to spread from livestock farms (Ghosh and LaPara 2007).

The only way to comprehend the temporal variability of AMR is through surveillance. To determine the frequency and rate of resistance persistence, point prevalence surveys (PPSs) are employed. Several human PPS studies have regularly assessed antibiotic prescriptions for critically ill hospital patients based on clinical symptoms. Animal PPSs differ from human PPSs as only healthy animals are sampled before slaughtering, and sick and weak animals are often neglected. Because the microorganisms and degree of resistance differ greatly between healthy animals and ill humans, it is challenging to relate human and animal PPSs to look for any AMR correlations (Moore 2019).

To effectively combat global antibiotic resistance, improving contact to clean water, maintaining proper hygiene, guaranteeing strong governance, and growing public healthcare must be addressed. According to a study by Collignon et al., inadequate cleanliness was found to be significantly linked to AMR levels, but antibiotic use was not. The strong correlation between the infrastructure index and *E. coli* resistance is supported by the fact that *E. coli* are fecal bacteria, and their presence in water indicates fecal contamination in humans or animals (McEwen and Collignon 2018). In dairy operations, drugs are used to treat diseases that dairy cattle can get. According to estimates, 41% of all medications used in livestock are antimicrobials. Total antimicrobial consumption varies greatly by nation, ranging from 5,393 tons in Europe to 9,300 tons in the United States, according to several studies. Antimicrobials are used in mature cattle and calves for both preventive and therapeutic purposes (González Pereyra et al. 2015). By providing resistant strains of bacteria with a competitive edge, these practices can increase the number of AMR bacteria. Additionally, if resistance genes are carried on horizontally transmissible elements such as plasmids and conjugative transposons, they can successfully spread to new bacterial hosts. Resistance-granting chromosomal changes that enable a relatively quick change in the target molecule or decrease antibiotic absorption result in resistant strains like fluoroquinolones (Ahmed et al.2024).

The growth of chickens given crude *Streptomyces aureofaciens* fermentations was higher than that of poultry feed containing purified B12. These findings led to the discovery of an unidentified growth factor – later identified as chlortetracycline – in crude fermentations. According to Mellon et al., livestock are treated with about 70% of the antibiotics produced. Some fungi and bacteria create antimicrobial substances that aid in their survival in soil and biofilms, which are teeming with microorganisms. Since organisms that produce antibiotics require strategies to counteract the effects of their products in order to survive in the presence of them, it is also likely that antibiotic resistance evolved naturally before humans started using drugs. Selection pressure increased the benefit of preserving resistance genes in various bacterial groups as the use of antibiotics in human medicine and animal feed production increased. Bacterial evolutionary progress eventually included mechanisms to retain, accumulate, and disperse resistance genes among bacterial populations. Starr and Reynolds (1951) were the first to propose that antibiotics usage in animals induced resistance, observing resistance to streptomycin in coliform bacteria from turkeys given the antibiotic. Since then, numerous studies have discovered a link between the use of antibiotics in livestock and the rise in the prevalence of organisms resistant to those antibiotics in livestock, agricultural products, and the farm environment. Foodborne pathogens that cattle can carry and spread to humans through the food chain include *Salmonella, Yersinia, Listeria, Campylobacter,* and some strains of *E. coli.* Regardless of whether antibiotics are administered, animal stresses such as transport and weaning have been shown to enhance the incidence of bacteria that shows resistance in pigs (Wang and Li 2024). Transposons are genetic components that induce self-excision and reinsertion into new hosts' genetic code, allowing for horizontal gene transfer. Because transposons frequently include resistance genes, including introns, it is speculated that they play a key role in the evolution of antimicrobial resistance (Mathew et al. 2007).

In various species, respiratory and gastrointestinal infections are among the most prevalent, and mastitis is common in milk-producing animals, such as cows, goats, sheep, and buffaloes. Breeding pigs, adult cattle, and buffaloes are normally treated separately with injectable formulations, whereas fattening pigs, poultry, and fish are mostly treated orally by group medicine through feed or water (Bengtsson and Greko 2014). Soon after, antibiotic resistance was discovered, and Fleming warned in a 1945 interview with *The New York Times* that using penicillin incorrectly could cause resistant bacteria to be selected. Antibiotic-resistant "mutant forms" of *Staphylococcus aureus* make the host more infected with the pathogen and increase the risk of transmission to other hosts. When a resistant bacterium does not respond to treatment, there are serious morbidity and mortality consequences as well as higher economic costs. The development of enzymes that deactivate the antimicrobial, the presence of efflux systems that lower intracellular antibiotic concentrations, the absence of the target, or other characteristics of a microorganism's general anatomy or physiology can all be considered inborn (inherent) resistance to antimicrobials. However, because of this natural form of resistance, the health of both humans and animals is not the primary concern. Rather, the great majority of bacteria that are resistant to antibiotics have developed genetic alterations as a result of horizontal genetic material transfer from other bacterial strains or mutations (vertical evolution). Antimicrobials function by either killing or preventing the production of proteins (tetracyclines), quinolones (quinolones), bacterial cell walls (penicillin), and metabolic pathways (tetracyclines, sulfonamides). Biofilm formation, surface permeability changes, efflux or enzymatic inactivation of the compound before it reaches its target site, alteration or overproduction of the target site, and acquisition of alternative metabolic pathways to those inhibited by the drug are some of the microbial strategies for evading the effects of antimicrobials (Capita and Alonso-Calleja 2013).

Antimicrobials are constantly supplied to drinking water or supplemented to daily feed for an extensive duration of time at minor doses as prophylactic measures while antimicrobial growth promoters are supplemented to feed or drinking water on a routine basis at a minimal dosage (Rahman 2019). Two significant biological processes that contribute to the evolution and dissemination of resistance are vertical gene transfer and horizontal gene transfer (HGT) (Hedman et al. 2020). Studies show *E. coli* was resistant to carbapenems as well as an extended range of β-lactams, which could have spread to the environment when litter was employed as manure (Sebastian et al. 2021). As a reservoir for opportunistic, multidrug-resistant extraintestinal diseases, foodborne pathogens can either cause acute illness or persist in the gut microbiota asymptomatically (Skandalis et al. 2021).

2.8 HUMAN HEALTH CONCERNS

The use of antibiotics in animal research, especially as antimicrobial growth promoters, has a significant effect on the development of antibiotic resistance in microorganisms and increases the likelihood that human pathogenic organisms will also develop antibiotic resistance. Animal-derived antibiotic-resistant bacteria, including *Salmonella, E. coli, Enterococcus,* and *Campylobacter* species, can infect and harm people through contact or the food chain (Bengtsson and Greko 2014). Additionally, pathogens in the human gut flora can transfer resistance genes from animals to humans. Although food products of animal origin are the most common source of human infections, humans can contract *Salmonella* isolates from animals by coming into direct contact with sick animals or animal feces. Animal-based foods are a common source of carriers and symptoms of salmonella infections in intensive livestock farming. Meat and meat products are contaminated during slaughter, processing, and packaging by salmonella found in the digestive tracts of these animals. Meat and meat products, eggs, and other foods can then spread the infection to humans. According to Han et al. (2024), human illness is not always caused by salmonella infection.

The three mechanisms that have been examined are as follows: direct contact with bacteria that are resistant to antibiotics that come from an animal source, disruptions in the species barrier that result in the ongoing spread of resistant strains from cattle to humans, and transfer of resistance genes from animals to humans. Millions of cases of salmonellosis and campylobacteriosis brought on by tainted meat and other foods are reported each year in the United States alone (Chang et al. 2015). The link between animals and human health remains direct human-to-animal contact, intake of contaminated animal-sourced commodities, and drinking of contaminated water and vegetables. Even at low concentrations, the metabolites of antibiotics that are administered to the animals are released in the urine and feces and remain in the environment long enough to produce resistant bacteria in the soil (Sweileh 2021).

Vancomycin-resistant enterococci (VRE) epidemiology varies significantly between the United States and Europe. In the intestinal microbiota of farm animals and healthy humans in Europe, *Enterococcus fecium* carrying the vanA resistance element that causes vancomycin resistance was common, but until 2008, the United States did not have any VRE prevalence in livestock or healthy people. The widespread agricultural use of a glycopeptide known as avoparcin, which has been widely used in Europe since the 1970s but has never received approval in the United States, has been blamed for this shift. It has been shown that avoparcin, which confers vancomycin cross-resistance, selects for VRE in animals. Numerous opportunities for human infection and the potential for resistant bacteria to colonize the human niche are made possible by the large reservoir of VRE in animals. Molecular epidemiological studies have revealed that the VRE strains isolated from humans and animals are identical (Chang et al. 2015). When genes are transferred between bacteria via mobile genetic elements like integrases, plasmids, and transposases, this process is known as horizontal gene transfer (HGT). HGT is therefore essential to the quick spread of resistant genes. Veterinarians, farm and slaughterhouse workers, and people who have direct contact with agricultural workers are at a higher risk of contracting resistant bacteria due to their daily exposure to infected animals (Ma et al. 2021). The human intestine, the food chain, and fish rearing are all examples of horizontal gene transfer. Geographical, ecological, and phylogenetic boundaries arc not always followed by antimicrobial resistance. According to Heuer et al. (2009), the use of antimicrobial medications in one ecological niche, like aquaculture, may therefore affect the prevalence of antimicrobial resistance in other ecological niches, like the human environment.

Infections with antibiotic-resistant strains are associated with poorer personal hygiene, relapse rates, chronicity, and pathological conditions of resistant organisms. People with compromised immune systems, such as undernourished children, cancer patients, and adults living with HIV, are particularly at risk because survival often depends on receiving effective treatment to prevent and treat serious infections (Ahmed et al. 2024). Since antibiotics are necessary for patient safety and to avoid complications in the future, antibiotic resistance also jeopardizes sophisticated medical procedures like organ transplants and prosthesis implants. In some infections, treatment failure is associated with an increased risk of death. For instance, it has been demonstrated that the mortality rate from methicillin-resistant *Staphylococcus aureus* bloodstream infections is two to three times higher than that of infections brought on by nonresistant bacteria. Researchers have found links between pathogenicity and antibiotic resistance. Genes linked to enterotoxin production or iron sequestering, which increase the invasiveness of their host strains, can be found in the same mobile genetic elements of genomic islands as genes that cause virulence, such as those found in several antibiotic resistance genes. As a result, the transfer and expression of these genes happen simultaneously (Capita and Alonso-Calleja 2013). AMR makes effective antibiotics ineffective and, more significantly, lengthens hospital stays, treatment durations, and dosages (Ur Rahman and Mohsin 2019). The main way that multidrug-resistant *E. coli* is spread is through food, especially meat. Although multidrug-resistant *E. coli,* which is commonly found in healthy people's stomachs, may not be the direct cause of disease, it could act as a breeding ground for the AMR gene. In addition to spreading through open defecation and inadequate sanitation, these resistant genes can spread to other Gram-negative pathogenic microbiota in the gut and become harmful, resulting in illness (Aworh et al. 2021).

The effects of eating meat products contaminated with resistant bacteria were demonstrated in a study by Sørensen et al. (2001), which found that animal-derived glycopeptide-resistant *Enterococcus fecium* that was consumed through chicken or pork remained in human stool for up to 14 days after consumption. According to Henning Sorum, resistance has typically spread and been established more quickly from aquatic bacterial sources to humans than from terrestrial animal bacterial sources to humans (Henning Sorum 2005). A study by Zhang and colleagues using Chinese agricultural workers revealed apramycin-resistant *E. coli* bacteria, which are utilized only in agriculture, not in human medicine. Workers on each farm in the study that used apramycin as an antimicrobial growth promoter were found to have apramycin resistance genes. The resistance gene, *aac(3)-IV,* was identical in all pig, poultry, and human isolates, and some resistance profiles were species specific Zhu et al. (2020). It has been demonstrated that antibiotics or resistant bacteria are present in farm dust, indoor and outdoor air currents from swine feeding facilities, groundwater associated with feeding operations, and food crops treated with antibiotic-containing manure. Innumerable microbial species are exposed to trace levels of antibiotics through their leakage into the environment, which is sufficient to select bacteria with resistance mutations and encourage the establishment and spread of resistance genes among bacteria of different kinds (Marshall and Levy 2011).

Human pathogens are tracked by several localized and international monitoring programs. In the United States, the National Antimicrobial Resistance Monitoring System has proven to be a valuable tool for monitoring resistance infections found in retail meats, human beings, and animal feed. Since advances in genetic analysis have made it possible to track the spread and emergence of horizontally transmissible genes, a new initiative called Reservoirs of Antibiotic Resistance (ROAR) aims to collect data on commensal and environmental strains that harbor antibiotic resistance genes. The ROAR project monitors the amount, diversity, and distribution of resistance genes by using geographic, phenotypic, and genotypic data from isolates from animal, water, plant, and soil sources worldwide, as well as commensals to track the evolution of resistance in human infections (Marshall and Levy 2011). In humans, resistance is managed through sanitary measures such as avoiding cross-contamination, reducing antibiotic use, and through hygiene practices such as avoiding contact between the mouth and feces. Reducing the use of antibiotics is ultimately the only way to reduce resistance in large animal populations. Changing animal husbandry practices, improving feed composition, and eliminating infectious diseases can all help. Additionally, the global use of antibiotics in animals will be cut in half if they are no longer used as feed additives to promote growth in animals raised for human consumption. This will reduce the public health risk of bacteria that are resistant or genes spreading from animals to people, but it would also be critical for maintaining antibiotic efficacy in veterinary care (Han et al. 2024).

2.9 ALTERNATIVES TO ANTIBIOTICS

A "Guidance for Industry" from the US Food and Drug Administration (FDA)'s Center for Veterinary Medicine 2013 describes guidelines and suggests limitations on the use of antibiotics in animal husbandry. Even though the new FDA guidance allows antibiotics to be used in animal husbandry to treat infections, they will no longer be allowed to be used for growth promotion, improved performance, or improved feed efficiency (Allen et al. 2013). In order to introduce new, safe, resistant-free, and affordable antibiotic alternatives, more innovative and proactive approaches to combat antibiotic resistance are urgently required (Yang et al. 2024). Hence, various novel strategies that have promising results as alternatives to antibiotics are discussed in the following sections.

2.9.1 Immunomodulators

Immunomodulators are compounds that can enhance immune function by altering the immune system's reaction to a stimulus in a positive way (Sonja Nodland

2020; Ruwali et al. 2018). Heat stress changes an animal's metabolism after birth and affects its immunologic development, which lowers fertility and, consequently, milk production. OmniGen-AF (OG), a feed additive that is patented, was created for the enhancement of immune function and has been shown to have immunomodulatory effects in ruminants and rodent models (McBride et al. 2020). Silicon dioxide, calcium and sodium aluminosilicate, brewer's dehydrated yeast, mineral oil, calcium carbonate, rice hulls, niacin supplement, biotin, D-calcium pantothenate, vitamin B12 supplement, choline chloride, thiamine mononitrate, pyridoxine hydrochloride, and riboflavin-5 are all included in the proprietary OmniGen-AF formulation (Skibiel et al. 2017). The innate immune response and daily milk production of dairy cows with an OG of 35 to 46 days relative to calving improved when they were challenged with a pathogen (Leiva et al. 2017). When OG was added to immunocompromised sheep's diet, the quantity of L-selectin on neutrophils and lymphocytes in circulation rose (Skibiel et al. 2017).

Global glitches and financial losses in the small-scale ruminant industry are due to blood-feeding parasites that cause dehydration anemia, loss of body weight, and demise in young animals (Ehsan et al. 2021). Various parasite species, including *Fasciola gigantica,* release calcium-binding proteins, which cause the hosts' immune system to respond with immunoglobulin E (IgE). The immunogenic potential of these proteins has been assessed, and primary immunization studies using certain tegumental proteins (TegPs) have shown that they offer a significant level of infection protection. Goat peripheral blood mononuclear cells (PBMCs) may bind HcESPs. PBMCs have immunomodulatory effects on a variety of factors that may help *Haemonchus contortus* evade the immune system (Ehsan et al. 2021). In poultry farming, STIMULAN, a commercially available immunomodulator, has been shown to be effective in treating acute and subchronic *Salmonella* infections (Khatab et al. 2021).

2.9.2 Probiotics

Probiotics are defined as "live bacteria that impose significant health benefits on the host when administered in suitable levels" and "direct-fed microbials" by the FAO and WHO (Ban and Guan 2021). The advantages of probiotics include the colonization and repair of the host's aberrant gut microbiome, the effective removal of pathogenic organisms and the generation of different antimicrobial molecules, cell adhesion, cell antagonism, and mucin expression, as well as control of the host's innate and acquired immunity (Gulbe et al. 2020; Li et al. 2021). Because of their biological activity and inhibitory qualities toward pathogenic microbes, probiotics have been proposed as an alternative to antibiotics. They have been demonstrated to produce a variety of nonviable metabolic by-products, including bacteriocins, organic acids, acetaldehydes, diacetyl, ethanol, and hydrogen peroxide, which are nontoxic, nonpathogenic, and resistant to enzyme systems in mammals (Azad et al. 2018; Fukuyama et al. 2020). All livestock sectors, including swine, poultry, cattle, lamb, and juvenile ruminants, have used probiotics during the past 20 years (Ban and Guan 2021; Johnson et al. 2019).

2.9.3 Prebiotics

Prebiotics are a "non-digestible food element that benefits the host by selectively encouraging the growth of one or a restricted number of bacteria in the colon, hence improving host health", according to Gibson and Roberfroid. Certain prebiotics prevent the colonization of pathogenic bacteria by binding to their type 1 fimbriae (Zmrhal and Slama 2020). Prebiotics also increase the number of goblet cells, which leads to increased mucin synthesis (Limbu et al. 2024). Beneficial microbiota can create organic acids and bacteriocins in response to prebiotic therapy, which prevents pathogenic bacteria from colonizing (Choi et al. 2017). Pathogen-associated molecular patterns can cause prebiotics to have a direct effect on the immune response. Pattern recognition receptors on immune system antigen-presenting cells are used to identify pathogen-associated molecular patterns (Zmrhal and Slama 2020).

2.9.4 Antimicrobial Peptide

The synthesis of peptides (host defense peptides [HDPs]), which is being studied as a possible antibiotic-free, disease-resistant, and inhibition technique, improves endogenous host defense. HDPs, or antimicrobial peptides, are short peptides that are primarily produced by phagocytes and mucosal epithelial cells. The immune system depends on HDPs, which have direct antibacterial activity against bacteria, fungi, protozoa, and viruses (Yang et al. 2021). Without the use of antibiotics, efforts to increase HDP accumulation on mucosal surfaces may improve animal immunity, disease resistance, and overall health (Robinson et al. 2018). HDPs are also known to have a variety of regulatory effects on immunity, including endotoxin neutralization, immune cell chemotaxis, phagocyte activation, and the onset of adaptive immunity. Some examples are butyrate, forskolin, and lactose that act in synergy to boost innate immunity by upregulating the expression of HDPs of the host, which was reported in poultry and can be employed for other animal species (Yang et al. 2021).

2.9.5 Host Defense Peptide Classification

Most host defense peptides (HDPs) in vertebrates belong to the defense or cathelicidin families. Defensins are classified based on α, β, and θ defensins and the explicit spacing pattern of cysteine residues. Although θ defensins have only been identified in primates, β-defensins are present in almost all vertebrates, whereas α-defensins are only found in a small number of mammalian species. The presence of cathelin domain, which is highly conserved, distinguishes cathelicidins, CAMP (cyclic adenosine monophosphate), or LL–37. In chickens, four cathelicidin genes have been found, notably fowlicidins 1–3 (or CATH1–3) and cathelicidin-B1, while cattle have 10 cathelicidin genes, and pigs have 11 cathelicidin genes (Robinson et al. 2018).

2.9.6 Egg Yolk Antibodies

In order to combat infectious diseases, egg yolk antibodies, also known as immunoglobulin Y (IgY), are emerging as a promising passive immunologic supplement and a substitute for antibiotics (Karamzadeh-Dehaghani et al. 2021). Egg yolk oral administration IgY is the most suitable method for suppressing enteric infections. IgY (chicken immunoglobulin Y) can move from serum to yolk during the egg-production process and performs similar tasks to IgG (mammalian immunoglobulin G). The yolk of an egg includes a substantial amount of IgY, which can provide passive protection to chicks against infections at both the embryonic and post-hatching stages (Tetik 2024). In order to produce a greater quantity of specific egg yolk IgY, which may be utilized as a potential antibiotic substitute in human and veterinary medicine, chickens are immunized with a specific antigen (Wang et al. 2021). One of the most common infections treated in calf diarrhea is enterotoxigenic *E. coli* (ETEC) K99 (Wang et al. 2019). *Campylobacter* infections, which are primarily associated with poultry products, are the most common bacterial foodborne disease that is treated with this method (Vandeputte et al. 2019).

2.9.7 Bacteriophage

The use of "phage therapy", or the application of lytic phages against pathogenic bacteria, which results in their lysis and eventual destruction—one of the crucial characteristics of phages that has garnered renewed attention—is an appealing alternative to antibiotic treatment due to the negative effects and ineffectiveness of antibiotics as well as the emergence of novel antibiotic-resistant strains (Kaur et al. 2021). Conventional phage therapy has shown encouraging results in human clinical trials. In vitro studies and animal models both use phages as therapeutic agents (Zhang et al. 2021b). Potential uses of phage therapy include reducing the amount of *Salmonella* bacteria that infiltrate the digestive tract, preventing the development of colibacillosis, which starts in the respiratory tract and air sacs, combating *Campylobacter* in the digestive tract of birds, and combating *E. coli* strains (Wernicki et al. 2017).

2.9.8 Endolysins

Endolysins are peptidoglycan hydrolases derived from phages that cause osmolysis and the release of replicating viral particles by breaking down the peptidoglycan layer in microorganisms (Barnas et al. 2024). According to Liu et al. (2024), bacteriophages are viruses that infect bacteria and produce lytic enzymes called endolysins, also known as lysins. LysKP213, another phage protein, facilitates endolysins' passage through the inner membrane and onto the peptidoglycan of the cell wall (Chu et al. 2024). Gram-positive bacteria do not have an outer membrane, so the peptidoglycan of the cell wall is vulnerable to the effects of externally applied endolysins (Tung et al. 2024). Because endolysins act outside of the pathogen, they can lyse these microorganisms and circumvent a number of intracellular drug resistance mechanisms, including efflux pumps. According to Swift et al. (2019), endolysins used in the production of animal feed pellets ought to be either thermostable or thermotolerant.

2.9.9 Antibacterial Vaccine

A vaccine that is safe, effective, and offers sustained defense against a variety of infections is the perfect one for pets. In addition, it needs to be affordable, easy to implement, often on a large scale (Hoelzer et al. 2018). There are two kinds of traditional vaccinations: live attenuated and inactivated/killed. A suspension of weakened or killed bacteria makes up the vaccine known as Bacterin. Live attenuated bacteria can replicate temporarily in the host antigen repertoire and exhibit a wide variety of functions. Furthermore, the fact that live strains stay in the animal body longer and are more likely to revert to full virulence is one of their main drawbacks. The dead vaccines are less dangerous than the live ones. In order to increase immune responses, the vaccine is produced by killing in vitro–generated bacterial cultures and loading them with adjuvants based in oil (Cheng et al. 2014).

2.9.10 Feed Enzymes

The enzymes are biologically active proteins that help break down nutrients chemically into smaller molecules for easier digestion and absorption. A number of enzymes produced by fermentation from microorganisms (fungi and bacteria) have been used in swine and poultry diets in recent years, and their importance in enhancing growth and feed conversion has long been acknowledged. Among the most frequently utilized enzymes are proteases, carbohydrases (including xylanase, cellulase, α-galactosidase, β-mannanase, α-amylase, and pectinase), and phytotase (Gadde et al. 2017; Yang et al. 2024).

2.9.11 Virion-Associated Peptidoglycan Hydrolases

The use of virion-associated peptidoglycan hydrolases (VAPGHs) to treat infectious disorders has piqued concern. These enzymes kill bacteria by degrading peptidoglycans and then osmotically lysing. Glycosidases, amidases, and endopeptidases are three types of VAPGHs based on the cleaving location on peptidoglycan. The phage tail and baseplate are the most prevalent places in the virion where VAPGH is found to help with peptidoglycan breakdown after phage adsorption. It has been discovered that VAPGH kills both Gram-negative bacteria with permeabilized outer membranes, such as *Pseudomonas aeruginosa*, and Gram-positive bacteria, such as *S. aureus* (Yuhui Yang et al. 2018).

2.9.12 Quorum Sensing Inhibitors

The density of the bacterial population determines the effectiveness of the quorum sensing (QS) communication mechanism. Numerous genes that regulate a variety of processes, such as bioluminescence, pathogenicity, biofilm formation, and sporulation, are activated by small diffusible signaling molecules. Since QS causes virulence in therapeutically relevant bacteria, inhibiting it appears to be a feasible method of managing these harmful bacteria. Uncontrolled antibiotic use has led to an alarming increase in the prevalence of bacteria resistant to antibiotics. To treat diseases brought on by organisms that are resistant to multiple drugs, new antibiotics and/or innovative approaches must be developed because medications are no longer the panacea. QS inhibition, also known as quorum quenching, has been one of these innovative strategies. Quorum sensing inhibitors (QSIs), or quorum quenchers, decrease the pathogenicity of bacteria instead of killing them or slowing their growth. The number of reports of treatment issues and failures in all known infections in humans, animals, and plants has increased due to multiple drug resistance. Alternatives to antibiotics for the treatment of bacterial infections are desperately needed in this situation, as are medications that can be used in combination with antibiotics to reduce the dosage of conventional antibiotics required for treatment.

QSIs are less likely to cause bacteria to develop a resistant phenotype because they do not place them in a life-and-death scenario. The bacterial mechanism of QS makes sure that there are enough bacterial cells for an external stimulus to cause a biological reaction. Numerous studies have been conducted in *Vibrio* since the QS mechanism was initially identified in *Vibrio fischeri* (Bhardwaj et al. 2013).

Biofilm is a complex microbial aggregation that is distinguished by the presence of an extracellular matrix composed of proteins, polysaccharides, and nucleic acids that is protective, dynamic, and adhesive. It is believed that the biofilm in which microbes live and communicate with one another (QS), which enables them to survive in hostile environments, is responsible for the vast majority of human illnesses. Autoinducers (AI) are extracellular signaling molecules released during QS, a type of cell-cell interaction that sets off a number of downstream cellular processes. While quorum quenching enzymes are enzymes that deactivate the QS signal, QS inhibitors are chemicals that interfere with the QS pathway. There are three methods to achieve quorum quenching: (I) blocking the production of the signal molecule, (II) blocking the degradation of the signal molecule, and (III) blocking the conduction of the signal molecule or its binding to receptors (Carradori et al. 2020).

2.10 CONCLUSION AND PROSPECTS

There has been a revolutionary change in the food-producing animal and poultry industry. As a result, antibiotics became essential because they aid in the treatment of bacterial and fungal infections in livestock. Therefore, their use in animal feed is done with utmost care. Currently, 40% of the antibiotics produced are utilized as feed additives. Antibiotics used are as follows: 1.5 million kilograms, 0.5 million kilograms, 1.4 million kilograms for the cattle, poultry, and swine industries, respectively, and 0.4 million kilograms for different animals. Antibiotics are used to improve growth rates, enhance feed conversion, and in prevention or recovery from infectious diseases such as mastitis, arthritis, respiratory problems, gastrointestinal infections, as well as bacterial infections, which include *Campylobacter fetus, Salmonella, Streptococcus suis, Leptospira, E. coli,* and *Enterococcus*. Antibiotics used in animal feed resulted in antibiotic resistance, which can result in life-threatening diseases in the human population. Thus, the public health concern is now more focused on identifying alternative control agents that could be supplemented in animal feeds. More attention and extensive research are being directed to identify potent antimicrobial substances that are resistance free. Strict legal regulations to control antibiotic abuse are required. The use of antibiotics in animal husbandry is rapidly rising worldwide, regardless of a country's geographic location, economic system, or legal system. Government agencies need to take a closer look at drugs used in animal farming and set important guidelines, such as dose, route of administration, and withdrawal time, and switch to safer, alternative antimicrobial agents that can reduce adverse effects on human life.

Acknowledgment: The authors would like to thank the Deanship of Scientific Research, Vice Presidency for Graduate Studies and Scientific Research, King Faisal University, Saudi Arabia, for its assistance (Grant No. KFU251718).

REFERENCES

Abdelhay Kaoud, H. 2019. "Introductory Chapter: Bacterial Cattle Diseases - Economic Impact and Their Control." In *Bacterial Cattle Diseases*. IntechOpen. https://doi.org/10.5772/intechopen.83635.

Ahmed, S K, S Hussein, K Qurbani, R H Ibrahim, A Fareeq, K A Mahmood, and M G Mohamed. 2024. "Antimicrobial Resistance: Impacts, Challenges, and Future Prospects." *Journal of Medicine, Surgery, and Public Health* 2: 100081.

Allen, H K, U Y Levine, T Looft, M Bandrick, and T A Casey. 2013. "Treatment, Promotion, Commotion: Antibiotic Alternatives in Food-Producing Animals." *Trends in Microbiology* 21 (3): 114–119. https://doi.org/10.1016/j.tim.2012.11.001.

Argudín, M A, A Deplano, A Meghraoui, M Dodémont, A Heinrichs, O Denis, C Nonhoff, and S Roisin. 2017. "Bacteria from Animals as a Pool of Antimicrobial Resistance Genes." *Antibiotics* 6 (2): 12. https://doi.org/10.3390/antibiotics6020012.

Awad, Y M, S C Kim, S A M Abd El-Azeem, K H Kim, K R Kim, K Kim, C Jeon, S S Lee, and Y S Ok. 2014. "Veterinary Antibiotics Contamination in Water, Sediment, and Soil near a Swine Manure Composting Facility." *Environmental Earth Sciences* 71 (3): 1433–1440. https://doi.org/10.1007/s12665-013-2548-z.

Aworh, M K, O Abiodun-Adewusi, N Mba, B Helwigh, and R S. Hendriksen. 2021. "Prevalence and Risk Factors for Faecal Carriage of Multidrug Resistant *Escherichia coli* Among Slaughterhouse Workers." *Scientific Reports* 11 (1): 13362. https://doi.org/10.1038/s41598-021-92819-3.

Azad, Md A K, M Sarker, and D Wan. 2018. "Immunomodulatory Effects of Probiotics on Cytokine Profiles." *BioMed Research International*: 1–10. https://doi.org/10.1155/2018/8063647.

Bacanlı, M and N Başaran. 2019. "Importance of Antibiotic Residues in Animal Food." *Food and Chemical Toxicology* 125: 462–466. https://doi.org/10.1016/j.fct.2019.01.033.

Balami, E Y, H I Abdulrahman, M M Gashua, H B Galadima, and I A Gulani. 2018. "Occurrence of Multi Drug Resistant Commensal *Escherichia coli* in Apparently Healthy Lambs and Kids from Maiduguri, Northeastern Nigeria." *Asian Journal of Research in Animal and Veterinary Sciences* 2: 1–10.

Ban, Y and L L Guan. 2021. "Implication and Challenges of Direct-Fed Microbial Supplementation to Improve Ruminant Production and Health." *Journal of Animal Science and Biotechnology* 12 (1): 109. https://doi.org/10.1186/s40104-021-00630-x.

Barnas, M R, W D Attuquayefio, D M Donovan, C D Skory, R W Hammond, G R Siragusa, and J R Timmons. 2024. "Yeast Expressing a Phage Endolysin Reduces Endogenous Clostridium perfringens Ex Vivo in 21-Day-Old Broiler Chicken Intestinal Fluids." *Avian Diseases* 68 (2): 129–133.

Barton, M D. 2000. "Antibiotic Use in Animal Feed and Its Impact on Human Health." *Nutrition Research Reviews*. 13 (2): 279–299. https://doi.org/10.1079/095442200108729106.

Bengtsson, B and C Greko. 2014. "Antibiotic Resistance-Consequences for Animal Health, Welfare, and Food Production." *Upsala Journal of Medical Sciences* 119 (2): 96–102. https://doi.org/10.3109/03009734.2014.901445.

Bhardwaj, K A, K Vinothkumar, and N Rajpara. 2013. “Bacterial Quorum Sensing Inhibitors: Attractive Alternatives for Control of Infectious Pathogens Showing Multiple Drug Resistance.” *Recent Patents on Anti-Infective Drug Discovery* 8 (1): 68–83. https://doi.org/10.2174/157489113805290809.

Bist, R B, K Bist, S Poudel, D Subedi, X Yang, B Paneru, S Mani, D Wang, and L Chai. 2024. “Sustainable Poultry Farming Practices: A Critical Review of Current Strategies and Future Prospects.” *Poultry Science*: 104295.

Boeckel, T P V, C Brower, M Gilbert, B T Grenfell, S A Levin, T P Robinson, A Teillant, and R Laxminarayan. 2015. “Global Trends in Antimicrobial Use in Food Animals.” *Proceedings of the National Academy of Sciences of the United States of America* 112 (18): 5649–5654. https://doi.org/10.1073/pnas.1503141112.

Boerlin, P. 2004. “Molecular Epidemiology of Antimicrobial Resistance in Veterinary Medicine: Where Do We Go?” *Animal Health Research Reviews* 5 (1): 95–102. https://doi.org/10.1079/AHRR200465.

Brown, K, R R E Uwiera, M L Kalmokoff, S P J Brooks, and G D Inglis. 2017. “Antimicrobial Growth Promoter Use in Livestock: A Requirement to Understand Their Modes of Action to Develop Effective Alternatives.” *International Journal of Antimicrobial Agents* 49 (1): 12–24. https://doi.org/10.1016/j.ijantimicag.2016.08.006.

Buret A G, 2010. “Immuno-Modulation and Anti-Inflammatory Benefits of Antibiotics: The Example of Tilmicosin.” *Canadian Journal of Veterinary Research* 74 (1): 1–10.

Bush, K and P A Bradford. 2016. “β-Lactams and β-Lactamase Inhibitors: An Overview.” *Cold Spring Harbor Perspectives in Medicine* 6 (8): a025247. https://doi.org/10.1101/cshperspect.a025247.

Butaye, P, L A Devriese, and F Haesebrouck. 2003. “Antimicrobial Growth Promoters Used in Animal Feed: Effects of Less Well Known Antibiotics on Gram-Positive Bacteria.” *Clinical Microbiology Reviews* 16(2): 175–188. https://doi.org/10.1128/CMR.16.2.175–188.2003.

Call, D R., M A Davis, and A A. Sawant. 2008. “Antimicrobial Resistance in Beef and Dairy Cattle Production.” *Animal Health Research Reviews* 9 (2): 159–167. https://doi.org/10.1017/S1466252308001515.

Capita, R and C Alonso-Calleja. 2013. “Antibiotic-Resistant Bacteria: A Challenge for the Food Industry.” *Critical Reviews in Food Science and Nutrition* 53 (1): 11–48. https://doi.org/10.1080/10408398.2010.519837.

Capstaff, N M and A J Miller. 2018. “Improving the Yield and Nutritional Quality of Forage Crops.” *Frontiers in Plant Science* 9: 535. https://doi.org/10.3389/fpls.2018.00535.

Carradori, S, N D Giacomo, M Lobefalo, G Luisi, C Campestre, and F Sisto. 2020. “Biofilm and Quorum Sensing Inhibitors: The Road so Far.” *Expert Opinion on Therapeutic Patents* 30 (12): 917–930. https://doi.org/10.1080/13543776.2020.1830059.

Cháfer-Pericás, C, Á Maquieira, and R Puchades. 2010. “Fast Screening Methods to Detect Antibiotic Residues in Food Samples.” *TrAC - Trends in Analytical Chemistry* 29 (9): 1038–1049. https://doi.org/10.1016/j.trac.2010.06.004.

Chandrasekara, A and T J Kumar. 2016. “Roots and Tuber Crops as Functional Foods: A Review on Phytochemical Constituents and Their Potential Health Benefits.” *International Journal of Food Science*: 1–15. https://doi.org/10.1155/2016/3631647.

Chang, Q, W Wang, G Regev-Yochay, M Lipsitch, and W P Hanage. 2015. “Antibiotics in Agriculture and the Risk to Human Health: How Worried Should We Be?” *Evolutionary Applications* 8 (3): 240–247. https://doi.org/10.1111/eva.12185.

Chattopadhyay, M K. 2014. “Use of Antibiotics as Feed Additives: A Burning Question.” *Frontiers in Microbiology* 5: 334. https://doi.org/10.3389/fmicb.2014.00334.

Chen, Q, X D Pan, B F Huang, J L Han, and B Zhou. 2019. "Screening of Multi-Class Antibiotics in Pork Meat by LC-Orbitrap-MS with Modified QuEChERS Extraction." *RSC Advances* 9 (48): 28119–28125. https://doi.org/10.1039/c9ra04853g.

Cheng, G, H Hao, S Xie, X Wang, M Dai, L Huang, and Z Yuan. 2014. "Antibiotic Alternatives: The Substitution of Antibiotics in Animal Husbandry?" *Frontiers in Microbiology* 5: 217. https://doi.org/10.3389/fmicb.2014.00217.

Cherian, G. 2020. "A Guide to the Principles of Animal Nutrition." Oregon State University, Web address: A Guide to the Principles of Animal Nutrition – Open Textbook Library. https://open.umn.edu/opentextbooks/textbooks/952.

Choi, H, S K Jung, J S Kim, K W Kim, K B Oh, P Y Lee, and S J Byun. 2017. "Effects of Dietary Recombinant Chlorella Supplementation on Growth Performance, Meat Quality, Blood Characteristics, Excreta Microflora, and Nutrient Digestibility in Broilers." *Poultry Science* 96 (3): 710–716. https://doi.org/10.3382/ps/pew345.

Chu, D, J Lan, L Liang, K Xia, L Li, L Yang, H Liu, and T Zhang. 2024. "The Antibacterial Activity of a Novel Highly Thermostable Endolysin, LysKP213, against Gram-Negative Pathogens is Enhanced When Combined with Outer Membrane Permeabilizing Agents." *Frontiers in Microbiology* 15: 1454618.

Collignon, P. 2005. "Fluoroquinolone Use in Food Animals." *Emerging Infectious Diseases* 11 (11): 1789–1792. https://doi.org/10.3201/eid1111.040630.

Cromwell, G L. 2002. "Why and How Antibiotics Are Used in Swine Production." *Animal Biotechnology* 13 (1): 7–27. https://doi.org/10.1081/ABIO–120005767.

Cruz-Rincon, S. 2021. "Common Bacterial Diseases in Backyard Chickens." https://veterinarypartner.vin.com/default.aspx?pid=19239&catId=102911&id=10048768 (Accessed on 14-06-2022).

Darwish, W S, E A Eldaly, M T El-Abbasy, Y Ikenaka, S Nakayama, and M Ishizuka. 2013. "Antibiotic Residues in Food: The African Scenario." *Japanese Journal of Veterinary Research* 61: S13–S22.

Duggar, B M. 1948. "Aureomycin: A Product of the Continuing Search for New Antibiotics." *Annals of the New York Academy of Sciences* 51 (2): 177–181. https://doi.org/10.1111/j.1749-6632.1948.tb27262.x.

Ehsan, M, R S Hu, J L Hou, H M. Elsheikha, X D Li, P H Liang, and X Q Zhu. 2021. "Fasciola Gigantica Tegumental Calcium-Binding EF-Hand Protein 4 Exerts Immunomodulatory Effects on Goat Monocytes." *Parasites and Vectors* 14 (1): 276. https://doi.org/10.1186/s13071-021-04784-5.

Erickson, P S and K F Kalscheur. 2019. "Nutrition and Feeding of Dairy Cattle." In *Animal Agriculture: Sustainability, Challenges and Innovations*, 157–180. Elsevier. https://doi.org/10.1016/B978-0-12-817052-6.00009-4.

FAO, 2015. "Responsible Use of Antibiotics in Aquaculture." https://agris.fao.org/agris-search/search.do?recordID=XF2006426547.

FAO, 2021. "Gateway to Dairy Production and Products." *Food and Agriculture Organization of the United Nations.*

Fejzic, N, M Begagic, S Šerić-Haračić, and M Smajlovic. 2014. "Beta Lactam Antibiotics Residues in Cow's Milk: Comparison of Efficacy of Three Screening Tests Used in Bosnia and Herzegovina." *Bosnian Journal of Basic Medical Sciences* 14 (3): 155–159. https://doi.org/10.17305/bjbms.2014.3.109.

Finlay, A C, and G L Hobby. 1950. "Terramycin, a new antibiotic." *Science*, New York, 111(2874), 85. https://doi.org/10.1126/science.111.2874.85

Floron C F, Jr. and V Fajt. 2018. "Proper Use of Sulfonamides in Market Show Animals." November 11. https://agrilifeextension.tamu.edu/.

Fukuyama, K, M A Islam, M Takagi, W Ikeda-Ohtsubo, S Kurata, H Aso, G Vignolo, J Villena, and H Kitazawa. 2020. "Evaluation of the Immunomodulatory Ability of

Lactic Acid Bacteria Isolated from Feedlot Cattle against Mastitis Using a Bovine Mammary Epithelial Cells in Vitro Assay." *Pathogens* 9 (5): 410. https://doi.org/10.3390/pathogens9050410.

Gadde, U, W H Kim, S T Oh, and H S Lillehoj. 2017. "Alternatives to Antibiotics for Maximizing Growth Performance and Feed Efficiency in Poultry: A Review." *Animal Health Research Reviews* 18 (1): 26–45. https://doi.org/10.1017/S1466252316000207.

Gallo, G F, and J L Berg. 1995. "Efficacy of a Feed-Additive Antibacterial Combination for Improving Feedlot Cattle Performance and Health." *The Canadian Veterinary Journal = La Revue Veterinaire Canadienne* 36 (4): 223–229.

Garcia-Galicia, I A., J A Arras-Acosta, M Huerta-Jimenez, A L Rentería-Monterrubio, J L Loya-Olguin, L M Carrillo-Lopez, J M. Tirado-Gallegos, and Alma D. Alarcon-Rojo. 2020. "Natural Oregano Essential Oil May Replace Antibiotics in Lamb Diets: Effects on Meat Quality." *Antibiotics* 9 (5): 248. https://doi.org/10.3390/antibiotics9050248.

Ghosh, S and T M LaPara. 2007. "The Effects of Subtherapeutic Antibiotic Use in Farm Animals on the Proliferation and Persistence of Antibiotic Resistance among Soil Bacteria." *ISME Journal* 1 (3): 191–203. https://doi.org/10.1038/ismej.2007.31.

González Pereyra, V, M Pol, F Pastorino, and A Herrero. 2015. "Quantification of Antimicrobial Usage in Dairy Cows and Preweaned Calves in Argentina." *Preventive Veterinary Medicine* 122 (3): 273–279. https://doi.org/10.1016/j.prevetmed.2015.10.019.

Gouws, R. 2019. *The Potential of Buffers (Acid Buf) or Plant Extracts (Xtract 7035) as Natural Alternatives for Ionophores in Lamb Feedlot Diets.* MSc (Agric) Animal Science Thesis: Animal Nutrition, Faculty of Natural & Agricultural Sciences University of Pretoria.

Gulbe, G, M Pilmane, V Saulīte, S Doniņa, J Jermolajevs, L Peškova, and A Valdovska. 2020. "Cells and Cytokines in Milk of Subclinically Infected Bovine Mammary Glands after the Use of Immunomodulatory Composition GLP 810." *Mediators of Inflammation*: 1–12. https://doi.org/10.1155/2020/8238029.

Gustafson, R H, and R E Bowen. 1997. "Antibiotic Use in Animal Agriculture." *Journal of Applied Microbiology* 83: 531–541.

Han, D, T Ren, Y Yang, Z Li, X Du, C Zhang, Q Pu, L He, K Zhao, R Guo, and J Xin. 2024. "Application and Substitution of Antibiotics in Animal Feeding." *Medycyna Weterynaryjna-Veterinary Medicine-Science and Practice* 80 (1): 5–11.

Hedman, H D., K A Vasco, and L Zhang. 2020. "A Review of Antimicrobial Resistance in Poultry Farming within Low-Resource Settings." *Animals* 10 (8): 1264. https://doi.org/10.3390/ani10081264.

Heuer, O E, H Kruse, K Grave, P Collignon, I Karunasagar, and F J Angulo. 2009. "Human Health Consequences of Use of Antimicrobial Agents in Aquaculture." *Clinical Infectious Diseases* 49 (8): 1248–1253. https://doi.org/10.1086/605667.

Ho, Y B, M P Zakaria, P A Latif, and N Saari. 2014. "Occurrence of Veterinary Antibiotics and Progesterone in Broiler Manure and Agricultural Soil in Malaysia." *Science of the Total Environment* 488–489 (1): 261–267. https://doi.org/10.1016/j.scitotenv.2014.04.109.

Hoelzer, K, L Bielke, D P Blake, E Cox, S M Cutting, B Devriendt, E Erlacher-Vindel, et al. 2018. "Vaccines as Alternatives to Antibiotics for Food Producing Animals. Part 1: Challenges and Needs." *Veterinary Research* 49 (1): 64. https://doi.org/10.1186/s13567-018-0560-8.

Hosain, Z, S M Lutful Kabir, and M Kamal. 2021. "Antimicrobial Uses for Livestock Production in Developing Countries." *International Journal of One Health* 14 (1): 210–221. https://doi.org/10.14202/vetworld.2021.210–221.

Hughes, P and J Heritage. 2013. "Antibiotic Growth-Promoters in Food Animals." *FAO Animal Production and Health Paper* 160: 129–152.

Ioannis Mavromichalis. 2017. "Understanding Oligosaccharides in Young Animal Nutrition." September 14. https://www.wattagnet.com/articles/31830-understanding-oligosaccharides-in-young-animal-nutrition?v=preview.

Johnson, C N., M H Kogut, K Genovese, H He, S Kazemi, and R J. Arsenault. 2019. "Administration of a Postbiotic Causes Immunomodulatory Responses in Broiler Gut and Reduces Disease Pathogenesis Following Challenge." *Microorganisms* 7 (8): 268. https://doi.org/10.3390/microorganisms7080268.

Karamzadeh-Dehaghani, A, A Towhidi, M Zhandi, N Mojgani, and A Fouladi-Nashta. 2021. "Combined Effect of Probiotics and Specific Immunoglobulin Y Directed against Escherichia Coli on Growth Performance, Diarrhea Incidence, and Immune System in Calves." *Animal* 15 (2): 100124. https://doi.org/10.1016/j.animal.2020.100124.

Karthik, K and M Prabhu. 2021. :"Bacterial Diseases of Goat and Its Preventive Measures." In *Goat Science - Environment, Health and Economy*. https://doi.org/10.5772/intechopen.97434

Kaur, S, A Kumari, A K Negi, V Galav, S Thakur, M Agrawal, and V Sharma. 2021. "Nanotechnology Based Approaches in Phage Therapy: Overcoming the Pharmacological Barriers." *Frontiers in Pharmacology* 12: 699054. https://doi.org/10.3389/fphar.2021.699054.

Kelly, A M. 2013. *Medicated Feed for Food Fish*. Southern Regional Aquaculture Center.

Khatab, S A, S A Hemeda, A F El-Nahas, W S.H Abd El Naby, S Hassan, J A Alorabi, and M A O Dawood. 2021. "Intra- and Inter-Breed Variation in Immune Response to Acute and Sub-Chronic Salmonella Infection and Commercial Immune-Stimulant in Two-Layer Breeds." *Veterinary Medicine and Science*: 1–9. https://doi.org/10.1002/vms3.621.

Kohn, H and W Widger. 2005. "The Molecular Basis for the Mode of Action of Bicyclomycin." *Current Drug Targets Infectious Disorders* 5 (3): 273–295. https://doi.org/10.2174/1568005054880136.

Lagha, A B, B Haas, M Gottschalk, and D Grenier. 2017. "Antimicrobial Potential of Bacteriocins in Poultry and Swine Production." *Veterinary Research* 48 (1): 22. https://doi.org/10.1186/s13567-017-0425-6.

Landers, T F, B Cohen, T E Wittum, and E L Larson. 2012. "A Review of Antibiotic Use in Food Animals: Perspective, Policy, and Potential." *Public Health Reports (Washington, DC:1974)* 127(1): 4–22. https://doi.org/10.1177/003335491212700103

Landfried, L K, E K Barnidge, P Pithua, R D Lewis, J A Jacoby, C C King, and C R Baskin. 2018. "Antibiotic Use on Goat Farms: An Investigation of Knowledge, Attitudes, and Behaviors of Missouri Goat Farmers." *Animals* 8 (11): 198. https://doi.org/10.3390/ani8110198.

Leiva, T, R F Cooke, A P Brandão, K M Schubach, L F D Batista, M F Miranda, E A Colombo, et al. 2017. "Supplementing an Immunomodulatory Feed Ingredient to Modulate Thermoregulation, Physiologic, and Production Responses in Lactating Dairy Cows under Heat Stress Conditions." *Journal of Dairy Science* 100 (6): 4829–4838. https://doi.org/10.3168/jds.2016–12258.

Li, Y, D Jia, J Wang, H Li, X Yin, J Liu, J Wang, et al. 2021. "Probiotics Isolated From Animals in Northwest China Improve the Intestinal Performance of Mice." *Frontiers in Veterinary Science* 8: 750895. https://doi.org/10.3389/fvets.2021.750895.

Limbu, D, B R Sarkar, and M D Adhikari. 2024. "Role of Probiotics and Prebiotics in Animal Nutrition." *Sustainable Agriculture Reviews: Animal Biotechnology for Livestock Production* 4: 173–204.

Lin, Z-N, L Ye, Z-W Li, X-S Huang, Z Lu, Y-Q Yang, H-W Xing, J-Y Bai, and Z-Y Ying. 2020. "Chinese Herb Feed Additives Improved the Growth Performance, Meat Quality, and Nutrient Digestibility Parameters of Pigs." *Animal Models and Experimental Medicine* 3 (1): 47–54. https://doi.org/10.1002/ame2.12104.

Lisa, S and R Robert. 2012. "FDA Order Limits Cephalosporin Use in Food Animals." https://www.cidrap.umn.edu.

Liu, B, Z Chang, Z Li, R Liu, and X Liu. 2024. "Prediction of Key Amino Acids of Salmonella Phage Endolysin LysST-3 and Detection of Its Mutants' Activity." *Archives of Microbiology* 206(4): 151.

López-Alonso, M. 2012. "Trace Minerals and Livestock: Not Too Much Not Too Little." *ISRN Veterinary Science*, 1–18. https://doi.org/10.5402/2012/704825.

Ma, F, S Xu, Z Tang, Z Li, and L Zhang. 2021. "Use of Antimicrobials in Food Animals and Impact of Transmission of Antimicrobial Resistance on Humans." *Biosafety and Health*. 3 (1): 32–38. https://doi.org/10.1016/j.bsheal.2020.09.004.

Macdiarmid, S C. 1978. "Antibacterial Drugs Used against Mastitis in Cattle by the Systemic Route." *New Zealand Veterinary Journal* 26 (12): 290–295. https://doi.org/10.1080/00480169.1978.34574.

Mallick, P, K Muduli, J N Biswal, and J Pumwa. 2020. "Broiler Poultry Feed Cost Optimization Using Linear Programming Technique." *Journal of Operations and Strategic Planning* 3 (1): 31–57. https://doi.org/10.1177/2516600x19896910.

Malomo, G A and N E Ihegwuagu. 2017. "Some Aspects of Animal Feed Sampling and Analysis." In *Ideas and Applications Toward Sample Preparation for Food and Beverage Analysis*. InTechopen. https://doi.org/10.5772/intechopen.70856.

Marshall, B M, and S B Levy. 2011. "Food Animals and Antimicrobials: Impacts on Human Health." *Clinical Microbiology Reviews* 24 (4): 718–733. https://doi.org/10.1128/CMR.00002–11.

Mathew, A G, R Cissell, and S Liamthong. 2007. "Antibiotic Resistance in Bacteria Associated with Food Animals: A United States Perspective of Livestock Production." *Foodborne Pathogens and Disease* 4 (2): 115–133. https://doi.org/10.1089/fpd.2006.0066.

Matsushima J. K. 1979. "Classification of Feeds." In *Feeding Beef Cattle. Advanced Series in Agricultural Sciences* 7. Heidelberg: Springer-Verlag Berlin. https://doi.org/10.1007/978-3-642-67199-9 e

McBride, M L, N C Burdick Sanchez, J A Carroll, P R Broadway, X A Ortiz, J L Collier, J D Chapman, et al. 2020. "Response to Adrenocorticotropic Hormone or Corticotrophin-Releasing Hormone and Vasopressin in Lactating Cows Fed an Immunomodulatory Supplement under Thermoneutral or Acute Heat Stress Conditions." *Journal of Dairy Science* 103 (7): 6612–6626. https://doi.org/10.3168/jds.2019–17548.

McEwen, S A, and P J Collignon. 2018. "Antimicrobial Resistance: A One Health Perspective." *Microbiology Spectrum* 6 (2). ARBA–0009. 1–26. https://doi.org/10.1128/microbiolspec.arba-0009–2017.

McGuire, J M, R L Bunch, R C Anderson, H E Boaz, E H Flynn, H M Powell, and J W Smith. 1952. "Ilotycin, a New Antibiotic." *Antibiotics & Chemotherapy (Northfield)* 2 (6): 281–283. https://pubmed.ncbi.nlm.nih.gov/24541924/.

Mehdi, Y, M P Létourneau-Montminy, M L Gaucher, Y Chorfi, G Suresh, T Rouissi, S K Brar, C Côté, A A Ramirez, and S Godbout. 2018. "Use of Antibiotics in Broiler Production: Global Impacts and Alternatives." *Animal Nutrition* 4 (2): 170–178. https://doi.org/10.1016/j.aninu.2018.03.002.

Menkem, Z E, B L Ngangom, S S A Tamunjoh, and F F Boyom. 2019. "Antibiotic Residues in Food Animals: Public Health Concern." *Acta EcologicaSinica* 39 (5): 411–415. https://doi.org/10.1016/j.chnaes.2018.10.004.

Mingeot-Leclercq, M P, Y Glupczynski, and P M Tulkens. 1999. "Aminoglycosides: Activity and Resistance." *Antimicrobial Agents and Chemotherapy* 43(4): 727–737. https://doi.org/10.1128/AAC.43.4.727.

Miroshnik, A S, M I Slozhenkina, I F Gorlov, M v Frolova, D v Nikolaev, and N I Mosolova. 2021. "The Use of Biologically Active Supplements as an Alternative for Application of Antibiotics in Production of Pork." In *IOP Conference Series:*

Earth and Environmental Science 677: 032009. IOP Publishing Ltd. https://doi.org/10.1088/1755–1315/677/3/032009.

Mithuna, R, R Tharanyalakshmi, I Jain, S Singhal, D Sikarwar, S Das, J Ranjitha, D Ghosh, M M Rahman, and B Das. 2024, "Emergence of Antibiotic Resistance Due to the Excessive Use of Antibiotics in Medicines and Feed Additives: A Global Scenario with Emphasis on the Indian Perspective." *Emerging Contaminants* 10 (4): 100389. https://doi.org/10.1016/j.emcon.2024.100389.

Mohr, K I. 2016. "History of Antibiotics Research." *Current Topics in Microbiology and Immunology* 398: 237–272. https://doi.org/10.1007/82_2016_499.

Monger, X C., A A Gilbert, L Saucier, and A T Vincent. 2021. "Antibiotic Resistance: From Pig to Meat." *Antibiotics* 10 (10): 1209. https://doi.org/10.3390/antibiotics10101209.

Moore, C E. 2019. "Changes in Antibiotic Resistance in Animals." *Science (New York, N.Y.)* 365 (6459): 1251–1252. https://doi.org/10.1126/science.aay9652.

Muloi, D, M J Ward, A B Pedersen, E M Fèvre, M E J. Woolhouse, and B A D. van Bunnik. 2018. "Are Food Animals Responsible for Transfer of Antimicrobial-Resistant Escherichia Coli or Their Resistance Determinants to Human Populations? A Systematic Review." *Foodborne Pathogens and Disease* 15 (8): 467–474. https://doi.org/10.1089/fpd.2017.2411.

Mutua, F, G Sharma, D Grace, S Bandyopadhyay, B Shome, and J Lindahl. 2020. "A Review of Animal Health and Drug Use Practices in India, and Their Possible Link to Antimicrobial Resistance." *Antimicrobial Resistance and Infection Control* 9 (1): 103. https://doi.org/10.1186/s13756-020-00760-3.

National Research Council (US) Committee to Study the Human Health Effects of Subtherapeutic Antibiotic Use in Animal Feeds. (1980). *The Effects on Human Health of Subtherapeutic Use of Antimicrobials in Animal Feeds*. Washington (DC): National Academies Press (US).

Nodland, S. 2020. "What Is an Immunomodulator, and How Does It Help Support Immunity?" https://wellmune.com/2020/05/07/what-is-an-immunomodulator-and-how-does-it-help-support-immunity/.

Ojo, O E, E Fabusoro, A A Majasan, and M A Dipeolu. 2016. "Antimicrobials in Animal Production: Usage and Practices among Livestock Farmers in Oyo and Kaduna States of Nigeria." *Tropical Animal Health and Production* 48 (1): 189–197. https://doi.org/10.1007/s11250-015-0939-8.

Oliver, S P, S E Murinda, and B M Jayarao. 2011. "Impact of Antibiotic Use in Adult Dairy Cows on Antimicrobial Resistance of Veterinary and Human Pathogens: A Comprehensive Review." *Foodborne Pathogens and Disease* 8(3): 337–355. https://doi.org/10.1089/fpd.2010.0730.

Pormohammad, A, M J Nasiri, and T Azimi. 2019. "Prevalence of Antibiotic Resistance in *Escherichia coli* Strains Simultaneously Isolated from Humans, Animals, Food, and the Environment: A Systematic Review and Meta-Analysis." *Infection and Drug Resistance* 12: 1181–1197. https://doi.org/10.2147/IDR.S201324.

Putnam, D H and S B Orloff. 2014. "Forage Crops." *Encyclopedia of Agriculture and Food Systems*: 381–405. https://doi.org/10.1016/B978-0-444-52512-3.00142-X.

Pyörälä, S, K E Baptiste, B Catry, E van Duijkeren, C Greko, M A Moreno, M C M F Pomba, M Rantala, M Ružauskas, P Sanders, E J Threlfall, J Torren-Edo, and K Törneke. 2014. "Macrolides and Lincosamides in Cattle and Pigs: Use and Development of Antimicrobial Resistance." *Veterinary Journal* 200(2): 230–239. https://doi.org/10.1016/j.tvjl.2014.02.028.

Quintanilla, P, E Doménech, I Escriche, M C Beltrán, and M P Molina. 2019. "Food Safety Margin Assessment of Antibiotics: Pasteurized Goat's Milk and Fresh Cheese." *Journal of Food Protection* 82 (9): 1553–1559. https://doi.org/10.4315/0362–028X.JFP-18–434.

Rahman, S. 2019. "The Under Reported Issue of Antibiotic-Resistance in Food-Producing Animals in Pakistan." *Pakistan Veterinary Journal* 39 (03): 323–328. https://doi.org/10.29261/pakvetj/2019.037.

Rizaldi, A, D W Lukman, and H Pisestyani. 2019. "Antibiotic Resistance of Escherichia Coli in Pork Sold at TamiangLayang Market, East Barito District." *Advances in Animal and Veterinary Sciences* 7 (9): 791–797. https://doi.org/10.17582/journal.aavs/2019/7.9.791.797.

Robinson, K, X Ma, Y Liu, S Qiao, Y Hou, and G Zhang. 2018. "Dietary Modulation of Endogenous Host Defense Peptide Synthesis as an Alternative Approach to In-Feed Antibiotics." *Animal Nutrition* 4 (2): 160–169. https://doi.org/10.1016/j.aninu.2018.01.003.

Roth, N, A Käsbohrer, S Mayrhofer, U Zitz, C Hofacre, and K J Domig. 2019. "The Application of Antibiotics in Broiler Production and the Resulting Antibiotic Resistance in *Escherichia coli*: A Global Overview." *Poultry Science* 98 (4): 1791–1804. https://doi.org/10.3382/ps/pey539.

Ruwali, P, T K Ambwani, and P Gautam. 2018. "In Vitro Immunomodulatory Potential of Artemisia Indica Willd. In Chicken Lymphocytes." *Veterinary World* 11 (1): 80–87. https://doi.org/10.14202/vetworld.2018.80–87.

Saha, U K, L S Sonon, D W Hancock, N S Hill, L Stewart, G L Heusner, and D E Kissel. (2010). "Common Terms Used in Animal Feeding and Nutrition." https://extension.uga.edu/.

Salvatore, M J and S E Katz. 1993. "Unified Procedure for the Determination of Antibiotics in Animal Feeds." *Journal of AOAC International* 76 (3): 514–525.

Sapkota, A R., L Y Lefferts, S McKenzie, and P Walker. 2007. "What Do We Feed to Food-Production Animals? A Review of Animal Feed Ingredients and Their Potential Impacts on Human Health." *Environmental Health Perspectives* 115 (5): 663–670. https://doi.org/10.1289/ehp.9760.

Sawant, A A, L M Sordillo, and B M Jayarao. 2005. "A Survey on Antibiotic Usage in Dairy Herds in Pennsylvania." *Journal of Dairy Science* 88 (8): 2991–2999. https://doi.org/10.3168/jds.S0022–0302(05)72979–9.

Schar, D, E Y Klein, R Laxminarayan, M Gilbert, and T P van Boeckel. 2020. "Global Trends in Antimicrobial Use in Aquaculture." *Scientific Reports* 10 (1): 21878. https://doi.org/10.1038/s41598-020-78849-3.

Sebastian, S, A A Tom, J A Babu, and M Joshy. 2021. "Antibiotic Resistance in *Escherichia coli* Isolates from Poultry Environment and UTI Patients in Kerala, India: A Comparison Study." *Comparative Immunology, Microbiology and Infectious Diseases* 75: 101614. https://doi.org/10.1016/j.cimid.2021.101614.

Sepe, L and A Argüello. 2019. "Recent Advances in Dairy Goat Products." *Asian-Australasian Journal of Animal Sciences* 32 (8): 1306–1320. https://doi.org/10.5713/ajas.19.0487.

Sharma M, A B Shrivastav, Y P Sahni, and G Pandey. 2012. "Overviews of the Treatment and Control of Common Fish Diseases." *International Research Journal of Pharmacy.* 3 (7): 123–127.

Siregar, R S, Y Widiyastuti, M S Siregar, and M Silalahi. 2024. "Ginger as an Animal Feed Additive: An Overview." *Journal of Animal and Plant Sciences* 34 (1): 31–49.

Sivagami, K, V J Vignesh, R Srinivasan, G Divyapriya, and I M Nambi. 2020. "Antibiotic Usage, Residues and Resistance Genes from Food Animals to Human and Environment: An Indian Scenario." *Journal of Environmental Chemical Engineering* 8 (1). https://doi.org/10.1016/j.jece.2018.02.029.

Sivaraman, S. 2018. *Antibiotic Use in Food Animals: India Overview.* ReAct Asia-Pacific.

Skandalis, N, M Maeusli, D Papafotis, S Miller, B Lee, I Theologidis, and B Luna. 2021. "Environmental Spread of Antibiotic Resistance." *Antibiotics* 10 (6): 640. https://doi.org/10.3390/antibiotics10060640.

Skibiel, A L., T F Fabris, F N Corrá, Y M Torres, D J McLean, J D Chapman, D J Kirk, G E Dahl, and J Laporta. 2017. "Effects of Feeding an Immunomodulatory Supplement to Heat-Stressed or Actively Cooled Cows during Late Gestation on Postnatal Immunity, Health, and Growth of Calves." *Journal of Dairy Science* 100 (9): 7659–7668. https://doi.org/10.3168/jds.2017–12619.

Sørensen, T L, M Blom, D L Monnet, N Frimodt-Møller, R L Poulsen, and F Espersen. 2001. "Transient Intestinal Carriage after Ingestion of Antibiotic-Resistant *Enterococcus faecium* from Chicken and Pork." *New England Journal of Medicine* 345 (16): 1161–1166. https://doi.org/10.1056/nejmoa010692.

Sorum, H. 2005. *Antimicrobial Resistance in Bacteria of Animal Origin*. Frank M. ed, Aarestrup. Washington, DC, USA: ASM Press. https://doi.org/10.1128/9781555817534.

Starr, M P, and D M Reynolds. 1951. "Streptomycin Resistance of Coliform Bacteria from Turkeys Fed Streptomycin." *American Journal of Public Health and the Nation's Health* 41 (11): 1375–1380. https://doi.org/10.2105/ajph.41.11_pt_1.1375.

Sweileh, W M. 2021. "Global Research Activity on Antimicrobial Resistance in Food-Producing Animals." *Archives of Public Health* 79 (1): 49. https://doi.org/10.1186/s13690-021-00572-w.

Swift, S M, K P Reid, D M Donovan, and T G Ramsay. 2019. "Thermophile Lytic Enzyme Fusion Proteins That Target *Clostridium pTarkerfringens*." *Antibiotics* 8 (4): 214. https://doi.org/10.3390/antibiotics8040214.

Tang, K L, N P Caffrey, D B Nóbrega, S C Cork, P E Ronksley, H W Barkema, A J Polachek, et al. 2017. "Restricting the Use of Antibiotics in Food-Producing Animals and Its Associations with Antibiotic Resistance in Food-Producing Animals and Human Beings: A Systematic Review and Meta-Analysis." *The Lancet Planetary Health* 1 (8): e316–e327. https://doi.org/10.1016/S2542–5196(17)30141–9.

Tark, D S, D C Moon, H Y Kang, S R Kim, H M Nam, H S Lee, S C Jung, and S K Lim. 2017. "Antimicrobial Susceptibility and Characterization of Extended-Spectrum β-Lactamases in *Escherichia coli* Isolated from Bovine Mastitic Milk in South Korea from 2012 to 2015." *Journal of Dairy Science* 100 (5): 3463–3469. https://doi.org/10.3168/jds.2016–12276.

Tetik, K, 2024. "Avian IgY Antibodies and Its Immunotherapeutic Applications." *Journal of Istanbul Veterinary Sciences* 8 (1): 64–74.

Tufarelli, V, M Ragni, and V Laudadio. 2018. "Feeding Forage in Poultry: A Promising Alternative for the Future of Production Systems." *Agriculture* 8 (6): 81. https://doi.org/10.3390/agriculture8060081.

Tung, C W, D Julianingsih, C Canagarajah, G Sellers, A Scriba, Z Alvarado-Martínez, Z Tabashsum, and D Biswas, 2024. The Effectiveness of Endolysin ENDO-1252 from Salmonella Bacteriophage-1252 against Nontyphoidal Salmonella Enterica." *FEMS Microbiology Letters*, 371: fnae051.

Ur Rahman, S and M Mohsin. 2019. "The under Reported Issue of Antibiotic-Resistance in Food-Producing Animals in Pakistan." *Pakistan Veterinary Journal* 39 (3): 323–328. https://doi.org/10.29261/pakvetj/2019.037.

Van Duijkeren, E, C Greko, M Pringle, K E Baptiste, B Catry, H Jukes, M A Moreno, M C M F Pomba, S Pyörälä, M Rantala, M Ružauskas, P Sanders, C Teale, E J Threlfall, J Torren-Edo, and K van Törneke. 2014. "Pleuromutilins: Use in Food-Producing Animals in the European Union, Development of Resistance and Impact on Human and Animal Health." *The Journal of Antimicrobial Chemotherapy* 69(8): 2022–2031. https://doi.org/10.1093/jac/dku123.

Vandeputte, J, A Martel, S Canessa, N van Rysselberghe, L de Zutter, M Heyndrickx, F Haesebrouck, F Pasmans, and A Garmyn. 2019. "Reducing Campylobacter Jejuni

Colonization in Broiler Chickens by In-Feed Supplementation with Hyperimmune Egg Yolk Antibodies." *Scientific Reports* 9 (1): 8931. https://doi.org/10.1038/s41598-019-45380-z.

Varadaraj, M. C. 2010. "Capacity Building: Building Analytical Capacity for Microbial Food Safety." In Boisrobert C, Stjepanovic A, Oh S, Lelieveld H, eds, *Ensuring Global Food Safety – Exploring global Harmonization*, 151–176. London, UK: Elsevier Inc.

Wang, H, X Zeng, and J Lin. 2021. "Ex Vivo Evaluation of Egg Yolk IgY Degradation in Chicken Gastrointestinal Tract." *Frontiers in Immunology* 21 (12): 746831. https://doi.org/10.3389/fimmu.2021.746831.

Wang, J, L Deng, M Chen, Y Che, L Li, L Zhu, G Chen, and T Feng. 2024. "Phytogenic Feed Additives as Natural Antibiotic Alternatives in Animal Health and Production: A Review of the Literature of the Last Decade." *Animal Nutrition* 22 (17): 244–264. https://doi.org/10.1016/j.aninu.2024.01.012.

Wang, L and D Li. 2024. "Current Status, Challenges and Prospects for Pig Production in Asia." *Animal Bioscience* 37 (4): 742. https://doi.org/10.5713/ab.23.0303

Wang, Z, J Li, J Li, Y Li, L Wang, Q Wang, L Fang, X Ding, P Huang, J Yin, Y Yin and Huansheng Yang. 2019. "Protective Effect of Chicken Egg Yolk Immunoglobulins (IgY) against Enterotoxigenic Escherichia Coli K88 Adhesion in Weaned Piglets." *BMC Veterinary Research* 15 (234): 1–12.

Washer, P and H Joffe. 2006. "The 'Hospital Superbug': Social Representations of MRSA." *Social Science and Medicine* 63 (8): 2141–2152. https://doi.org/10.1016/j.socscimed.2006.05.018.

Wernicki, A, A Nowaczek, and R Urban-Chmiel. 2017. "Bacteriophage Therapy to Combat Bacterial Infections in Poultry." *Virology Journal* 14 (1): 179. https://doi.org/10.1186/s12985-017-0849-7.

WHO. 2015. *WHO Library Cataloguing-in-Publication Data Global Action Plan on Antimicrobial Resistance*. www.paprika-annecy.com.

WHO. 2019. "Critically Important Antimicrobials for Human Medicine 6th Revision 2018 Ranking of Medically Important Antimicrobials for Risk Management of Antimicrobial Resistance Due to Non-Human Use."

Wu X, X Shen, X Cao, R Nie, H Zhang, C Tang, and W Wang. 2021. "Simultaneous Determination of Amphenicols and Metabolites in Animal-Derived Foods Using Ultrahigh-Performance Liquid Chromatography-Tandem Mass Spectrometry". *International Journal of Analytical Chemistry* 3: 3613670. https://doi.org/10.1155/2021/3613670.

Yadav, M K, A Khati, R Chauhan, P Arya, and A Semwal. 2021. "A Review on Feed Additives Used in Fish Diet." *International Journal of Environment, Agriculture and Biotechnology* 6 (2): 184–190. https://doi.org/10.22161/ijeab.62.21.

Yang, Q, M A Whitmore, K Robinson, W Lyu, and G Zhang. 2021. "Butyrate, Forskolin, and Lactose Synergistically Enhance Disease Resistance by Inducing the Expression of the Genes Involved in Innate Host Defense and Barrier Function." *Antibiotics* 10 (10): 1175. https://doi.org/10.3390/antibiotics10101175.

Yang, W, J Li, Z Yao, and M Li. 2024. "A Review on the Alternatives to Antibiotics and the Treatment of Antibiotic Pollution: Current Development and Future Prospects." *Science of The Total Environment*: 171757. https://doi.org/10.1016/j.scitotenv.2024.171757

Yang, Y, A J. Ashworth, C Willett, K Cook, A Upadhyay, P R Owens, S C Ricke, J M DeBruyn, and P A Moore. 2019. "Review of Antibiotic Resistance, Ecology, Dissemination, and Mitigation in U.S. Broiler Poultry Systems." *Frontiers in Microbiology* 10: 2639. https://doi.org/10.3389/fmicb.2019.02639.

Yang, Y, S Le, W Shen, Q Chen, Y Huang, S Lu, Y Tan, M Li, F Hu, and Y Li. 2018. "Antibacterial Activity of a Lytic Enzyme Encoded by Pseudomonas Aeruginosa Double

Stranded RNA Bacteriophage PhiYY." *Frontiers in Microbiology* 9: 1778. https://doi.org/10.3389/fmicb.2018.01778.
Yuan, X, Z Lv, Z Zhang, Y Han, Z Liu, and H Zhang, 2023. "A Review of Antibiotics, Antibiotic Resistant Bacteria, and Resistance Genes in Aquaculture: Occurrence, Contamination, and Transmission." *Toxics* 11 (5): 420.
Zhang, H, Y Luo, L Wu, Y Huang, and P Christie. 2015. "Residues and Potential Ecological Risks of Veterinary Antibiotics in Manures and Composts Associated with Protected Vegetable Farming." *Environmental Science and Pollution Research* 22 (8): 5908–5918. https://doi.org/10.1007/s11356-014-3731-9.
Zhang, L, L Shi, Q He, and Y Li. 2021a. "A Rapid Multiclass Method for Antibiotic Residues in Goat Dairy Products by UPLC-Quadrupole/Electrostatic Field Orbitrap High-Resolution Mass Spectrometry." *Journal of Analytical Science and Technology* 12 (1): 14. https://doi.org/10.1186/s40543-021-00268-4.
Zhang, Y, B Meng, X Wei, Y Li, X Wang, Y Zheng, C Wang, L Cui, and X Zhao. 2021b. "Evaluation of Phage Therapy for Pulmonary Infection of Mouse by Liquid Aerosol-Exposure *Pseudomonas aeruginosa*." *Infection and Drug Resistance* 14: 4457–4469. https://doi.org/10.2147/IDR.S326230.
Zhu, X., Li, P., Qian, C., Liu, H., Lin, H., Zhang, X., Li, Q., Lu, J., Lin, X., Xu, T., Zhang, H., Hu, Y., Bao, Q., & Li, K. (2020). Prevalence of aminoglycoside resistance genes and molecular characterization of a novel gene, aac(3)-IIg, among clinical isolates of the Enterobacter cloacae complex from a Chinese teaching hospital. *Antimicrobial Agents and Chemotherapy*, 64(9). https://doi.org/10.1128/AAC.00852-20'
Zmrhal, V and P Slama. 2020. "Immunomodulation of Avian Dendritic Cells under the Induction of Prebiotics." *Animals* 10 (4): 698. https://doi.org/10.3390/ani10040698.
"A Review of the Literature on Antibiotic Use and Antibiotic-Resistant Bacteria in Food Animals. In Antibiotic Use and Resistance in Food Animals Current Policy and Recommendations." Center For Disease Dynamics, Economics and Policy, USA, 2006.
"Center for Veterinary Medicine." 2021. *Arsenic-Based Animal Drugs and Poultry*. March 5.
"Effect of Antibiotics on Animal Feed." n.d. www1.udel.edu/chem/C465/senior/fall97/feed/present.html
"Introduction to Nutrition." 2019.
"Steve Craaig, 2017." https://fisheries.tamu.edu/files/2019/01/FST–269.pdf
"Use of Antibiotics in Dairy Cows." 2014. Health Products Regulatory Authority.
"World's Best Meat Producing Countries Based on FAO." 2021. *The Science Agriculture*.

3 Antimicrobial Resistance in Agriculture

Human and Animal Health

Amro Abd Al Fattah Amara,
Mounir Mohamed Salem-Bekhit,
Khairy Mohamed Abdallah Zoheir, Arti Gupta,
and Mohamed Morsi Mohamed Ahmed

3.1 INTRODUCTION

High use of antimicrobial agents (AmAs), for different purposes drives increased drug resistance, as microbes are exposed more to the AmAs. The existence of the different AmAs everywhere brings some comfort and safety. When antibiotics showed success against humans' pathogens, their use was extended to other fields. Antibiotics improve the economy worldwide by raising feed production's safety processes and enabling bulk production for livestock (LS), crops, fishes, and so on. After decades of their use, today most antibiotics have lost their power because of antibiotic resistance (AR). This has become a critical problem because AR affects consumers. In the agriculture system, the problem is complicated, as mega producers use the crowded fast farming style (CFFS) and cannot stop their use of antibiotics without dramatically affecting business. LS farms can serve as reservoirs for pathogenic zoonotic bacteria such as *Staphylococcus aureus* (*S. aureus*), *Salmonella* spp., *Campylobacter* spp., *Escherichia coli,* and *Listeria* spp., which can be transmitted into the environment through air, slurry, and feces from LS (Jėżak and Kozajda 2022). Why do we increasingly need to use AmAs? Why are we focused on one type, antibiotics? Why do we neglect the use of other alternatives? Did we wisely use antibiotics? Should we change our attitude and use more planned and controlled strategies? Did we understand the microbes and the AmAs? What do old procedures and experiences tell us? To understand the side effects of our misuse of antibiotics, in this chapter, we try to get a better understanding of these points and answer these questions. Agricultural applications use AmAs as microbicides, food preservatives, in poultry and LS farming, etc. (Amara 2023; Jėżak and Kozajda 2022). Nerveless, antibiotics are still the most used AmAs in agriculture. Although most of the targeted bacteria are killed by a dose of antibiotics, some escape death; bacteria survival can initiate the acquiring or development of an escape mechanism to the antibiotic. This resistance trait can be passed onto their offspring and other microbes. Structure modification and

DOI: 10.1201/9781003269847-3

readjustment could also turn sensitive microbes to resistance. In CFFS, antibiotics are used against different kinds of bacterial infections, and for growth promotion. The agricultural sector is reported as a major route for antibiotic-resistant genes (ARGs) to transfer to humans (Duro and Cook 2014). Agricultural activity that includes the use of antibiotics can be categorized into the following sections: (1) LS, (2) aquaculture, and (3) crops (O'Neill 2015). One should consider some issues that could influence the amount of antibiotics used, hot regions, population, polluted areas, and microbial biodiversity. Interventions will no doubt include improvements in infection control, better LS husbandry practices, use of vaccines, and adoption of diagnostic devices to ensure better targeted and appropriate veterinary prescribing. The subject of antibiotics and antibiotic use problems is highly scattered. For this reason, this chapter will classify their problems based on the expected solutions in the form of strategies. They depend on control and reduction of the use of antibiotics and introduce different ways to solve the AR problem. Individuals, farmers, as well as governments could benefit from the ideas and discussions included in the proposed strategies. Many links between each strategy and their effect on humans and animals are discussed. Other issues are also included and discussed.

3.2 DIFFERENT STRATEGIES TO CONTROL ANTIBIOTIC USE AND ANTIBIOTIC RESISTANCE

3.2.1 Strategy Number 1: Recall the Traditional Slow Farming Style

The success of the CFFS negatively affects small producers who have better-quality products but cannot compete with the low prices of the megaproducer. In a cascade step, the small producers find their products face many restrictions. As a consequence, respective amounts of fresh products are omitted from markets. Villages become consumers instead of being producers. Even the small producer is a built-in part of the economy of different countries, but a real conversion is happening. The trials to put them under the umbrella of the large producer have not stopped. Antibiotics enable the CFFSs' products to dominate the market. On the consumer side, Egyptian farmers who were early practicing the CFFS were wary of their own products, especially poultry, and invented a simple strategy. After poultry production (usually 45 days), the producers gave some of them to their family members and neighbors. Then most of them did something interesting, they kept the chicken to eat with their own chickens, which were grown according to TSFS for nearly 3 weeks to 40 days. They believed that this period was enough to get rid of any unwanted gradient(s). After years, hardly anyone still follows this practice. In some cases, this is because they are no longer growing chickens in their house areas. This TSFS has some unique properties, including use of fewer chemicals, hormones, and antibiotics, and good quality and fresh foods. Collectively, with antibiotics, the CFFS flourished and the TSFS declined, the health system (for human and animal) changed to be totally dependent on antibiotics, other alternatives were ignored, etc. The TSFS became part of the past for some countries. The consumer was forced to see an attractive well-packaged food, and sometimes precooked food, but there is no guarantee that the food does not contain antibiotic residues, genes from the vaccination process, preservatives, etc. After decades of using antibiotics in agriculture, many are interested in turning back to the TSFS for food production. Apparently, we are over-reliant on ourselves and on antibiotics, without thinking of the losses resulting from damaging a system that was established at the beginning of human life with all the experience gained over the ages. Nowadays, one can hardly find a farmer (or a farmer's wife) who knows the traditional way to make a chicken or duck lay on eggs until they can be hatched. In fact, non-fully-modernized countries have more chances to combine traditional experiences and modern technology. The inability to raise a small number of birds due to the losses of traditional knowledge will be finally covered by using antibiotics. This is not an invitation to return to the TSFS in its entirety but to preserve the important related knowledge and to support its practice. In a country with 50 million farmers, if 10^4 farmers are able to produce the product of one company, that means we have the product of 5,000 companies. One could calculate the amount of antibiotics that can be saved by practicing TSFS. In addition, this will be a hidden support for the economy of any country. In fact, TSFS mostly guarantees healthy and naturally immunized animals, which eat less, grow normally, are happier, and are safer if utilized as food. AR does not seriously affect TSFS because it is not dependent on antibiotics. Today, our focus should be to find alternatives to antibiotics to save ourselves and to reduce AR for the best of the CFFS, while a reduction in the AR improves the farming process. For that, the TSFS should be considered as one of the most important alternatives to the use of antibiotics.

3.2.2 Strategy Number 2: Avoid the Use of Antibiotics with Human Interest

Many reports highlight the side effects of antibiotics with human interest in agriculture applications (Saiful Islam, Kumar Sarkar, and Shiraj-Um-Mahmuda 2020). Antibiotics were used in LS farming since the mid-20th century (Woods 2019) as therapeutics, to treat infections, and for prophylactic purposes (Salem et al. 2023). Selective pressure was applied because of excessive and improper use of antibiotics in human and veterinary medicine, LSs' food, fish production, and aquaculture farming (McEwen and Collignon 2018). The overlap between antibiotics in human interests and those used in the agricultural sector led to different kinds of problems, like activating opportunistic pathogens, reducing the activity of the immune system, putting resistance genes under stress (prevent plasmids curing), spreading the ARGs, dominating the AR pathogens, etc. For example, penicillin and tetracyclines are widely employed in the food of the LS (Mathers, Flick, and Cox 2011). Macrolides, polymyxins, aminoglycosides, and third-generation cephalosporins are applied to LS (Barton 2000). Ionophores and sulfonamide are employed as prophylaxes against chicken coccidiosis (Chapman and Rathinam 2022). During the Second World War, penicillin was used to treat the wounded. As a result, veterinarians were granted access to lyophilized versions of the antibiotic, which marked the start of using antibiotics in LSs' food. Compared to previous nonantibiotic treatments, penicillin proved to be more effective (Koike, Mackie, and Aminov 2017). For example, mastitis is treated with oxytetracycline (Mathers, Flick, and Cox 2011).

Tetracycline was used in large quantities in various agricultural applications and has human interest. According to the US Food and Drug Administration (FDA 2015), tetracycline is one of the most important antibiotics in the United States. The quantity used reached 6,514,779 kg of active ingredients (FDA 2015), which was about 71% of all antibiotics sold for use in food-producing LS. The use of tetracyclines is extended to aquaculture and horticulture (Chopra and Roberts 2001). Some practices in agriculture (e.g., corded avian and herd farming) and aquaculture (fishers) use antibiotics with human interest. The problem can be summarized in that out of the 41 commonly used antibiotics in the producing LS (FDA report) sector, 31 are medically important for humans (O'Neill 2015). The antibiotics that are classified as medically important for humans and used in agriculture are amoxicillin, ampicillin, ceftiofur, cephapirin, chlortetracycline, cloxacillin, danofloxacin, dihydrostreptomycin, enrofloxacin, erythromycin, florfenicol, gamithromycin, gentamicin, lincomycin, neomycin, ormetoprim, oxytetracycline, penicillin, pirlimycin, spectinomycin, sulfadimethoxine, sulfamerazine, sulfamethazine, sulfaquinoxaline, tetracycline, tildipirosin, tilmicosin, tulathromycin, tylosin, tylvalosin, and virginiamycin. The other ten that are not medically important for humans are bacitracin, bambermycins, carbadox, laidlomycin, lasalocid, monensin, narasin, novobiocin, salinomycin, and tiamulin (O'Neill 2015).

3.2.3 Strategy Number 3: Stop the Improper Use of Antibiotics

Antimicrobial resistance (AMR) threatens the effective prevention and treatment of an increasing range of infections, including blood poisoning, pneumonia, diarrhea, gonorrhea, tuberculosis, HIV/AIDS, and malaria. One should differentiate between treatment because infection or illness could lead to a second microbial infection and treatment to protect against expected infection (prophylaxes). The misuse of antibiotics will increase the rate of AR. Subtherapeutic levels of antibiotics are usually used as growth promoters to improve feed efficiency and accelerate weight gain in LS (Niranjan et al. 2023). Reports tell that numerous developing nations lack effective antibiotic use control measures (Ayukekbong, Ntemgwa, and Atabe 2017; Lu and Wang 2023), which cause AR for human pathogenic bacteria (Salam et al. 2023). When the proper amount of antibiotic is not applied, resistance microbes could be elevated (Harbarth and Samore 2005; Chantziaras et al. 2013; Almansour et al. 2023). As discussed above, during the beginning of the 1950s, utilizing antibiotics in LS became prevalent (Woods 2019) and increased their therapeutic application (Page and Gautier 2012; Kitandu and Juranová 2006). Antibiotics employed as growth promoters are frequently administered at dosages lower than therapeutic dosages (Gonzalez Ronquillo and Angeles Hernandez 2017). In 2006, the European Union enacted a prohibition on utilizing antibiotics to promote growth in LS. The United States implemented regulations on this matter (Katakweba et al. 2018). On the human side, global antibiotic consumption increased by 36% between 2000 and 2010. One-half of this increased use is regarded as unneeded. In many countries, antibiotics can be bought without a prescription (Van Boeckel et al. 2014). Improbable personal human activities could also contribute in transferring AR to the agricultural sector. Human activities concerning antibiotic use include our activities in homes, hospitals, the place of production, pharmacies, and how we get rid of unused/expired antibiotics.

3.2.4 Strategy Number 4: Give Nature a Chance to Solve the Problems of Antibiotic Resistance

Antibiotics and AR affect both human and LS health (Economou and Gousia 2015). Many of the microbes that produce antibiotics have ARGs. Microbes that produce antibiotics could be found everywhere, including our microflora (e.g., bacitracin production by skin microbes). For that, ARGs could be found in different human living environments (microflora, LS, plants, water, and soil) and can spread from human-to-human or through zoonotic transmission from LS (or LS products) to humans or vice versa (El-Kafrawy et al. 2023). One point that should be highlighted is that nature's response is spontaneous. For that, it has different hydrocarbons and AR genes. Those genes are of great importance in curing nature and returning it to its normal balance. So, one should understand when to act against nature's activity and when to interact against their own activity. In our case, we should intervene against our own activity concerning contaminating the environment by leaving huge amounts of different antibiotic residues. Interfering with nature's own activity will delay the clearance of antibiotic residues.

3.2.5 Strategy Number 5: Conceptualize Our Counterattack

If antibiotics or AmAs are used to control/prevent pathogens (e.g., bulk LS farming), an excess of antibiotic residues is expected to be present and to contaminate the surroundings. Avoiding the transmission of ARGs to the beneficial microbes either naturally or artificially is crucial, while even though such genes are beneficial and even increase chances of survival against any treatment with antibiotics, they could transfer those genes to the pathogenic microbes. In other words, both normal and pathogenic microbes that carry ARGs should be under investigation. ARGs by any means should not be introduced voluntarily to microbes used by living structures, but they can be used in a controlled fermentation process that guarantees getting rid of them after the success of the production steps. For better achievement of any goal concerning the control of antibiotic use and AR, one should understand some important facts about antibiotics, AR, the resistance as a phenomenon, resistance gene transfer, etc. Antibiotics are naturally produced and exist in different ecosystems as part of the microbial balance and microbial antagonism. Humans have purified some and succeeded in solving their chemical structures and are able to produce them as synthetic antibiotics. In addition, non-natural structures or modified natural structures are produced synthetically. Finally, we load the original structure with extra amounts of AmAs. One important question should be addressed, why did AR at such a level not exist before? Our focus on antibiotics as the most successful AmAs is leading to the neglect of many other antimicrobial compounds. By recalling the traditional safe method for using AmAs, one can reduce dependence on antibiotics. There were many different traditional practices to fight harmful microbes in different fields (e.g., ethanol, salt, ash, drying, smoking, fermentation). Most of these practices were ignored after the discovery of the antibiotics. The few who resist widespread antibiotic use and use it selectively keep other traditional practices and alternatives going. Resistance genes protect microbes from antibiotics. They protect the resistant microbes from the producer and *vice versa* if the proper resistance gene(s) is existed; as well as from the antibiotics around. During the natural antagonism process, the antibiotic producers keep other microbes away. But the ARGs give the chance for more contact, exchange, and transfer of harmful genes, including the new ARGs. Yes, AR is one of the main claimed agents for spreading more harmful genes within microbes. Resistance harmful pathogens are responsible for exchanging and transferring additional other harmful genes. Reducing, controlling, and regulating antibiotic use will limit this kind of risk and keep them available. For controlling AR, one should trust in the scientific fact that we cannot stop resistance genes from being transferred here and there, but we can control the amount of antibiotics. The best choice is to properly use antibiotics. In each single step, one should use unique criteria that guarantee recovery from certain infections without contaminating the environment with antibiotic residues or resistance microbes. Good practice for using antibiotics should be known by everyone in the field. Alternative AmAs should be able to reduce our dependence on antibiotics. The gap between scientists and producers, political entities, consumers, etc., must disappear. Contaminated natural environments must be given enough chance to degrade these residues. One should respect the inevitability of balanced life, where all of its components react spontaneously in harmonic reactions (Amara 2013). Sometimes the big picture will be missed because we are so involved in the details. Chances must be given to the thinkers, philosophers, etc., to share ideas, suggestions, opinions, etc., in solving such scientific problems. Sometime laypeople have the key. In fact, they were producing safe and healthy foods for hundreds of years!

3.2.6 Strategy Number 6: Reduce the Used Amount of Antibiotics in Agriculture

As discussed above, antibiotics are used not only for treating human infections but also in many other applications, such as in food preservation, as AmAs, in agricultural applications, for growing poultry and LS, etc. (Podlacha et al. 2021; Amara 2023). The excessive use of antibiotics began early after antibiotic discovery. Logically, it was expected that antibiotics used in agricultural applications would be different from those of human interest or that at least one or more classes would be kept safe for emergency use. Some countries have observed and applied these uses early. The most important points about the use of antibiotics in agriculture are their used quantities and types. The quantity of the used antibiotics could be reduced either by installing an effective regulation system or by spreading information to the public and the user about appropriate use. Laws can help in regulating the total produced and used amounts. There is no accurate account of used antibiotics per year worldwide. The reported estimate for the used amount is said to range from 63,000 tons (Van Boeckel et al. 2015) to over 240,000 tons (Grace 2015; Manyi-Loh et al. 2018). Negative factors in management of antibiotic use include shortage management plans, development of unstable and inflexible strategies, underperformance in husbandry zone planning, and insignificant hygienic practices in crowded LS husbandry (Saiful Islam, Kumar Sarkar, and Shiraj-Um-Mahmuda 2020). Here, one should give exception to the TSFS, where LS are observed under continuous care, and antibiotics are only used if there is infection. In aquaculture, antibiotics are supplied with the feed. The uneaten feed will contaminate the water, and the eaten feed, or at least part of it, will be excreted into the water. Meek et al. (2015) suggested that about 70% to 80% of antibiotics come from fish excreta (Meek, Vyas, and Piddock 2015). The smallest amount of antibiotic used in the agricultural sector is for growing crops, where only about 0.2% to 0.4% of the total antibiotic is used (Smalla and Tiedje 2014).

3.2.7 Strategy Number 7: Conduct National and International Studies

Because of the complication of connecting the health of humans, animals, and the environment, AMR can be controlled through applying a collective strategy. This refers to a cooperative effort by diverse healthcare specialists including experts in human health, domestic LS, wild animals, plants, and the environment (Bassetti et al. 2022). The Swiss AR Report describes the use of antibiotics and occurrence of AR in Switzerland. It was published in 2022 by collaborative work of "The Swiss Centre for Antibiotic Resistance", ARCH-Vet, and Informations system Antibiotika in der Veterinärmedizin (Thanner, Drissner, and Walsh 2016). Many consider that AR in LS, the environment, and humans is "one health perspective" (Pandey et al. 2024). Multiple reports have indicated a 35% increase in global antibiotic consumption during the first decade of the 21st century (Van Boeckel et al. 2014). A study conducted by Mulchandani et al. (2023) using statistical models estimated antimicrobial usage of 99,502 tons in 2020 and projected an 8% increase to 107,472 tons based on the latest trend (Mulchandani et al. 2023). Alternatives were proposed. In 2002, the United Nations Food and Agriculture Organization/World Health Organization (FAO/WHO) published the "Guidelines for evaluating probiotics in food". These guidelines established safety and efficacy standards for probiotics, systematizing their discovery and selection (2002) (FAO/WHO 2002). de Melo Pereira et al. (2018) pointed different criteria that should be used to evaluate the selected probiotics. In 2019, the WHO identified AMR as a problem and one of the top ten threats to global public health that humanity faces. In the same year, the United Nations issued a report declaring that alarming resistance levels have been reported in countries of all income levels, which is resulting in the spread of common diseases (WHO 2019a). The European Food Safety Authority and the European Medicines Agency published a joint scientific opinion on how to reduce the need for antimicrobial use in food-producing LS. The opinion focused on how to prevent the spread of bacteria resistant to antimicrobials (Use et al. 2017). The European Union Commission issued guidelines for the prudent use of antimicrobials in veterinary medicine to its members (Watson 2016). The major LS production types (such as pigs, chattels, poultry, aquaculture, and rabbits) were covered, and these guidelines also cover other species (including pets, for LS, and non-food-producing species).

3.2.8 Strategy Number 8: Build a Surveillance and Management System

The European Food Safety Agency and the European Centers for Disease Control strongly recommend the use of whole-genome sequencing and the building of platforms for surveillance (FOPH-FSVO 2022). Food-producing LS are reported as an expected source of AR issues concerning humans (Moudgil et al. 2017). The National Farmed Animal Health and Welfare Council pointed out approaches including clinical microbiology, infection control (biosecurity), regulations, surveillance on antibiotic use and resistance, LS management, husbandry, and alternatives to antibiotics (Council 2016). Appropriate use of antimicrobials is discussed in different conferences and including a coordinated network, LS producers, pharmacists, veterinary pharmaceutical industries, and regulatory authorities is necessary to enforce antibiotic use control (Van Vuuren 2014). The use of a surveillance system will help in monitoring the used type, quantity of the antibiotic, the applications used, and final amount of the discharge that contaminated the environment, the LS, and human food. There is no unique surveillance system for AR over the entire world. In the United States, the National Antimicrobial Resistance Monitoring System uses the Clinical and Laboratory Standards Institute interpretation criteria. In Europe, EARS-Net uses the European Committee on Antimicrobial Susceptibility Testing criteria. The surveillance system could help in giving early alarms about the spread of resistance. It usually needs labs and statistical systems to analyze data. There are other types of surveillance systems concerned with monitoring new ideas, policies, regulations, etc. Some kinds are responsible for monitoring AR in particular agricultural products, such as dairy animals (Sharma et al. 2018). It is important to install good surveillance systems for some issues, like population number, patient number, animal number, age, gender, resistance gene sequences, genome sequence, amount of produced antibiotics, etc.

3.2.9 Strategy Number 9: Develop Institutional and Governmental Regulations

Effective action on tackling AMR requires a coordinated response from agriculture, industry, and governments and international agencies. It also requires collaboration between pharmacists, farmers, veterinarians, physicians, and patients (Xiong, Sun, and Zeng 2018). AMR is a complex global issue requiring a shared responsibility for all countries (Van Boeckel et al. 2014). The US government issued a national action plan to combat antimicrobial-resistant bacteria (Guardabassi et al. 2018). The health and economic load of AMR is significant and has the potential to become catastrophic if appropriate action is not taken. The government activists also promote enhancing coordination between countries to develop a true global action plan to tackle AMR. One report prepared by the WHO is titled "Global action plan on antimicrobial resistance" (2015). Some interesting ideas concerning governmental activities include the following:

1. "Whole-of-society engagement including a one health approach" (Pandey et al. 2024; WHO 2015).

Antimicrobial resistance will affect everybody, regardless of where they live, their health, economic circumstances, lifestyle or behavior. It will affect sectors beyond human health, such as animal health, agriculture, food security and economic development. Therefore, everybody – in all sectors and disciplines – should be engaged in the implementation of the action plan, and in particular in efforts to preserve the effectiveness of antimicrobial medicines through conservation and stewardship programmes (WHO 2019b).

2. "Prevention first".

Every infection prevented is one that needs no treatment. Prevention of infection can be cost effective and implemented in all settings and sectors, even where resources are limited. Good sanitation, hygiene and other infection prevention measures that can slow the development and restrict the spread of difficult-to-treat antibiotic-resistant infections are a "best buy" (WHO 2015).

3. "Access" to the existing and new antimicrobial medicines, health facilities, healthcare professionals, veterinarians, preventive technologies, diagnostic tools, knowledge, education, information etc. (WHO 2015).
4. "Sustainability" for every positive activity, where all countries should have a national plan against the misuse of the antibiotic and the AR (WHO 2015).
5. "Incremental targets for implementation". The general concept is to allow each country to determine the priority actions that it needs because each country has different priorities, conditions, facilities, etc. (WHO 2015). This problem of AR and AMRs requires a communal responsibility for all countries to protect citizen health and combat the international AMR threat (Van Boeckel et al. 2014). A standard for a certain level of antimicrobial use is needed for producers, users, and final product consumers. It should be a regulation that addresses concerns if any supplier does not respect the standard criteria for antibiotics storage. Globally, like the laypeople response, governments should respond to scientists' advice and start to control the antibiotics that contaminate food products everywhere. Global awareness and collaboration to resolve this problem should introduce many ideas to help bypass this situation. Taking steps and transferring ideas, opinions, experiences, knowledge, etc., are helping in installing international policies to reduce the use of antibiotics.

3.2.10 Strategy Number 10: Use Other Antimicrobial Agents

Natural antimicrobial compounds widely exist in the environment (Sharaf et al. 2013; Holmes et al. 2016) and can be summarized in the following points: (1) microbes that produce antibiotics to antagonize each other in their natural habitat; (2) microbes that produce secondary metabolites with antimicrobial activities during their normal life span; (3) plant and LS's natural products with antimicrobial activities (Amara,

El-Masry, and Bogdady 2008); (4) natural metals, atoms, compounds, and any chemical structures that have antimicrobial activities (i.e., acids, alkalis, salts, metals, CO_2, H_2O_2, etc.); and (5) microbials that react against other by direct attacks, such as the bacterial and algal viruses (phages). It might be important to highlight that laypeople have their own protocols for handling some illnesses. For example, if they observe an elevation in temperature, they first use any resources available at home like lemon, onion, garlic, honey, tea, coffee, etc. For example, if there is diarrhea, they use lemon, honey, concentrated tea (or a spoon of powdered tea), lemon juice alone or mixed with starch, or coffee (as a powder) mixed with starch. The concentrated boiled tea has fast action, but only a small amount (about 5 ml) should be used. There is a complete Arabic practice called "Teriaq", which means to equilibrate the effect of one toxin or one illness with another one (which might not be toxic). The people who are expert in this field found that the concentrated tea is one part of the "Teriaq" that could equilibrate the effect of other toxins. In fact, these practices could be explained scientifically; for example, the salt inactivates the enzymes.

3.3 SOME ALTERNATIVES TO ANTIBIOTICS

3.3.1 Plant Products

Today, there is a global demand to reuse the AmAs that are traditionally used as well as the modern ones. Laypeople always have their own practices, which are handed down through memory and beliefs. For example, tea was not consumed in the Madill East until British colonization. But nowadays tea is used in many applications. A small volume (10 ml) of concentrated boiled tea (with or without sugar) or consumed directly as a powder is used to cure diarrhea. The burned dry plants and woody waste are usually in the form of ashes that have antimicrobial properties and were used traditionally for injuries, in leather treatment, and as a cleaning agent. Wood ashes are a source of macronutrients for plants which are present in the following series of decreasing values: Ca > C > K > Mg > P > S > N. Ash is used in agriculture as a liming agent (Symanowicz et al. 2018). Sands mixed with ash are also used for general cleaning purposes. Part of plants can be used in different antimicrobial activities. Arabs brush their teeth with *Salvadora presica* branches and use it as a purgative and to cure wounds. Volatile oils derived from several plants can heal microbial diseases and are used to clean certain instruments. Lemon, which is known for its acidity, is a potent agent against *Streptococcus pyrogenes* and *Staphylococcus aureus*. A fast and a highly efficient folk practice is forcing the child to absorb (not drink) lemon syrup where the mother presses the cut-half part of the lemon in the mouth and lets it be taken by the child little by little. During the treatment process, the lemon is also added to the child's food (e.g., yogurt).

3.3.2 Vaccines

Vaccines are still one of the most important protective measures for LS raised in the CFFS (Yegoraw et al. 2021). Factors can influence the choice of vaccination method, including type, location, labor force, type of production, production cycle, health status, innate immunity, and vaccination costs (Rauw et al. 2009). In some countries (Belgium, the Netherlands, and Germany), vaccination of farm LS against some diseases is mandatory. In France, only long-lived poultry (laying and breeding poultry) are vaccinated (Rauw et al. 2009). Some bacterial pathogens produced as vaccines are *Salmonella* vector vaccines delivering three *Clostridium perfringens* antigens to protect poultry from necrotic enteritis (Wilde et al. 2019), *Campylobacter jejuni* vaccine (Zeng, Xu, and Lin 2010), attenuated *Salmonella typhimurium* live vaccines, and many others. A new method called sponge-like protocol is showing increasing interest in microbe evacuation and control of pathogens (El-Baky and Amara 2014; Amara, Salem-Bekhit, and Alanazi 2013; Sheweita et al. 2022).

3.3.3 Related to Microbes' Antagonism and Their Products

3.3.3.1 Acids

The most common acids produced as by-products of fermentation processes are lactic acid, acetic acid, and propionic acid (Amara 2023). Commercially, acids such as lactic acid are produced by some *Lactobacillus* spp., such as *L. sakei* and *L. curvatus* (Zhang et al. 2018; Amara 2023), acetic acid by *Acetobacter aceti,* and propionic acid by dairy *Propionibacterium* spp., such as *Propionibacterium freudenreichii, P. jensenii, P. acidipropionici* (Piwowarek et al. 2018). Acetic acid has more antimicrobial activity than lactic acid. Acids produced from certain fermentation processes are preferable for their aroma (Bahuguna et al. 2020). Besides being used as AmAs, they are used as additives to enhance flavor and shelf life (Agriopoulou et al. 2020). Vinegar is produced by a bacterium group *Acetobacter.* Other organic acids, such as formic, benzoic, and phenyl lactic acids are produced. They appear to play an important role in inhibiting pathogenic bacteria (Tharmaraj and Shah 2009; Amara 2023).

3.3.3.2 Ethanol and Carbon Dioxide

Carbon dioxide (CO_2) has antimicrobial activity (Blatchford et al. 2019). The fermentation processes usually include CO_2 and one or more of the organic products, such as ethanol, organic acids, and other organic molecules. Microbes utilize sugar while producing alcohol and CO_2. Most commercial bread is made by CO_2 rise with the yeast (Katsi et al. 2021). CO_2 creates an anaerobic environment by replacing existent molecular oxygen. It can decrease pH and has destructive effects on cell membranes (Naidu, Bidlack, and Clemens 1999). Ethanol could kill its own producer. For that, their production is stopped at the critical point where no enzymes can react any more.

3.3.3.3 Bacteriocin

The bacteriocin is a ribosomal synthesized peptide and microbial products originally defined as proteinaceous compounds affecting the growth or viability of closely related organisms. They are variant in their types (Shyntum et al. 2019). The bacteriocin is applied in the control of foodborne pathogenic and spoilage bacteria (Pogány Simonová, Chrastinová, and Lauková 2020). They are largely used in food preservation and were reported as safe products to preserve food (Gautam and Sharma 2009).

3.3.3.4 Hydrogen Peroxide

Some microbes produce hydrogen peroxide (H_2O_2) under aerobic conditions of growth. They release it into the environment to protect themselves (Daeschel 1989). It was used in microbial and virus evacuations. H_2O_2 proved to have antimicrobial activity and the ability to degrade the genomic elements in bacteria, fungi, yeast, and viruses (Amara 2015; Amara, Salem-Bekhit, and Alanazi 2013). It is a strong oxidizing agent and can be used against bacteria, fungi, and viruses (also bacteriophages). It is able to degrade both RNA and DNA and affects other macromolecules as proven by several works concerning microbial cells' and viruses' evacuation using the sponge-like protocol. H_2O_2 shows antimicrobial activity against yeast and Gram-positive and Gram-negative bacteria (Suskovich et al. 2010).

3.3.4 Other Antimicrobial Agents

There are many other antimicrobial by-products produced by beneficial microbes and probiotics such as acetaldehyde (Nieminen et al. 2014), cyclic dipeptides (Jinendiran et al. 2020), and deconjugated bile acids (Oelschlaeger 2010). They are microbial by-products and can be produced during the food fermentation process and support the food as preservatives and agents that resist spoilage (Amara 2023).

3.3.4.1 Microbes against Microbes

3.3.4.1.1 Probiotic

Probiotics are defined as live microorganisms that confer health benefits to host organisms when administered in sufficient amounts. Basic research was applied to understand the mechanisms of action of probiotics in various environments. Probiotics influence the mammalian gut flora in a way that improves health. The mammalian gut microbial community is a complex community consisting of more than 500 different bacterial species. One common probiotic today is *Lactobacillus acidophilus* which can be found in yogurt, acidified milk, and dietary supplements. Probiotics given in combination with antibiotics can help prevent antibiotic-related diarrhea. Yeast *S. boulardii* and three *Lactobacillus* strains have proven useful in this situation (Amara 2023).

3.3.4.1.2 Bacteriophage

Bacteriophages can be used for the treatment of infections caused by various bacteria, ranging from *Staphylococcus aureus*, *Pseudomonas aeruginosa*, *Salmonella* spp., and *Shigella* spp. Human phage therapy studies have shown that bacteriophages are effective in treating patients for a variety of clinically important bacteria, and these therapies are already used effectively in some Eastern European countries (D'Accolti et al. 2021). Direct treatment with bacteriophages is applied to control different bacterial species (D'Accolti et al. 2021). Overall, bacteriophage therapy tends to be extremely time sensitive. For example, phage therapy had limited efficacy when administered more than 16 hours after experimental infection. Bacteriophages were used for disease prevention and treatment, with promising results. For example, they have protected chickens from respiratory disease after experimental infection with

E. coli (Huff et al. 2004). Salmonella infection in day-old broiler chicks was treated by a phage cocktail containing bacteriophages specific to *Salmonella enteritidis*. Bacteriophages have also been evaluated as treatments for colibacillosis in chickens. Phage therapy has also shown promising results in piglets and calves (Johnson et al. 2008). The major obstacles to using bacteriophages for disease treatment in LS include the lack of rapid and accurate diagnostics, which are needed because the phages typically are effective only against a very narrow range of bacterial strains. There is usually a risk of phage inactivation *via* the host immune response, and rapid emergence of resistant bacterial strains (Cheng et al. 2014). Phage cocktails that contain several different bacteriophage strains can help address these limitations (Amara 2024).

3.3.5 Related to Protein and Its Derived Structures

3.3.5.1 Lysozyme

Enzymes can have antimicrobial activities, such as lysozymes, which react on the bacterial cell wall. Lysozymes is an enzyme found in saliva and sweat and tears that destroys the cell walls of some bacteria. Plant-protective compositions comprising lysozyme are provided.

3.3.5.2 Antimicrobial Peptides

The advantages of antimicrobial peptides are their activity against a general spectrum, including most Gram-positive and Gram-negative bacteria, bactericidal and rapid action, targeted resistance, and low immunogenicity (El-Baky and Al Fattah Amara 2023). Different studies and early clinical symptoms have not led to advanced systemic treatments. In some instances, active ingredients used by spray can improve systemic therapy. Organized peptides of the defense (small, natural peptides) and regulation of the defense innate (small, synthetic peptides) have effects on the indirect antimicrobial. Some peptides increase anti-inflammatory chemokine and cytokine expression and reduce the expression of proinflammatory cytokines.

3.3.5.3 Direct Control to the Protein That Deactivates Antibiotics

A unique idea has been introduced, which is control of the enzymes that degrade antibiotics and need divalent cations by a chelating agent like EDTA (Hussein and Amara 2006; Amara 2022b). The idea is old and observed by many but re-highlighted by Hussain and Amara (2006) and Amara and Hussain (2006) (Hussein and Amara 2006). The first reinvestigation started in 2014, and the first published data was presented in the VAAM symposium 2005 proceedings (Amara 2005a), but the complete survey done on different strains of *Pseudomonas aeruginosa* was published in 2006 (Amara 2005a; Hussein and Amara 2006; Amara and Hussain 2006). Enzymes, active proteins, and active macromolecules can be ion dependent to complement their activity (and their charges) in catalysis (Amara 2015). The macromolecules do their function by the terminal charges in their active catalytic residue (e.g., the catalytic amino acids) or complement their work by the help of one or more than one cation(s) that exist around (Amara 2015, 2019). Naturally, ions are everywhere and even can localize in highly specific orientation around their specific target. The specific protein in such a case will work spontaneously. Meanwhile, if such ions did not exist or existed in non-free form (chelated), the protein will not be able to do its activity either completely or in a perfect way. Active atoms, ions, and their co-ordinance around the target are like a multistep machine where each part makes a step essential for the second part of a particular function. Interfering with such terminal atoms or ion activity or location could terminate their function or change it. Organization, coordination, step-fitting, and the function/structure/specificity are critical issues in the interaction of the macromolecules and their substrates. Eliminating single fin steps from the process (divalent cations) will corrupt all the processes (Amara 2005b, 2005a, 2011; Sheweita et al. 2022). Another example from nature could be giving different biofilm rigidity. The Ca^{+2} load could give either flexible or rigged alginate biofilm (Amara 2022a). There are many other types of AmAs that could not all be included in this chapter.

3.4 CONCLUSION

The chapter suggests ten strategies to control the use of antibiotics to control antibiotic resistance. The logical analysis of the antibiotic resistance problem includes searching for its origin, why did it not exist before? Why is it increasing and not decreasing over time? The role of the misuse of antibiotics is discussed in different applications but with more focus on the agriculture sector, particularly the crowded fast farming style. And why has the traditional slow farming style not depended for hundreds of years on antibiotics? The strategies include nearly all the involved factors starting with farming styles through the role of institutions and governments. Simple ideas and solutions are introduced, like re-encouraging the growth of the traditional farming style, reducing dependence on antibiotics by using similar or better alternatives, recalling successful folk practices, etc. Also introduced solid solutions and actions that can be considered as guides to the government to implement rules covering the use of antibiotics by individuals, houses, farmers, hospitals, countries, and anyone in need of antibiotics or in need to find alternatives. Although the chapter gives brief discussion, more details can be found in the literatures. Even a scientist could wonder, can lemons really cure a child's thorax infection and decrease a child's temperature? The proof can be obtained by trying this practice yourself. Yes, you will need to force the child to consume the syrup coming out from your hand pressing on the half-lemon. During the child's resistance to that, the child in fact increases the area that will receive the lemon. Trying to change such a practice by collecting the lemon syrup in a reservoir will not give the same result. In the folk remedies of any country, there are many wonderful practices that give good results. The misuse of antibiotics and the AR microbes and genes cause serious problems in all sectors that depend on the antibiotic. In this chapter, we focus on the agriculture sector and the human and the AR microbes that harm them both. It discusses different kinds of problems and introduces ten strategies to solve most of the problems. Logical analysis of the different problems that lead to elevation of AR is described. The misuse of antibiotics in non-human applications leads to contamination of farming products with their residues. That accelerates the emergence of resistance to the antibiotics. Today, there is 100% resistance to antibiotics for some bacterial species. The information and the suggested examples given in this chapter expand the scientific literature to some folk remedy experiences to highlight the importance of traditional tactics in farming practices. Meanwhile, it is advised that experts in microbiology should handle any of them in any place. Environmental changes (and other chemical, physical, and biological changes) could change the selection type, and experts must be consulted. Modern alternatives to antibiotics are introduced, including some of the traditional ones that are re-evaluated and proven successful, such as the probiotic and the microbial by-products that can be used as preservatives, such as the bacteriocin, organic acids, salt, ash, etc. The chapter also includes some management strategies like the use of surveillance systems within a country or worldwide. Nature, which is the most effective part in the antibiotic and antibiotic resistance story, is given special interest. Antibiotic resistance genes exist naturally but dominate nature through our misuse of antibiotics. Nature can help us but only if given a correct chance. This can occur simply because nature contains different kinds of hydrocarbon-degrading microbes and enzymes that could deactivate the antibiotics.

3.5 ACKNOWLEDGMENT

The authors acknowledge anyone who by any means supported this study. Special thanks to Dr Arti Gupta.

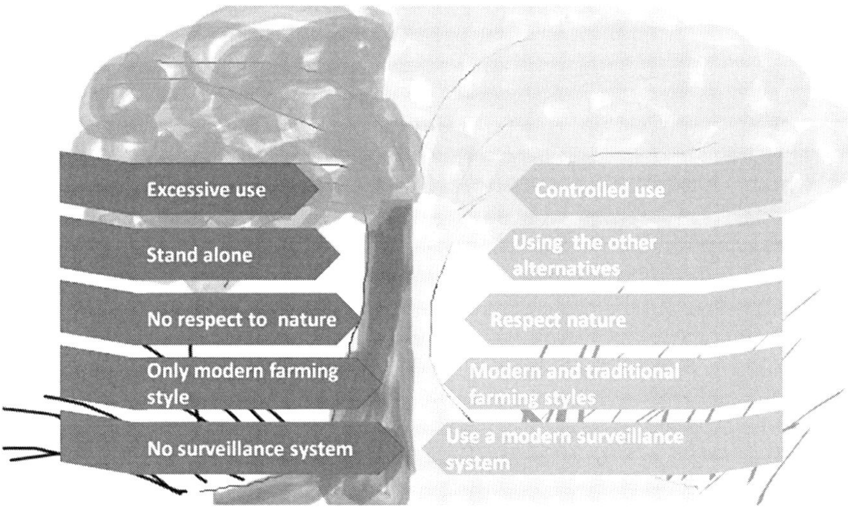

FIGURE 3.1 Nature symbolized as a tree: one-half is affected with the misuse of the antibiotic (the left), and the other half is the correct use of the antibiotic. In the left part, the tree is going to die, its green color is changed; in the right part, where there is a good use for the antibiotics and its alternatives, the tree is becoming green and healthy. The image symbolizes which choice one should select, and the final result if one selects an incorrect answer. (Created in this study.)

REFERENCES

Agriopoulou, S, E. Stamatelopoulou, M. Sachadyn-Król, and T. Varzakas. 2020. "Lactic Acid Bacteria as Antibacterial Agents to Extend the Shelf Life of Fresh and Minimally Processed Fruits and Vegetables: Quality and Safety Aspects." *Microorganisms* 8 (6). https://doi.org/10.3390/microorganisms8060952.

Almansour, A M., M A Alhadlaq, K O Alzahrani, L E Mukhtar, A L Alharbi, and S M Alajel. 2023. "The Silent Threat: Antimicrobial-Resistant Pathogens in Food-Producing Animals and Their Impact on Public Health." *Microorganisms* 11 (9):2127. https://doi.org/10.3390/microorganisms11092127.

Amara, A A. 2005a. "New Strategy for Controlling Resistant Strain of Pseudomonas Aeruginosa." *BioSpectrum* 74 In *Proceeding of Tagungsband zum 2. Gemeinsamen kongress der DGHM und VAAM Göttingen*, 25–28 September. Göttingen, Germany.

Amara, A A. 2005b. "Simple Slot Lysis and Plasmid Curing for Determination the Transfer of Resistant Factors Within the Pathogenic Microbial Strains." *BioSpectrum* 74 In *Proceeding of Tagungsband zum 2. Gemeinsamen kongress der DGHM und VAAM Göttingen*, 25–28 September. Göttingen, Germany.

Amara, A A. 2011. "Opportunistic Pathogens and Their Biofilm "Food for Thought." In Edit by Mendez-Vilas, *Science Against Microbial Pathogens: Communicating Current Research and Technology Advances*, Vol. 2, 813–824. FORMATEX Microbiology Series No 3.

Amara, A A. 2015. *Kostenlose viral Ghosts, Bacterial Ghosts, Microbial Ghosts and More*. Vol. Book. Germany: Schüling Verlage.

Amara, A A. 2019. "An Overview About Our Understanding of Protein Activity, Specificity, Structure and Function." In Edited by V K Gupta and M G Tuohy. Vol. 1, Chapter 14. *Microbial Catalysts*. Nova Publishers.

Amara, A A. 2023. "Fermented Food Microbes and Their Preservative Byproducts." In Edited by A Gupta and R Prasad. *Antimicrobials in Food Science and Technology*. CRC Press.

Amara, A A. 2024. "Phags and Their Role in Natural and Directed Microbial Control." Edited by S Varjani, I Haq, A Pandey, V K Gupta, and X T Bui. *Waste Management in Climate Change and Sustainability Perspectives*. CRC, Press.

Amara, A A A F. 2013. "The Inevitability of Balanced Lives: Genes and, Foods in Action and, Interactions." *IIOAB Journal* 4 (2):1–27.

Amara, A A A F. 2022a. "Natural Polymer Types and Applications." In Edited by V K Gupta, S D Sarker, M Sharma, M E Pirovani, Z Usmani, and C Jayabaskaran. *Biomolecules from Natural Sources*. Wiley.

Amara, A A A F. 2022b. "The Role of Divalent Cations in Antibiotic Sensitivity." Edited by V K Gupta, S D Sarker, M Sharma, M E Pirovani, Z Usmani, and C Jayabaskaran, *Biomolecules from Natural Sources*, Wiley.

Amara, A A, M H El-Masry, and H H. Bogdady. 2008. "Plant Crude Extracts Could Be the Solution: Extracts Showing in Vivo Antitumorigenic Activity." *Pakistan Journal of Pharmaceutical Sciences* 21 (2):159–171.

Amara, A A and Z M Hussain. 2006. "Offline Controlling of Pseudomonas Aeruginosa Resistant to Protein Inhibitor Antibiotics Using Combination of EDTA and Na-Citrate or Disinfectant(s)." *Microbiology (China)* 100 (33): 95.

Amara, A A, M M Salem-Bekhit, and F K Alanazi. 2013. "Sponge-Like: A New Protocol for Preparing Bacterial Ghosts." *The Scientific World Journal* 2013. https://doi.org/10.1155/2013/545741.

Ayukekbong, J A, M Ntemgwa, and A N Atabe. 2017. "The Threat of Antimicrobial Resistance in Developing Countries: Causes and Control Strategies." *Antimicrobial Resistance and Infection Control* 6: 47. https://doi.org/10.1186/s13756-017-0208-x.

Bahuguna, A, I G Jo, J S Lee, and M Kim. 2020. "Effects of Radishes, Apples, and Pears on the Lactic Acid Bacteria and Nutritional and Functional Qualities of Flavored Soy Sauce." *Foods* 9 (11). https://doi.org/10.3390/foods9111562.

Barton, M D. 2000. "Antibiotic Use in Animal Feed and Its Impact on Human Healt." *Nutrition Research Reviews* 13 (2):279–299. https://doi.org/10.1079/095442200108729106.

Bassetti, S, S Tschudin-Sutter, A Egli, and M Osthoff. 2022. "Optimizing Antibiotic Therapies to Reduce the Risk of Bacterial Resistance." *European Journal of Internal Medicine* 99: 7–12. https://doi.org/10.1016/j.ejim.2022.01.029.

Blatchford, P A, S G Parkar, W Hopkins, J R Ingram, and K H Sutton. 2019. "Dose-Dependent Alterations to In Vitro Human Microbiota Composition and Butyrate Inhibition by a Supercritical Carbon Dioxide Hops Extract." *Biomolecules* 9 (9). https://doi.org/10.3390/biom9090390.

Chantziaras, I, F Boyen, B Callens, and J Dewulf. 2013. "Correlation Between Veterinary Antimicrobial Use and Antimicrobial Resistance in Food-Producing Animals: A Report on Seven Countries." *Journal of Antimicrobial Chemotherapy* 69 (3):827–834. https://doi.org/10.1093/jac/dkt443.

Chapman, H D and T Rathinam. 2022. "Focused Review: The Role of Drug Combinations for the Control of Coccidiosis in Commercially Reared Chickens." *International Journal for Parasitology. Drugs and Drug Resistance* 18: 32–42. https://doi.org/10.1016/j.ijpddr.2022.01.001.

Cheng, G, H Hao, S Xie, X Wang, M Dai, L Huang, and Z Yuan. 2014. "Antibiotic Alternatives: The Substitution of Antibiotics in Animal Husbandry?" *Frontiers in Microbiology* 5: 217–217. https://doi.org/10.3389/fmicb.2014.00217.

Chopra, I, and M Roberts. 2001. "Tetracycline Antibiotics: Mode of Action, Applications, Molecular Biology, and Epidemiology of Bacterial Resistance." *Microbiology and Molecular Biology Reviews: MMBR* 65 (2): 232–260. https://doi.org/10.1128/MMBR.65.2.232-260.2001.

Council., National Farmed Animal Health and Welfare. 2016. *Antimicrobial Stewardship in Food Animals in Canada*. National Farmed Animal Health and Welfare Council. Bluevale, ON, Canada.

D'Accolti, M, I Soffritti, S Mazzacane, and E Caselli. 2021. "Bacteriophages as a Potential 360-Degree Pathogen Control Strategy." *Microorganisms* 9 (2). https://doi.org/10.3390/microorganisms9020261.

Daeschel, M A. 1989. "Antimicrobial Substances from Lactic Acid Bacteria for Use as Food Preservatives." *Food Technology* 43 (1): 164s.

de Melo Pereira, G V, B de Oliveira Coelho, A I Magalhães Júnior, V Thomaz-Soccol, and C R Soccol. 2018. "How to Select a Probiotic? A Review and Update of Methods and Criteria." *Biotechnology Advances* 36 (8): 2060–2076. https://doi.org/10.1016/j.biotechadv.2018.09.003.

Duro, L M and K L Cook. 2014. "Impact of Antibiotics Use IJN Agriculture. What Are the Benefits and Risks?" *Current Opinion in Microbiology* 19: 37–44.

Economou, V and P Gousia. 2015. "Agriculture and Food Animals as a Source of Antimicrobial-Resistant Bacteria." *Infection and Drug Resistance* 8: 49–61. https://doi.org/10.2147/IDR.S55778.

El-Baky, N A, and A A Amara. 2014. "Newcastle Disease Virus (LaSota Strain) as a Model for Virus Ghosts Preparation Using H2O2 Bio-Critical Concentration." *International Science and Investigation Journal* 3 (5): 38–50.

El-Baky, N A and A A A F Amara. 2023. "Natural Antimicrobial Therapeutic Peptides: Milk Lactoferricin and *Spirulina platensis* Peptides." In Edited by A Gupta and R Prasad. *Antimicrobials in Pharmaceutical and Medicinal Research*, 19–48.

El-Kafrawy, S A, A T Abbas, C Oelkrug, M Tahoon, S Ezzat, A Zumla, and E I Azhar. 2023. "IgY Antibodies: The Promising Potential to Overcome Antibiotic Resistance." *Frontiers in immunology* 14: 1065353–1065353. https://doi.org/10.3389/fimmu.2023.1065353.

FAO/WHO. 2002. *Guidelines for the Evaluation of Probiotics in Food; Report of a Joint FAO/WHO Working Group on Drafting Guidelines for the Evaluation of Probiotics in Food.* London, Ontario, Canada.

FDA. 2015. *Summary Report on Antimicrobials Sold or Distributed for Use in Food-Producing Animals.* Available at: https://docs.google. com/viewer?url=http%3A%2F%2Fwww.fda.gov%2Fdownloads%2FF orIndustry%2FUserFees%2FAnimalDrugUserFeeActADUFA%2FUC M440584.pdf.

FOPH-FSVO. Federal Office of Public Health and Federal Food Safety and Veterinary Office. 2022. *Swiss Antibiotic Resistance Report 2022. Usage of Antibiotics and Occurrence of Antibiotic Resistance in Switzerland.*

Gautam, N., and N. Sharma. 2009. "Bacteriocin: Safest Approach to Preserve Food Products." *Indian Journal of Microbiology* 49 (3):204–211. https://doi.org/10.1007/s12088-009-0048-3.

Gonzalez Ronquillo, M and J C A Hernandez. 2017. "Antibiotic and Synthetic Growth Promoters in Animal Diets: Review of Impact and Analytical Methods." *Food Control* 72: 255–267. https://doi.org/10.1016/j.foodcont.2016.03.001.

Grace, D. 2015. *Review of Evidence on Antimicrobial Resistance and Animal Agriculture in Developing Countries.* Evidence on Demand.

Guardabassi, L, M Apley, J E Olsen, P-L Toutain, and S Weese. 2018. "Optimization of Antimicrobial Treatment to Minimize Resistance Selection." *Microbiology Spectrum* 6 (3). https://doi.org/10.1128/microbiolspec.arba-0018–2017.

Harbarth, S and M H Samore. 2005. "Antimicrobial Resistance Determinants and Future Control." *Emerging Infectious Diseases* 11 (6):794–801. https://doi.org/10.3201/eid1106.050167.

Holmes, A H., L S P Moore, A Sundsfjord, M Steinbakk, S Regmi, A Karkey, P J Guerin, and L J V Piddock. 2016. "Understanding the Mechanisms and Drivers of Antimicrobial Resistance." *The Lancet* 387 (10014): 176–187. https://doi.org/10.1016/s0140–6736(15)00473–0.

Huff, W E, G R Huff, N C Rath, J M Balog, and A M Donoghue. 2004. "Therapeutic Efficacy of Bacteriophage And Baytril (enrofloxacin) Individually and in Combination to Treat Colibacillosis in Broilers." *Poultry Science* 83 (12).1944–1947. https://doi.org/10.1093/ps/83.12.1944.

Hussein, M Z, and A A Amara. 2006. "Case-by-Case Study Using Antibiotic-EDTA Combination to Control *Pseudomonas aeruginosa*." *Pakistan Journal of Pharmaceutical Sciences* 19 (3): 236–243.

Jeżak, K, and A Kozajda. 2022. "Occurrence and Spread of Antibiotic-Resistant Bacteria on Animal Farms and in Their Vicinity in Poland and Ukraine-Review." *Environmental Science and Pollution Research International* 29 (7):9533–9559.

Jinendiran, S, W Teng, H U Dahms, W Liu, V K Ponnusamy, C C C Chiu, B S D Kumar, and N Sivakumar. 2020. "Induction of Mitochondria-Mediated Apoptosis and Suppression of Tumor Growth in Zebrafish Xenograft Model by Cyclic Dipeptides Identified from Exiguobacterium Acetylicum." *Scientific Reports* 10. https://doi.org/10.1038/s41598-020-70516-x.

Johnson, R P, C L Gyles, W E Huff, S Ojha, G R Huff, N C Rath, and A M Donoghue. 2008. "Bacteriophages for Prophylaxis and Therapy in Cattle, Poultry and Pigs." *Animal Health Research Reviews* 9 (2): 201–215. https://doi.org/10.1017/s1466252308001576.

Katakweba, A A S, A P. Muhairwa, A M Lupindu, P Damborg, J T Rosenkrantz, U M Minga, M M A Mtambo, and J E Olsen. 2018. "First Report on a Randomized Investigation of Antimicrobial Resistance in Fecal Indicator Bacteria from Livestock, Poultry, and Humans in Tanzania." *Microbial Drug Resistance* 24 (3): 260–268. https://doi.org/10.1089/mdr.2016.0297.

Katsi, P, I S Kosma, S Michailidou, A Argiriou, A V Badeka, and M G Kontominas. 2021. "Characterization of Artisanal Spontaneous Sourdough Wheat Bread from Central Greece: Evaluation of Physico-Chemical, Microbiological, and Sensory Properties in Relation to Conventional Yeast Leavened Wheat Bread." *Foods* 10 (3). https://doi.org/10.3390/foods10030635.

Kitandu, A, and R Juranová. 2006. "Progress in Control Measures for Chicken Coccidiosis." *Acta Veterinaria Brno* 75 (2): 265–276. https://doi.org/10.2754/avb200675020265.

Koike, S, R Mackie, and R Aminov. 2017. "Agricultural Use of Antibiotics and Antibiotic Resistance Chapter 8." In Edited by M L Pérez and S Mirete. *Cell Biology Research Progress*. New York, USA: Nova Science Publishers, Inc., 217–250.

Lu, C and K Wang. 2023. "Natural Resource Conservation Outpaces and Climate Change: Roles of Reforestation, Mineral Extraction, and Natural Resources Depletion." *Resources Policy* 86: 104159. https://doi.org/10.1016/j.resourpol.2023.104159.

Manyi-Loh, C, S Mamphweli, E Meyer, and A Okoh. 2018. "Antibiotic Use in Agriculture and Its Consequential Resistance in Environmental Sources: Potential Public Health Implications." *Molecules (Basel, Switzerland)* 23 (4):795. https://doi.org/10.3390/molecules23040795.

Mathers, J J, S C. Flick, and L A Cox. 2011. "Longer-Duration Uses of Tetracyclines and Penicillins in U.S. Food-Producing Animals: Indications and Microbiologic Effects." *Environment International* 37 (5): 991–1004. https://doi.org/10.1016/j.envint.2011.01.014.

McEwen, S A, and P J Collignon. 2018. "Antimicrobial Resistance: A One Health Perspective." *Microbiology Spectrum* 6 (2). https://doi.org/10.1128/microbiolspec.arba-0009–2017.

Meek, R W, H Vyas, and L J V Piddock. 2015. "Nonmedical Uses of Antibiotics: Time to Restrict Their Use?" *PLoS Biology* 13 (10): e1002266–e1002266. https://doi.org/10.1371/journal.pbio.1002266.

Moudgil, P, J S Bedi, A D Moudgil, J P S Gill, and R S Aulakh. 2017. "Emerging Issue of Antibiotic Resistance from Food Producing Animals in India: Perspective and Legal Framework." *Food Reviews International* 34 (5): 447–462. https://doi.org/10.1080/87559129.2017.1326934.

Mulchandani, R, Y Wang, M Gilbert, and T P Van Boeckel. 2023. "Global Trends in Antimicrobial Use in Food-Producing Animals: 2020 to 2030." *PLOS Global Public Health* 3 (2): e0001305–e0001305. https://doi.org/10.1371/journal.pgph.0001305.

Naidu, A S, W R Bidlack, and R A Clemens. 1999. "Probiotic Spectra of Lactic Acid Bacteria (LAB)." *Critical Reviews in Food Science and Nutrition* 39 (1):13–126.

Nieminen, M T, L Novak-Frazer, V Rautemaa, R Rajendran, T Sorsa, G Ramage, P Bowyer, and R Rautemaa. 2014. "A Novel Antifungal Is Active against Candida albicans Biofilms and Inhibits Mutagenic Acetaldehyde Production In Vitro." *PLoS One* 9 (5). https://doi.org/10.1371/journal.pone.0097864.

Niranjan, D, N B. Sridhar, U S. Chandra, S S. Manjunatha, A Borthakur, M H Vinuta, and B R Mohan. 2023. "Recent Perspectives of Growth Promoters in Livestock: An Overview." *Journal of Livestock Science* 14 (1): 53–64. https://doi.org/10.33259/jlivestsci.2023.53–64.

O'Neill, J. 2015. "Antimicrobials in Agriculture and the Environment: Reducing Unnecessary Use and Waste the Review on Antimicrobial Resistance." 1–44. http://amrreview.org/.

Oelschlaeger, T A. 2010. "Mechanisms of Probiotic Actions A Review." *International Journal of Medical Microbiology* 300: 57–62.

Page, S W, and P Gautier. 2012. "Use of Antimicrobial Agents in Livestock." *Revue Scientifique et Technique de l'OIE* 31 (1): 145–188. https://doi.org/10.20506/rst.31.1.2106.

Pandey, S, H Doo, G B Keum, E S Kim, J Kwak, S Ryu, Y Choi, J Kang, S Kim, N R Lee, K K Oh, J-H Lee, and H B Kim. 2024. "Antibiotic Resistance in Livestock, Environment and Humans: One Health Perspective." *Journal of Animal Science and Technology* 66 (2): 266–278. https://doi.org/10.5187/jast.2023.e129.

Piwowarek, K, E Lipińska, E Hać-Szymańczuk, M Kieliszek, and I Ścibisz. 2018. "Propionibacterium spp.—Source of Propionic Acid, Vitamin B12, and Other Metabolites Important for the Industry." *Appl Microbiol Biotechnol* 102 (2): 515–538. https://doi.org/10.1007/s00253-017-8616-7.

Podlacha, M, Ł Grabowski, K Kosznik-Kawśnicka, K Zdrojewska, M Stasiłojć, G. Węgrzyn, and A. Węgrzyn. 2021. "Interactions of Bacteriophages with Animal and Human Organisms-Safety Issues in the Light of Phage Therapy." *International Journal of Molecular Sciences* 22 (16): 8937.

Pogány Simonová, M, Ľ Chrastinová, and A Lauková. 2020. "Autochtonous Strain *Enterococcus faecium* EF2019(CCM7420), Its Bacteriocin and Their Beneficial Effects in Broiler Rabbits—A Review." *Animals (Basel)* 10 (7). https://doi.org/10.3390/ani10071188.

Rauw, F, Y Gardin, B T van den, and B Lambrecht. 2009. "La Vaccination contre la maladie de newcastle chez le poulet (Gallus gallus)." *Societe et Environnement* 13: 587.

Saiful Islam, K B M, S K Sarkar, and S Shiraj-Um-Mahmuda. 2020. "Studies on Veterinary Antibiotic Residues in Chicken Eggs and Their Public Health Significance in Bangladesh Perspective." *Veterinary Medicine and Public Health Journal* 1 (3):115–119. https://doi.org/10.31559/vmph2020.1.3.9.

Salam, Md A, Md Y Al-Amin, M T Salam, J S Pawar, N Akhter, A A Rabaan, and M A A Alqumber. 2023. "Antimicrobial Resistance: A Growing Serious Threat for Global Public Health." *Healthcare (Basel, Switzerland)* 11 (13): 1946. https://doi.org/10.3390/healthcare11131946.

Salem, H M, A M Saad, S M Soliman, S Selim, W F A Mosa, A E Ahmed, S K Al Jaouni, M S Almuhayawi, M E Abd El-Hack, K A El-Tarabily, and M T El-Saadony. 2023. "Ameliorative Avian Gut Environment and Bird Productivity Through the Application of Safe Antibiotics Alternatives: A Comprehensive Review." *Poultry Science* 102 (9): 102840–102840. https://doi.org/10.1016/j.psj.2023.102840.

Sharaf, M, A Amara, A Aboul-Enein, S Helmi, A Ballot, and P Schnitzler. 2013. "Antiherpetic Efficacy of Aqueous Extracts of the Cyanobacterium *Arthrospira fusiformis* from Chad." *Pharmazie* 68 (5): 376–380. https://doi.org/10.1691/ph.2013.2807.

Sharma, C, N Rokana, M Chandra, B P Singh, R D Gulhane, J P S Gill, P Ray, A K Puniya, and H Panwar. 2018. "Antimicrobial Resistance: Its Surveillance, impact, and Alternative Management Strategies in Dairy Animals." *Frontiers in Veterinary Science* 4: 237.

Sheweita, S A, A A Amara, H Gamal, A A Ghazy, A Hussein, and M Bahey-el-din. 2022. "Bacterial Ghosts of Pseudomonas aeruginosa as a Promising Candidate Vaccine and Its Application in Diabetic Rats." *Vaccines* 10 (6). https://doi.org/10.3390/vaccines10060910.

Shyntum, D Y., N P Nkomo, N L Shingange, A R Gricia, D Bellieny-Rabelo, and L N Moleleki. 2019. "The Impact of Type VI Secretion System, Bacteriocins and Antibiotics on Bacterial Competition of *Pectobacterium carotovorum* subsp. brasiliense and the Regulation of Carbapenem Biosynthesis by Iron and the Ferric-Uptake Regulator." *Frontiers in Microbiology* 10. https://doi.org/10.3389/fmicb.2019.02379.

Smalla, K and J M Tiedje. 2014. "Editorial Overview: Ecology and Industrial Microbiology." *Current Opinion in Microbiology* 19: v–vii. https://doi.org/10.1016/j.mib.2014.06.011.

Suskovich, J, B Kos, J Beganovich, A Pavunc, K Habjanic, and S Matosich. 2010. "Antimicrobial Activity – The Most Important Property of Probiotic and Starter Lactic Acid Bacteria." *Food Technology and Biotechnology* 48 (3):296–307.

Symanowicz, B, M Becher, D Jaremko, and K Skwarek. 2018. "Possibilities for the Use of Wood Ashes in Agriculture." *Journal of Ecological Engineering* 19 (3):191–196. https://doi.org/10.12911/22998993/86156.

Thanner, S, D Drissner, and F Walsh. 2016. "Antimicrobial Resistance in Agriculture." *mBio* 7 (2):e02227. https://doi.org/10.1128/mBio.02227–15.

Tharmaraj, N, and N P Shah. 2009. "Antimicrobial Effects of Probiotics Against Selected Pathogenic and Spoilage Bacteria in Cheese-Based Dips." *International Food Research Journal* 16: 261–276.

Use, E M A, Committee for Medicinal Products for Veterinary, Efsa Panel on Biological Hazards, D Murphy, A Ricci, Z Auce, J G Beechinor, H Bergendahl, R Breathnach, J Bureš, J P D Da Silva, J Hederová, P Hekman, C Ibrahim, E Kozhuharov, G Kulcsár, E L Persson, J M Lenhardsson, P Mačiulskis, I Malemis, L Markus-Cizelj, A Michaelidou-Patsia, M Nevalainen, P Pasquali, J-C Rouby, J Schefferlie, W Schlumbohm, M Schmit, S Spiteri, S Srčič, L Taban, T Tiirats, B Urbain, E-M Vestergaard, A Wachnik-Święcicka, J Weeks, B Zemann, A Allende, D Bolton, M Chemaly, P S F Escamez, R Girones, L Herman, K Koutsoumanis, R Lindqvist, B Nørrung, L Robertson, G Ru, M Sanaa, M Simmons, P Skandamis, E Snary, N Speybroeck, B T Kuile, H Wahlström, K Baptiste, B Catry, P S Cocconcelli, R Davies, C Ducrot, C Friis, G Jungersen, S More, C M Madero, P Sanders, M Bos, Z Kunsagi, J T Edo, R Brozzi, D Candiani, B Guerra, E Liebana, P Stella, J Threlfall, and H Jukes. 2017. "EMA and EFSA Joint Scientific Opinion on Measures to Reduce the Need to Use Antimicrobial Agents in Animal Husbandry in the European Union, and the Resulting Impacts on Food Safety (RONAFA)." *EFSA Journal. European Food Safety Authority* 15 (1): e04666–e04666. https://doi.org/10.2903/j.efsa.2017.4666.

Van Boeckel, T P, C Brower, M Gilbert, B T Grenfell, S A Levin, T P Robinson, A Teillant, and R Laxminarayan. 2015. "Global Trends in Antimicrobial Use in Food Animals." *Proceedings of the National Academy of Sciences of the United States of America* 112 (18): 5649–5654. https://doi.org/10.1073/pnas.1503141112.

Van Boeckel, T P, S Gandra, A Ashok, Q Caudron, B T Grenfell, S A Levin, and R Laxminarayan. 2014. "Global Antibiotic Consumption 2000 to 2010: An Analysis of National Pharmaceutical Sales Data." *The Lancet Infectious Diseases* 14 (8): 742–750. https://doi.org/10.1016/s1473–3099(14)70780–7.

Van Vuuren, M. 2014. "Appropriate Use of Antimicrobials in Animal Health. . ." In *Proceedings of the Antimicrobial Resistance Summit*. Birchwood Hotel, Sandton, South Africa.

Watson, R. 2016. "EU to Launch New Antimicrobial Resistance Action Plan." *BMJ*: i6328. https://doi.org/10.1136/bmj.i6328.

WHO. 2015. "Global Action Plan on Antimicrobial Resistance." 1–28. https://iris.who.int/bitstream/handle/10665/193736/9789241509763_eng.pdf?sequence=1 ISBN 978 92 4 150976 3.

WHO. 2019a. "New Report Calls for Urgent Action to Avert Antimicrobial Resistance Crisis." https://www.who.int/news/item/29-04-2019-new-report-calls-for-urgent-action-to-avert-antimicrobial-resistance-crisis.

WHO. 2019b. "The Journal of Global Antimicrobial Resistance meets the World Health Organization (WHO)." *Journal of Global Antimicrobial Resistance* 18: 305–308. https://doi.org/10.1016/j.jgar.2019.07.022.

Wilde, S, Y Jiang, A M Tafoya, J Horsman, M Yousif, L A Vazquez, and K L Roland. 2019. "Salmonella-Vectored Vaccine Delivering Three Clostridium Perfringens Antigens

Protects Poultry Against Necrotic Enteritis." *PLoS One* 14 (2). https://doi.org/10.1371/journal.pone.0197721.

Woods, A. 2019. "Decentring Antibiotics: UK Responses to the Diseases of Intensive Pig Production (ca. 1925–65)." *Palgrave Communications* 5 (1). https://doi.org/10.1057/s41599-019-0246-5.

Xiong, W, Y Sun, and Z Zeng. 2018. "Antimicrobial Use and Antimicrobial Resistance in Food Animals." *Environmental Science and Pollution Research* 25 (19): 18377–18384. https://doi.org/10.1007/s11356-018-1852-2.

Yegoraw, A A, A M Assen, P F Gerber, and S W Walkden-Brown. 2021. "Transmission of Infectious Laryngotracheitis Virus Vaccine and Field Strains: The Role of Degree of Contact and Transmission by Whole Blood, Plasma and Poultry Dust." *Veterinary Research* 52. https://doi.org/10.1186/s13567-021-00959-1.

Zeng, X, F Xu, and J Lin. 2010. "Development and Evaluation of CmeC Subunit Vaccine against *Campylobacter jejuni*." *J Vaccines Vaccin* 1 (3). https://doi.org/10.4172/2157–7560.1000112.

Zhang, Y, L Zhu, P Dong, R Liang, Y Mao, S Qiu, and X Luo. 2018. "Bio-Protective Potential of Lactic Acid Bacteria: Effect of *Lactobacillus sakei* and *Lactobacillus curvatus* on Changes of the Microbial Community in Vacuum-Packaged Chilled Beef." *Asian-Australasian Journal of Animal Sciences* 31 (4): 585–594. https://doi.org/10.5713/ajas.17.0540.

4 Role of β-Lactams, β-Lactamases Inhibitors, and Efflux Pump Inhibitors in the Battle Against Antibiotic Resistance

*Kiran R Marathe, Sunil H Koli, Athrava S Patil, and Satish V Patil**

* Corresponding Author

4.1 INTRODUCTION

Nowadays, antibiotics are a major part of modern medicine and are extensively used to treat infectious diseases in humans and in veterinary healthcare. However, due to extensive use of antibiotics, microorganisms have developed resistance, reducing the effectiveness of treatment as bacteria become less susceptible to these drugs. The failure of antibiotic treatment leads to increases in morbidity and mortality. The development of resistance to antimicrobials is recognized as one of the most life-threatening risks to humans, accountable for millions of deaths worldwide each year (Murray et al. 2022). Furthermore, the rapid global increase in antibiotic-resistant microbes (Ventola 2015) has posed significant challenges for health professionals in treating microbial diseases due to the limited availability of effective drug options (Tadesse et al. 2017). Antimicrobial resistance causes about 700,000 deaths yearly around the world (Bojar et al. 2021), which is predicted to increase up to 10 million by 2050 (Pulingam et al. 2022).

The use of antibiotics in veterinary treatment has become a growing concern, as these drugs are relied upon more frequently to enhance animal health and productivity (Van et al. 2020). However, the World Organization for Animal Health reported a declined global use of antimicrobials in animals by 13% from 2017 to 2019 (World Organisation for Animal Health 2023). Improper antibiotic use and the disposal of animal waste containing antibiotic residues are major contributors to the emergence of resistance in microorganisms. The widespread use of antibiotics in companion animals, aquaculture, poultry, and livestock significantly accelerates the rise of

DOI: 10.1201/9781003269847-4

resistance (Palma et al. 2020). Antibiotic resistance originating from animals poses a direct or indirect threat to human health. As evidenced, the alarming prevalence of methicillin-resistant *Staphylococcus* isolated from companion animals, i.e., dogs and cats, has been reported in a 24-month survey study. This study reported that different species of *Staphylococcus,* including *S. schleiferi* (40%), *S. aureus* (35%), and *S. intermedius* (17%), showed elevated resistance to methicillin (Morris et al. 2006).

The situation becomes worse day by day; hence, there is an urgent need to study the problem and find out the actual cause of the fast development of resistance in bacteria. There is a possibility that no antibiotics will remain effective in the long term in the "post-antibiotic era" due to the rise in multidrug resistance (MDR) (Falagas and Karageorgopoulos 2008).

To address this issue, under the One Health approach, the European Medicine Agency (EMA) established guidelines categorizing antibiotics used in veterinary medicine to preserve their effectiveness for public health and to reduce risks to animals. This categorization divides antibiotics into four groups, A, B, C, and D, based on the impact their use in animals may have on human health, particularly regarding the development of antibiotic resistance and the necessity of their use in veterinary treatments (EMA/688114/2020) (European Medicines Committee 2020) (Table 4.1). Group D (prudence) includes antibiotics recommended as first-line treatments. Group C (caution) comprises antibiotics that have alternatives in human medicine and should be used only when no antibiotics in group D are effective. Group B (restrict) consists of highly restricted antibiotics that should be used sparingly to minimize public health risks. Group A (avoid) includes antibiotics that are prohibited for veterinary treatment (see the details of antibiotics categorization, https://www.ema.europa.eu/en/news/categorisation-antibiotics-used-animals-promotes-responsible-use-protect-public-and-animal-health).

However, antibiotic resistance is intrinsically associated with their use, especially bacteria that primarily developed resistance against various classes of antibiotics, either through intrinsic or acquired mechanisms. Bacteria can develop resistance to antibiotics through four major mechanisms: (1) production of antibiotic-inactivating enzymes, (2) modification of target sites for antibiotics, (3) alteration in membrane permeability to antibiotic, and (4) upregulation of efflux pumps to expel the antibiotics into the surrounding environment (Blair et al. 2015) (Figure 4.1). The bacterial pathogen uses different resistance mechanisms according to the classes of antibiotics and its mechanism of action to protect them against it. They emerge with resistance expressing either a single mechanism or a combination of two or more.

Among the various classes of antibiotics, β-lactams are considered safe, potent, and highly effective in both human and veterinary use due to their favorable pharmacokinetic and pharmacodynamic properties. Different β-lactams are currently authorized for the treatment of bacterial infections in veterinary medicine. β-Lactam antibiotics inhibit bacterial cell wall synthesis by targeting essential enzymes, specifically transglycosylase (which catalyzes the process of transglycosylation) and transpeptidases (which catalyze the process of cross-linking of peptidoglycan in bacterial cells). These enzymes are also known as penicillin-binding proteins (PBPs) (Bush and Bradford 2016).

TABLE 4.1
β-Lactam Antibiotics Classes and Substances Authorized for Veterinary Use in the European Union (EMA/688114/2020)

Class of β-lactams	A (Avoid)	B (Restrict)	C (Caution)	D (Prudence)
Penicillins	**Amdinopenicillins:** Mecillinam, pivmecillinam			**Aminopenicillin:** Amoxicillin ampicillin and metampicillin **Antistaphylococcal penicillins:** Cloxacillin, dicloxacillin, nafcillin, and oxacillin **Narrow spectrum penicillins:** Benzathine benzylpenicillin, Benzathine Phenoxymethylpenicillin, Benzylpenicillin penethamate hydriodide, Pheneticillin phenoxymethylpenicillin, Procaine benzylpenicillin
Monobactams	Aztreonam			
Carbapenems	Meropenem/ Doripenem, faropenem			
Cephalosporin's	Ceftobiprole, ceftaroline	Cefoperazone, cefovecin, cefquinome, and ceftiofur	Cefacetrile, cefadroxil, cefalexin, cealonium, cefalotin, cefapirin, and cefazolin	
Penicillin's-BLIs	Pepracillin-tazobactam		Amoxicillin-clavulanic acid and ampicillin-sulbactam	
Cephalosporin's-BLIs	Ceftolozane-tazobactam			

Abbreviations: **A**, not for food-producing animals only for individual companion animals under exceptional circumstances; **B**, restricted use in animals due mitigate the risk to public health; **C**, should only be used when none of the antibiotics in category D are active; **D**, antibiotics that should be used as first line treatments.

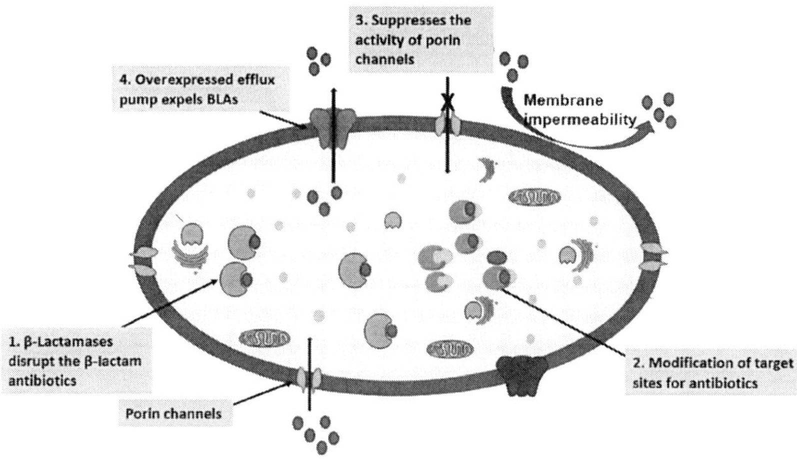

FIGURE 4.1 Schematic representation of four resistance mechanisms developed by bacteria.

However, even before the introduction of this class of antibiotic in clinical use, the bacteria had evolved the enzymes that hydrolyze the β-lactam antibiotics, called β-lactamase enzymes (Abraham and Chain 1940). Among the various resistance mechanisms, the β-lactamases play a significant role in β-lactam resistance in bacteria (Livermore 1995; Li et al. 2007). In this context, the rise of extended-spectrum β-lactamase (ESBL) (Woerde et al. 2023) and carbapenemase-producing bacteria in animals is a great concern regarding human health (Ramírez-Castillo et al. 2023). β-Lactamases are found across various bacterial species and are primarily classified in two different ways: (1) Ambler classification, in which the β-lactamases are categorized in four groups, i.e., A, B, C, and D, based on the amino acid motif in active site of enzymes (Ambler 1980) and (2) Bush-Jacobi-Medeiros classification system that categorizes β-lactamases into groups 1 to 4 based on the substrate and inhibitor profiles (Bush and Jacoby 2010). Among these, Ambler's classification is the most commonly used classification for β-lactamases, and more than 850 different β-lactamases are reported in bacterial pathogens.

In Ambler's classification, classes A, C, and D operate through a shared mechanism, hydrolyzing substrates via serine at the active site to form an acyl enzyme complex. Conversely, class B comprises metallo-β-lactamases (MBLs) that use zinc as a cofactor for substrate degradation. Even among these classes, some β-lactamases have narrow substrate specificity, such as penicillinases or cephalosporinases, while ESBLs have extended their spectrum and hydrolyzed penicillins, cephalosporins, and monobactams. Carbapenemases have been known for the additional hydrolysis of carbapenems, which are considered rescue therapy in humans. The most recently introduced MBLs are of primary concern due to their broad substrate specificity including all penicillins, cephalosporins, and carbapenems, except monabactams, and limited effective treatment available. Nevertheless, the emergence of certain broad-spectrum β-lactamases, including ESBLs, carbapenemases, and even MBLs in animal-origin bacteria is greatly alarming.

Although mechanisms exist to thoroughly track the prevalence and evolution of β-lactamases in bacteria of human origin (Livermore et al. 2007), limited efforts have been made for tracking in animals (Li et al. 2007). Considering these facts, this chapter highlights the β-lactamase-based mechanisms of resistance in bacteria of animals, including their types of β-lactamases and risks associated with human health. Also discussed are the various generations of β-lactamases inhibitors (BL/BLIs) authorized for veterinary and human use, the limitations of current β-lactam/BLIs combinations, the need for novel BLIs, and future strategies to combat this issue. Furthermore, the second section focuses on the efflux pump–mediated resistance to antibiotics and sources and types of different efflux pump inhibitors (EPIs) tested in vitro in research laboratories that can be future potent EPIs that may have clinical use.

4.2 β-LACTAM RESISTANCE IN ANIMAL ORIGIN BACTERIAL PATHOGENS

β-Lactams are among the most frequently prescribed class of antibiotics in human and veterinary medicine (Pleydell et al. 2012; Galarce et al. 2021). The World Organization for Animal Health (OIE) has published a list of key antibiotics essential for veterinary medicine, while the EMA has authorized various antibiotic classes for use in veterinary practices across the European Union (World Organization for Animal Health 2019). Both lists emphasize the inclusion of β-lactams and other antibiotics in treatments for animals. Table 4.1 depicts the β-lactams authorized by EMA. Similarly, various research publications list the β-lactams used in animal medicines and food production (Stegemann et al. 2006; Li et al. 2007; Galarce et al. 2021). In 2021, a survey for antimicrobial use in companion animals conducted in Chile reported the highest prescribed antibiotic class was β-lactams, which account for 51.1% penicillin and 34.4% cephalosporin. The use of β-lactams in animals has a long history and led to the observed acquired resistance in bacteria from animals and animal products. This resistance has significant implications for human health due to the potential transmission of infections to humans.

Resistance to β-lactams among bacteria from animals has been well documented in the literature, based on the antimicrobial resistance surveillance systems implemented by many countries, including DANMAP (Denmark), SVARM (Sweden), NORM-Vet (Norway), and MARAN (Netherlands) (Diallo et al. 2020).

According to DANMAP reports, in Denmark, the prevalence of ampicillin resistance among *Escherichia coli* from animals and meat between 2007 and 2012 ranged from 12% to 20% in broilers, with broiler meat at 11% to 20%, pigs at 19% to 29%, and pork at 29% to 33%. In 2022, ampicillin resistance in isolates from pigs increased to 35%, while resistance in broilers decreased to 9%. Ampicillin resistance in zoonotic *Salmonella* species from domestic pork increased from 38% in 2007 to about 56% in 2012, and further to 79% in 2022. No ampicillin resistance was detected among *Enterococci* isolates from broilers (Agersø et al. 2013; Duarte et al. 2023).

In Sweden, according to SVARM surveillance, ampicillin resistance in *E. coli* from pig clinical samples increased from 27% in 2012 to 38% in 2022. Similarly, in cattle fecal samples, ampicillin resistance rose from 48% in 2012 to 76% in 2022.

Resistance in *E. coli* from horses was 7% in 2012 and increased to 10% in 2022. *Pasteurella* spp. from calves showed no resistance to ampicillin in 2012, but resistance increased to 14% by 2022. The overall ampicillin resistance among *E. coli* (an indicator organism) from animals ranged from 3% to 30% during 2011–2012 and remained at 3% to 25% in 2022. Resistance to ampicillin in *Salmonella enterica* and *Salmonella typhimurium* from domestic animals was 11% and 16% in 2012, respectively, but dropped to 0% in 2022. The prevalence of ampicillin resistance among *Salmonella* has decreased, and the overall occurrence of *Salmonella* in animals remains low in Sweden (Hellman et al. 2013; Aspevall et al. 2022).

In Norway, *E. coli* isolates from poultry breeder flocks and broiler meat samples were found to harbor ESBLs and AmpC, with acquired resistance to third-generation cephalosporins reported at 7.3% and 32.2% respectively, in 2012 (NORM/NORM-VET 2012). Ampicillin resistance in *E. coli* isolates from animals ranged from 1.9% to 15% in 2012 and remained relatively constant at 7.5% to 14.5% in 2022. Ampicillin resistance in *Salmonella* spp. from animals decreased from 3% in 2012 to 1% in 2022. Additionally, resistance in *Enterococcus faecium* isolates from turkey rose to 6.4% in 2022 (NORM/NORM-VET 2022). In the Netherland, ampicillin resistance in *Salmonella typhimurium* isolated from cattle increased from 51.2% in 2012 to 73.2% in 2021. Moreover, ampicillin resistance in *E. coli* isolates of food-producing animals ranged from 0.7% to 36% based on 2021 surveillance data (MARAN 2022).

Data collected from various surveillance systems indicate different trends in β-lactam resistance among isolates derived from animals and food products in different regions. However, resistance to ampicillin remains high, particularly among Gram-negative bacterial pathogens. Additionally, the emergence of resistance to cephalosporins has been reported in some cases. This conclusion is based on limited surveillance data mentioned earlier, and further studies tracking resistance in pathogens from animal sources are necessary.

4.3 BACTERIAL PATHOGENS IN ANIMALS AND EMERGENCE OF β-LACTAMASES

Although the β-lactams are broad-spectrum antibiotics and are prescribed for both Gram-positive and Gram-negative bacterial infections, the β-lactamase-driven resistance is higher in Gram-negative bacteria. Therefore, for better understanding, we have categorized it into two parts, as presented in the following sections.

4.3.1 Gram-Positive Bacteria and β-Lactamases

The major Gram-positive bacteria pathogens found in various animals include *Staphylococcus aureus*, coagulase-negative *Staphylococcus* spp., *Streptococcus* spp. (Keane et al. 2013; Nocera et al. 2021), *Enterococcus faecalis*, and *Enterococcus faecium* (Torres et al. 2018). Among these, *Staphylococcus* spp. have been extensively studied and are responsible for causing a variety of infections in both human and animals (Nocera et al. 2021). In animals, mastitis in dairy animals is a major concern for the dairy industry, while pyoderma in pet animals (such as dogs) and various dermatologic infections in horses are also significant (Malik et al. 2007).

Staphylococcus has acquired the resistance to β-lactams primarily through alterations in PBPs, particularly in the PBP 2A protein encoded by the *mecA* gene (Fuda et al. 2005). However, β-lactamase-mediated resistance has also been reported in staphylococci (Watts and Salmon 1997; De Oliveira et al. 2020, Olsen et al. 2006). In 2007, Malik et al. conducted a study to evaluate β-lactamase-mediated resistance and detection of the *blaZ* gene in *Staphylococcus* spp. originating from animals. A total of 331 staphylococcal isolates were collected from dogs and cats in Australia. Among these, 23 isolates were resistant to penicillin, ampicillin, and cephalothin, and 13/23 isolates tested positive for the *blaZ* gene, which is responsible for resistance to these antibiotics in cats and dogs (Malik et al. 2007).

The β-lactamases produced by *Staphylococcus* are classified as Ambler class A and are narrow-spectrum penicillinases. Various species of *Staphylococcus*, including coagulase-negative *Staphylococcus, S. aureus, S. epidermidis,* and *S. haemolyticus* have been documented to possess the *blaZ* gene, which encodes the penicillin-hydrolyzing enzyme responsible for the development of penicillin resistance (Olsen et al. 2006; Nobrega et al. 2018; Bolte et al. 2020; Naranjo-Lucena and Slowey 2023). In 2011, Keserű et al. reported the membrane-bound and free forms of β-lactamases produced by *Staphylococcus aureus* isolated from human and cattle. These β-lactamases were able to hydrolyze methicillin and oxacillin, making the isolates borderline methicillin resistant. Interestingly, the enzyme is similar to the BlaZ penicillinase produced by these isolates (Keserű et al. 2011). The gene encoding the β-lactamase proteins may evolve independently from plasmid and chromosome encoded *blaZ* genes (Olsen et al. 2006).

Similarly, *Enterococcus* spp. exhibit resistance to β-lactams due to production of *blaZ*-encoded penicillinases (Patterson et al. 1990; Murray 1992). Although *Enterococcus* are capable of producing β-lactamases, high-level resistance to this class of antibiotics is mainly associated with increased production of PBP5 and polymorphism in the β subunit in this protein (Gagetti et al. 2019).

4.3.2 Gram-Negative Bacteria and β-Lactamases

In 2017, the World Health Organization published a list of "priority pathogens" that includes Gram-negative bacteria known as ESKAPE pathogens. These are a group of serious, life-threatening pathogens classified as multidrug-resistant organisms (Figure 4.2) (Breijyeh et al. 2020; Oliveira et al. 2023). Due to the diversity in membrane proteins and peptidoglycans, Gram-negative bacteria are able to acquire various mechanisms of resistance (Reygaert 2018). The β-lactamase-mediated resistance to β-lactams is the main mechanism, along with the downregulation of porin channels and overexpression of efflux pumps, contributing to increased resistance. In animals, *Campylobacter* spp., *E. coli*, and *Salmonella* spp. are the leading bacterial pathogens, along with some other Gram-negative pathogens.

The β-lactamase profile of Gram-negative organisms is vast, and many carry the chromosomal ampC gene for AmpC an Ambler class C β-lactamase, which is either constitutive or inducible (Bush et al. 1995). Intriguingly, in *E. coli,* this ampC is uninducible. Initially, though, it exerted a low level of resistance to penicillins and narrow-spectrum cephalosporins due to difficulties and weak promoter presence for

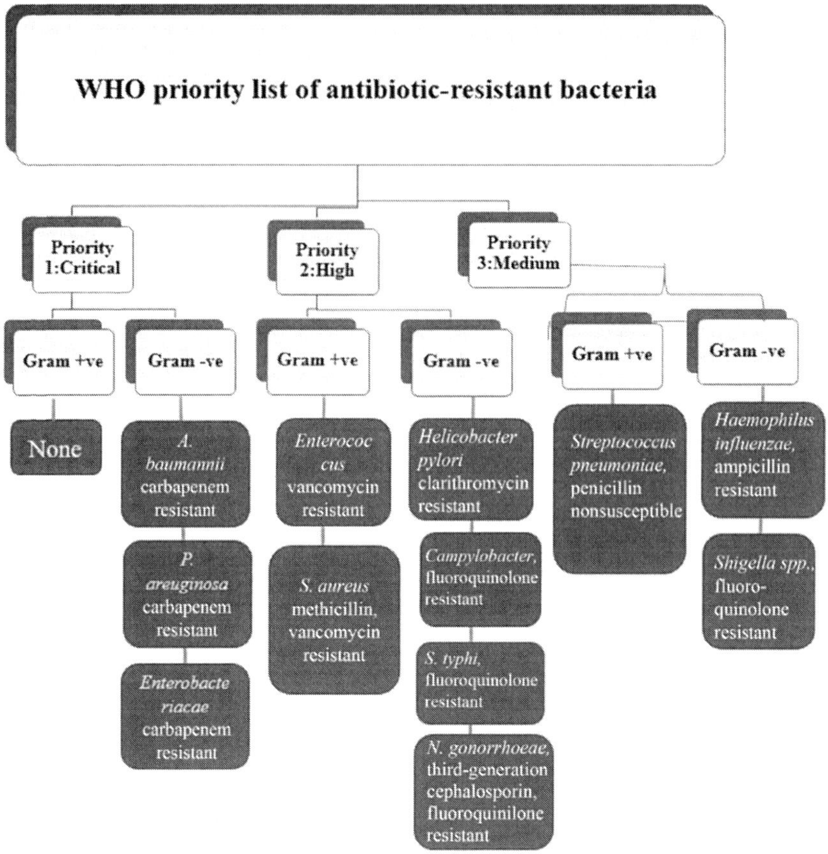

FIGURE 4.2 List of priority pathogens grouped under three priority categories according to their antibiotic resistance according to the World Health Organization (Breijyeh et al. 2020).

the ampC gene. Moreover, over time, the bacteria of animals acquired mutations in ampC genes, leading to higher production of AmpC, which is responsible for the resistance to major substances in the β-lactams (penicillins, early generations of cephalosporins and monobactams) (Li et al. 2007). In 1992–2005, Li et al. (2007) reviewed the emergence of AmpC and ESBLs among the *E. coli* and *Salmonella* spp., isolated from different animals, and the majority of studies reported the CMY2 type of β-lactamase enzymes. Additionally, both genera of bacteria produced the class A β-lactamase, including ESBLs; these enzymes mainly comprise plasmid-encoded TEM, SHV, and CTX-M. Furthermore, few reports highlighted the OXA type of β-lactamases, including OXA-30 and OXA-1 (Li et al. 2007). Although these ESBLs from bacteria derived from animals expand the hydrolytic spectrum of β-lactams, most of them are inhibited by the inhibitor clavulanic acid. Notably, Pomba-Féria and Caniça (2003) reported the emergence of inhibitor resistance TEM-1 in dogs, which bears resemblance to inhibitor-resistant TEM in *E. coli* isolated from humans.

Silva et al. (2019) attempted to report the prevalence of ESBLs and emerging carbapenemases producing Enterobacterales among the food-producing animals in Europe, including poultry, pigs, and cattle. They reported the highest percentage of CTX-M-1 with 38% in poultry, 59% in cattle, and 70% in pigs, followed by SHV-12 in poultry and pigs, and CTX-14 and CTX-15 in cattle. The predominant pathogen in all the tested animals was *E. coli*. The rise of ESBL-producing bacteria in companion animals has particular concerns because the *E. coli* and *K. pneumoniae* derived from dogs and cats during 2011–2021 express ESBLs ($bla_{\text{CTX-M-15}}$) and contribute resistance to multiple drugs, including penicillins, cephalosporins, monobactams, and some even to BL/BLIs (amoxicillin-clavulanic acid, ampicillin-sulbactam, and piperacillin-tazobactam), but are susceptible to carbapenems (Woerde et al. 2023). In 2023, a Korean study highlighted the high prevalence of β-lactamases harboring *E. coli* in poultry and that they are responsible for making the multidrug-resistant bacteria. The highest 81.8% of non-ESBLs gene $bla_{\text{TEM-1}}$, followed by 0.9% to 6.0% of ESBLs gene (SHV, Variety of CTX-M, and OXA-1) and 3.6% of AmpC gene $bla_{\text{CMY-2}}$ were represented among all (Jung et al. 2023).

Additionally, the carbapenemases, enzymes that hydrolyze the penicillins, cephalosporins, monobactams, and carbapenems, have been reported in *Acinetobacter* spp. from livestock in dairy cattle in 2012, which harbor the $bla_{\text{OXA-23}}$ carbapenemase gene (Poirel et al. 2012). Thereafter, MBL-type VIM-1 carbapenemase-producing *E. coli* was isolated from pig farms and *Salmonella* spp. from livestock farms in Germany (Fischer et al. 2012, 2013).

Falgenhauer and colleagues also published a similar report highlighting VIM-1 carbapenemase-encoding plasmids found in *E. coli* and *Salmonella* from livestock farms in Germany (Falgenhauer et al. 2017).

Several studies have been reviewed on carbapenem resistance and carbapenemase-producing bacteria in livestock, companion animals, and food-producing animals (Köck et al. 2018). Most of the studies reported *E. coli* as the main pathogen harboring the *bla* genes for NDM-1, NDM-5, NDM-9, NDM-17 (Wang et al. 2017a; Liu et al. 2017), OXA-48 (Melo et al. 2017), and VIM-1 (Roschanski et al. 2017), followed by *Klebsiella* spp., which harbor NDM-5 (He et al. 2016), OXA-48 (Stolle et al. 2013), and VIM-1 (González-Torralba et al., 2016), and *Salmonella* spp. that carry $bla_{\text{NDM-1}}$ and $bla_{\text{VIM-1}}$ (Wang et al. 2017b), with few other bacteria *Citrobacter, Enterobacter, E.coli, Morganella, Proteus,* and *Providencia* carry $bla_{\text{IMP-27}}$ (Mollenkopf et al. 2017) that confer carbapenem resistance to these isolates, especially to meropenem that is considered the last-resort drug for humans (Köck et al. 2018). The emergence of these carbapenemases is an alarming threat, and we cannot ignore the possibility of transfer of these genes to human bacteria that increased resistance to the last-resort carbapenems; therefore, there is an urgent need to formulate strategies to combat this issue.

4.4 β-LACTAMASE INHIBITORS

The enzymes that inactivate β-lactams by breaking the four-membered ring structure are called β-lactamases. Bacteria produce a wide range of β-lactamases to protect themselves from the bactericidal effects of β-lactams. The rising resistance to β-lactams, including third-generation cephalosporins and particularly the last-resort drugs known as carbapenems, is recognized as a global threat. Hence, there is an urgent need to tackle such a type of resistance using novel strategies. To combat resistance of Gram-negative bacteria, administration of BL/BLI combinations proved to be most effective (Zasowski et al. 2015).

The first BLI, clavulanic acid, was derived from *Streptomyces clavuligerus* in 1976. Initially, clavulanic acid was combined with amoxicillin to treat various Gram-negative infections (Reading et al. 1977). Thereafter, it was combined with ticarcillin, i.e., ticarcillin-clavulanate. Clavulanic acid is active against only class A (Serine) β-lactamases and is unable to inhibit classes B, C, and D (González-Bello 2017). Similarly, the next nontoxic BLI, tazobactam, is responsible for inhibition of class A β-lactamases and is used in combination with piperacillin, ceftolozane, and cefepime (Bassetti et al. 2020; Mushtaq et al. 2021). Sulbactam is one of the earlier BLIs that was combined with ampicillin. The sulbactam and clavulanic acids synthesized chemically and were the most commonly used BLIs, which act as irreversible competitive inhibitors of multidrug-resistant Enterobacteriaceae (Impey et al. 2020; Omollo et al. 2021; Parvaiz et al. 2021).

In the treatment of MDR *P. aeruginosa* infections, combinations of ceftolozane-tazobactam and ceftazidime-avibactam were used as BL/BLIs (Bassetti et al. 2020; Blomquist and Nix 2021). All three BLIs are β-lactam structure based inhibitors with narrow-spectrum inhibition potential. Eventually, the bacteria developed resistance against these narrow-spectrum BLIs through structural modifications, which led to an increase in the hydrolytic activity that hydrolyzed the BLIs. Over generations, the non-β-lactam structure BLIs are being discovered with a broader enzyme inhibition spectrum. The new class of diazabicyclooctane (DBO), avibactam, is a non-β-lactam structure BLI that effectively inhibits Ambler classes A, C, and D in various bacterial pathogens. Also, it is highly stable to all known β-lactamases. When combined with avibactam, the third-generation cephalosporin ceftazidime has been shown to be effective in treating Gram-negative infections.

Similarly, the relebactam was introduced with imipenem+cilastatin to selectively tackle the *K. pneumoniae* carbapenemase (KPC), with other class A and C β-lactamase-producing bacterial infections. Recently, the new boron-based BLIs were also introduced with carbapenems, i.e., meropenem+vaborbactam, to tackle the carbapenem-resistant Enterobacterales. Different combinations of BL/BLIs used for treatment of various bacterial infections and those in clinical trials are summarized in Table 4.2. Currently, the use of β-lactams combined with BLIs is recognized as a proven potential strategy to combat antibiotic-resistant bacteria, particularly those that produce β-lactamase. For this purpose, various combinations of β-lactams with BLIs are used to overcome resistance, e.g., a ceftolozane-tazobactam combination used to treat Enterobacterales and *P. aeruginosa* infections, which cause complicated intra-abdominal infections (Eckmann and Solomkin 2015). Similarly, ceftazidime-avibactam effectively treats urinary tract infection and intra-abdominal infections caused by ceftazidime-resistant Enterobacterales (Carmeli et al. 2016).

Still, currently approved combinations that undergo clinical trials do not fully solve the problem of antibacterial resistance because of alterations in amino acid sequences, mutations in target proteins, and activation of different efflux pumps

TABLE 4.2
List of some β-Lactamase Inhibitor, BL/BLIs Combinations Approved for Human Medicine and Under Clinical Development

Sr. No.	β-lactamase inhibitor	Source	Representative Combination	Mechanism of Action	Target Organisms	References
1.	Clavulanic acid	*Streptomyces clavuligerus*	Amoxicillin-clavulanic acid	Inhibits Class A type (Serine) lactamases	All β-lactamase producing organisms, *S. auerus.*	Reading et al. 1977; Gonzalez-Bello 2017
2.	Sulbactam	Derivative of penicillin	Amphicillin-sulbactum	Irreversible competitive inhibition of Class A type lactamases	Acinetobacter and Bacteroides. Enterobacteriaceae, Skin infections	Impey et al. 2020; Omollo et al. 2021
3.	Tazobactam	Synthetic penicillin sulfone	Piperacillin-tazobactam Ceftolozane-tazobactam	Effective against the OHIO-1, SHV-1, and TEM groups of beta-lactamases	bloodstream infections, febrile neutropenia, MDR *P. aeruginosa*	Hayashi et al. 2010
4.	Avibactam	Synthetic diazabicyclooctane	Ceftazidime-avibactam Aztreonam-avibactam (clinical development stage)	OXA-48 producers, Class A, KPC and Class C β-lactamase Inhibits class B and C β-lactamase	MDR Enterobacterales Enterobacteriaceae and *P. aeruginosa*	Longshaw et al. 2020 Biedenbach et al. 2015; Crandon and Nicolau 2013
5.	Relebactam	Synthetic Diazabicyclooctane	Imipenem-cilastatin-relebactam	Inhibit class A KPC and class C β-lactamase	Aerobic G- bacteria, Enterobacteriaceae, *Pseudomonas*, and *Acinetobacter*	Heo 2021
6.	Durlobactam	Synthetic Diazabicyclooctane	Sulbactum-durlobactam	Inhibits Class A type (Serine) and Class C type lactamases	Enterobacterales and *Pseudomonas aeruginosa*	Papp-Wallace et al. 2019, 2023; O'Donnell et al. 2024
7.	Vaborbactam	Synthetic cyclic boronate	Meropenem-vaborbactam	Class A and KPC	Carbapenem resistant *Enterobacterales*	Bhowmick and Weinstein 2020
6.	Taniborbactam	Synthetic bicyclic-boronate	Cefepime-taniborbactam	Inhibits all class A, C, D and Metallo β-lactamase	MDR *P. aeruginosa*	Daigle et al. 2018

(Continued)

TABLE 4.2 *(Continued)*
List of some β-Lactamase Inhibitor, BL/BLIs Combinations Approved for Human Medicine and Under Clinical Development

Sr. No.	β-lactamase inhibitor	Source	Representative Combination	Mechanism of Action	Target Organisms	References
8.	Funobactam	Synthetic Diazabicyclooctane	Imipenem-Funobactam	Effective against class A, C, and D enzyme particularly OXA-24/40-like β-lactamases	*A. baumannii*	Fratoni et al. 2023; Li et al. 2022
9.	Catechingallate	Plant flavonoid from green tea	Oxacillin-Catechin, Pen cillin-Catechin	B lactamase inhibitor	*S. aureus* NCTC 6571 and MTCC 737	Abreu et al. 2012
10.	Qurcetin	Plant alakloid	Piperacillin or imipenem- Qurcetin	Inhibit OXA-48 β-lactamase	Enterobacteriaceae	Zhang et al. 2022
11.	Shikimic acid	Plant aromatic compound	Oxacillin-Shikimic acid	Inhibits Class A type (Serine) lactamase	MDR *Staphyloccocus auerus*	Hou et al. 2024; Zhang et al. 2024a
12.	Baicalein	Active compound of *Scutellariabaicalensis*	Penicillin G/ amoxicillin-Baicalein	Inhibit penicillinase enzyme	MDR *Staphyloccocus auerus*	Qian et al. 2015
13.	Apigenin	Natural plant based poly phenolic flavonoid	Ampicillin-Apigenin and ceftriaxone- Apigenin		Methicillin-resistant *Staphylococcus aureus* (MRSA)	Akilandeswari and Ruckmani 2016
14.	Pantoprazole	Synthetic drug	Merophenm-Pantoprazole	Inhibit metalloβ-lactamase	Carbaphenem resistant *K. pneumoniae*	Abdulaal et al. 2024
15.	Tellimagrandin I and rugosin B	Rose red petals plant source	Tellimagrandin I and oxacillin, tetracycline	Inhibit β-lactamase	Methicillin-resistant Staphylococcus aureus (MRSA)	Shiota et al. 2004

(Wang et al. 2020). Therefore, designing new combinations with broad-spectrum new BLIs is essential, and this will depend on the selection of correct BLIs. The selection of BLIs will be carried out on the basis of specificity of inhibitor against β-lactamase, its pharmacokinetics properties, efficacy of penetration in cellular compartments, and less adverse effects (Zhang et al. 2024b).

A β-lactam antibiotic was derived from natural sources and further chemically modified for an increase spectrum of antibacterial activity. Nowadays, plant-derived BLIs are also used in combination with β-lactam antibiotics to tackle MDR pathogens. Various plant-derived compounds have the potential to inhibit β-lactamases. Among these compounds are catechin and epigallocatechin gallate, derived from the tea leaves of *Camellia sinensis* (L.) Kuntze, which effectively restore and enhance β-lactam activity against MRSA (Abreu et al. 2012). Furthermore, Baicalein (*Scutellaria baicalensis*) (Qian et al. 2015), various alkaloid compounds (*Cienfuegosia digitata Cav.*), (Konaté et al. 2012) corilagin and tellimagrandin I (red rose petals) (Shiota et al. 2000; Shiota et al. 2004), and Glycosmis-derived rugosin B, and 5-O-methylglovanon (Zhou et al. 2011) significantly enhanced the activity of β-lactams in combination against MDR strains of *S. aureus*. A combination of shikimic acid, oxacillin, and ceftiofur proved to be an excellent strategy for inhibiting MRSA (Hou et al. 2024; Zhang et al. 2024a). Similarly, Zhang et al. (2022) highlight the OXA-48 β-lactamase inhibitory potential of the plant-derived compounds, such as quercetin; luteolin; taxifolin; kaempferol; apigenin; 3′,4′,7-trihydroxyflavone; and fisetin in Enterobacterales. These BLIs significantly restored the activity of imipenem and piperacillin against the resistant *E. coli* (Zhang et al. 2022). The plant-derived compounds and their combinations with β-lactam antibiotics are summarized in Table 4.2.

Although there are a variety of BLIs available in combination with β-lactams with varying degrees of coverage, none of the clinically available BLIs or plant-derived BLIs showed effective inhibition of Ambler class B β-lactamases. The class B metallo-β-lactamases (MBL) are the major hurdle for treating the MDR Gram-negative bacteria due to the broad substrate profile, including all the β-lactams, except aztreonam. Therefore, class B β-lactamases, especially New Delhi Metallo- β-lactamases (NDM), are the major clinical threat in clinics globally. The in vitro MBL inhibition activity of EDTA is really attention grabbing; however, it has very narrow toxicity windows that limit its clinical utility. Nevertheless, there is one combination of ceftriaxone-sulbactam -disodium EDTA that is attempted and that is tested for MBL infections trough ramdomized control trials and found efficacy against MBL producer Enterobacterales in complicated urinary tract infections. This drug combination has approved by DCGI, India, only for extended-spectrum β-lactamases producer Gram-negative bacterial infections (Samantaray et al. 2023). Similarly, recently, a β-lactamases stable new siderophore-conjugated antibiotic, cefiderocol, was introduced by Shionagi, a Japanees company. Cefiderocol showed a significant efficacy against the NDM producers Gram-negative bacteria. However, some reports highlighted a high level of resistance for cefiderocol among the carbapenem-resistant *E. coli*. The emergence of resistance against cefiderocol is mainly observed in NDM-5 harboring isolates, which have sequence type 167, and KPC-ST131 producer *E. coli* (Wang et al. 2022; Castillo et al. 2024).

Additionally, a bicyclic boron-based BLI, taniborbactam, has the ability to inactivate several clinically relevant MBLs, and this will be available with a combination of cefepime. Unfortunately, this combination is also does not meet the full coverage for MBLs; it is unable to inhibit the imipenemases (IMP) (Matsumura et al. 2017).

The use of BL/BLIs for treating MDR infections is a promising strategy; however, in some bacteria, they use different mechanisms for developing antibiotic resistance, such as upregulation of efflux pumps. Therefore, we also focused on the efflux pumps present in MDR bacteria and its inhibitors to control expulsion of antibiotics.

4.5 ROLE OF EFFLUX PUMPS IN ANTIBIOTIC RESISTANCE AND ITS INHIBITORS

In bacteria, expression of active efflux pump proteins is one of the important mechanisms of the development of antibiotic resistance. Efflux pumps are membrane-embedded transporter proteins in bacteria that expel substrate from the cell. The substrates for the efflux pump are antibiotics, which impart the bacteria antibiotic-resistant phenotype (Piddock 2006). The first antibiotic resistance efflux pump was reported in the 1960s in MRSA. After that, a plethora of efflux pumps have been reported in both Gram-positive and Gram-negative bacteria. The Gram-positive bacteria, such as MRSA, *S. pneumoniae, Enterococcus* spp., *Listeria monocytogenes*, and *Clostridium difficile,* as well as Gram-negative species like *E. coli, K. pneumoniae, Salmonella* spp., *Steno maltophilia, P. aeruginosa, A. baumannii, Campylobacter jejuni, Neisseria gonorrhoeae,* and *Vibrio cholerae* (Prestinaci et al. 2015; Yang et al. 2021; Sharma et al. 2019). To date, six major families of efflux pumps have been identified: resistance-nodulation-division (RND), major facilitator superfamily (MFS), multidrug and toxic compound extrusion, ATP-binding cassette (ABC), small multidrug resistance, and proteome bacterial antimicrobial compound efflux families (Figure 4.3) (Zhang et al. 2024a).

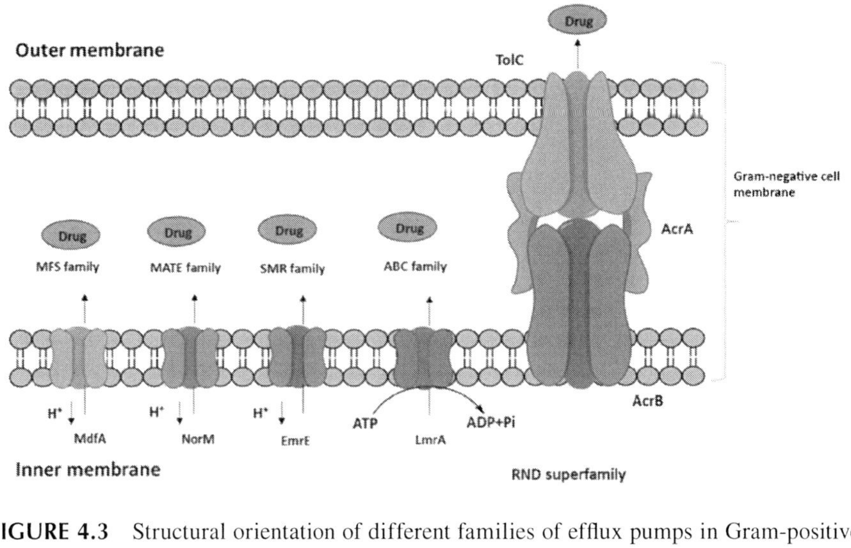

FIGURE 4.3 Structural orientation of different families of efflux pumps in Gram-positive and Gram-negative bacteria.

Among these efflux pumps, ABCs are the primary active transporters that draw energy from ATP-hydrolysis, and the remaining five families of efflux pumps are secondary active transporters that translocate the substrate using an ion or proton gradient to draw energy (Sharma et al. 2019). The efflux transporters located in cytoplasmic membranes, primarily of the MFS type, are a major cause of drug resistance in Gram-positive bacteria. Examples of efflux pumps from the MFS family include NorA (*S. aureus*), PmA (*S. pneumoniae*), and EmeA (*E. faecalis*) that are responsible for expelling the broad spectrum of antibiotics through them (Prestinaci et al. 2015; Yang et al. 2021). In contrast, RND family transporters are the major efflux pumps in Gram-negative bacteria that impart antimicrobial resistance. Its tripartite structure covers the inner and outer cytoplasmic membranes separated by periplasmic space in Gram-negative bacteria (Sharma et al. 2019). The RND family is a well-organized operon that plays a crucial role in antimicrobial resistance by enhancing resistance to a range of antibiotic classes, including penicillins, aminoglycosides, tetracyclines, and fluoroquinolones (Mueller 2023).

Efflux pumps are embedded intrinsic proteins, with their encoding genes regulated at the gene expression level. Mutations in regulatory proteins can lead to overexpression of these genes, upregulating the efflux pumps and contributing to drug resistance (Webber and Piddock 2003). Therefore, there is an urgent need to look for novel strategies for controlling the efflux system in MDR pathogens. To control efflux activity in bacteria, researchers have explored various mechanisms, such as (1) reducing the expression of efflux pump genes, (2) redesigning antibiotics so they are not recognized as substrates by the efflux pumps, and (3) inhibiting the assembly of active efflux pumps (Bhardwaj and Mohanty 2012). Here we focus on only inhibition of efflux pumps using the natural or synthetic EPIs on the basis of the efflux pump family, their types and mechanisms of action of resistance. The EPIs from natural and synthetic sources are listed in Table 4.3. It was found that the RND type of efflux pumps is most common and imparts resistance in Gram-negative bacteria due to its structural organization. Similarly, the efflux pumps of the MFS, multidrug and toxic compound extrusion, ABC, and small multidrug resistance families and their inhibitors and susceptible antibiotics are also listed in Table 4.3. It was found that the Nor A type of efflux pump, which belongs to the MFS family and is mostly present in *S. aureus,* was intensively studied, and many natural and synthetic EPIs were reported against it, like coumarins, reserpine, kaempferol, curcumin, capsaicin, ginsenoside, and conjugated α-bisabolol-β-cyclodextrin (Table 4.3). Reserpine and essential oils isolated from plant sources acted as inhibitors of MepA and Tet. A type of efflux pump belongs to the MATE family found in *S. aureus, E. coli,* and *Bacillus subtilis* (Stavri et al. 2007; Neyfakh et al. 1991; Gupta et al. 2016). Similarly, the inhibitors (EPIs) against the efflux pump of the ABC and small multidrug resistance families are also summarized in Table 4.3.

The use of EPIs to overcome antibiotic resistance is attractive and a scientifically excellent strategy, but as per the literature survey, the actual use of EPIs in preclinical and clinical studies is lacking. There are many limitations to the use of EPIs, such as the need for high optimum doses, substrate specificity, and the fact that the exact mechanisms of action of many EPIs remain unknown (Sharma et al. 2019). To overcome these limitations, thorough screening and optimization studies for EPIs are necessary.

TABLE 4.3
List of Efflux Pumps and Their Inhibitors

Sr. No.	Efflux pump	Mechanism of action	EPI	Bacterial strains	Substrate/Synergy	References
			Natural and Synthetic EPIs			
	RND Family					
1.	AcrAB-TO1C	Allowing efficient antibiotic efflux via energy derived from proton motive force (PMF). Secondary type of efflux pump	Cathinone (Plant)	*Salmonella typhimurium*	Ciprofloxacin	Piddock et al. 2010
			Theobromine (Plant)	*S. typhimurium, K. pneumoniae*	Ciprofloxacin, Tetracycline	Neyfakh et al. 1991
			Artesunate (Plant)	*E. coli*	Penicillin G; ampicillin, cefazolin, cefuroxime, cefoperazone	Li et al. 2011, 2015
2.	OqxAB	Overexpression leads to resistance	–	*K. pneumoniae, Enterobacter* spp.,	Quinolone and chloramphenicol	Ruiz 2019; Wong et al. 2015
3.	MexAB-OprM	constitutive expression confers intrinsic resistance	Phenyl-arginine beta-naphthylamide (PAbN) (Synthetic) Mechanism acting as an RND substrate that occupies an affinity site in the large substrate-binding pocket different from that occupied by a given antibiotic	*P. aeruginosa*	Fluoroquinolones, tetracycline, chloramphenicol, novobiocin and β-lactams, such as carbenicillin,	Kourtesi et al. 2013
			Pyridopyrimidine and pyranopyridine derivatives	*P. aeruginosa*	Fluoroquinolones	Mahmood et al. 2016
			EA-371a and EA-371d (Microbial source *Streptomyces* spp.)	*P. aeruginosa*	Levofloxacin	Lee et al. 2001
			Conessine (Plant)	*P. aeruginosa Acinetobacter baumannii*	Cefotaxime, levofloxacin, tetracycline, novobiocin and rifampicin	Siriyong et al. 2016, 2017
4.	MdtABC-TolC			*Escherichia coli*	Quinolones	Gauba and Rahman 2023

5.	AdeABC	Overexpressions efflux pump offers AMR in *A. baumannii*	NMP (1-(1naphthylmethyl)-piperazine) (Synthetic)	*A. baumannii, Enterobacter aerogenes, K. pneumoniae*	Fluoroquinolones, β-lactams, tigecycline, chloramphenicol, and aminoglycoside	Bohnert and Kern 2005 Dou et al. 2017
6.	CmpL1-4		–	*C. glutamicum*	Aminoglycosides, carbenicillin, fluoroquinolones, and novobiocin	Yang et al. 2014
	MFS family					
1.	Nor A	Chromosomally located proteins causing the overexpression, regulatory mutations in proteins	Coumarins (Plant)	*S. aureus*	Norfloxacin and Ciprofloxacin	Roy et al. 2013
			Reserpine (Plant)	*S. aureus, B. subtilis, Streptococcus pneumoniae*	Norfloxacin and ciprofloxacin,	Stavri et al. 2007; Neyfakh et al. 1991
			Curcumin (Plant)	*S. aureus*	Norfloxacin and ciprofloxacin	Joshi et al. 2014
			Kaempferol (Plant)	*S. aureus*	Norfloxacin and ciprofloxacin	Holler et al. 2012
			Diterpenes (Plant) (ferruginol)	*S. aureus, Mycobacterium* spp.	Tetracycline, erythromycin, norfloxacin, and isoniazid	Smith et al. 2007a
			Tannic acid (Plant)	*S. aureus*	Tetracycline, Norfloxacin	Tintino et al. 2017
			Baicalein(Plant)	*S. aureus, E. coli*	Ciprofloxacin and tetracycline	Chan et al. 2011

(Continued)

TABLE 4.3 *(Continued)*
List of Efflux Pumps and Their Inhibitors

Sr. No.	Efflux pump	Mechanism of action	EPI	Bacterial strains	Substrate/Synergy	References
			Natural and Synthetic EPIs			
			Piperine (Plant)	*S. aureus*	Norfloxacin, ciprofloxacin	Kumar et al. 2008
			Capsaicin	*S. aureus*	Ciprofloxacin	Kalia et al. 2012
			Ginsenoside20(S)-Rh2 (Rh2) (Chemically Synthesized)	*S. aureus*	Ciprofloxacin	Zhang et al. 2014
			α-Bisabolol, complexation with β-Cyclodextrin	*S. aureus*	Tetracycline and norfloxacin	Pereira da Cruz et al. 2020
2.	QacA/QacB	A proton motive forcedependent antiport mechanism, secondary type	(E)-1-(4-aminophenyl)-3-(4-ethoxyphenyl) prop-2-en-1-one (CPA-Etox) (Synthetic)	*S. aureus*	Chlorhexidine	De Sousa et al. 2024
3.	Blt	A proton motive force dependent mechanism	Roemerine (Plant)	*B. subtilis*	Norfloxacin, ciprofloxacin	Ahmed et al. 1995; Avci et al. 2019
	MATE Family					
1.	MepA	A proton motive force dependent mechanism	Reserpine	*S. aureus*, *B. subtilis*	Norfloxacin, ciprofloxacin, and tetracycline	Stavri et al. 2007; Neyfakh et al. 1991
			Clerodanediterpene 16α-hydroxycleroda-3,13 (14)-Z-dien-15,16-olide (Chemically synthesized)	*S. aureus*	Norfloxacin and ciprofloxacin	Gupta et al. 2016
2.	Tet(A) and (B)	–	Essential oils (Salvia species)	*A. baumannii* *S. epidermidis* *S. aureus*	Tetracycline	Gupta et al. 2016

			13-cyclopentylthio-5-OH-TC (13-CPTC), semisynthetic tetracycline (TC) analogs	*E. coli*	Tetracycline	Nelson and Levy 1999
			Totarol	*S. aureus, Mycobacterium spp.*	Erythromycin and isoniazid	Smith et al. 2007b
	ABC Family					
1.	MsrA	ATP dependant mechanism	Diosmetin	*S. aureus*	Erythromycin and norfloxacin	Chan et al. 2013
			Carnosic acid	*S. aureus*	Erythromycin	Oluwatuyi et al. 2004
			Carbonyl cyanide m-chlorophenyl hydrazine (CCCP) (Synthetic)	*K. pneumoniae*	Fluoroquinolone	Tang et al. 2020
2.	LmrA	Hydrolysis of ATP results in translocation substrate to the exterior and its release.	-	*L. lactis*	–	Van Veen et al. 2000
3.	Lmr S		Cumin seed oil, cuminaldehyde	*S. aureus*	EtBr and fusidic acid	Kakarla et al. 2017
	SMR family					
1.	MacB			*Acinetobacter baumannii*	Fluoroquinolones,	Gauba and Rahman 2023
2.	QacC/Smr (P)	A proton motive force dependent mechanism	1,8-Naphthyridines (Synthetic)	*S. aureus*	Tetraphenylphosphonium β-lactam antibiotics	Grinius et al. 1992; Oliveira-Tintino et al. 2023

Abbreviations: ABC, ATP-binding cassette; MATE, multidrug and toxic compound extrusion; MFS, major facilitator superfamily; RND, resistance-nodulation-division; SMR, small multidrug resistance

4.6 CONCLUSION

In summary, the emergence of multidrug resistance is a global threat and serious problem. The use of β-lactams with BLIs and EPIs is proven to be an effective and useful way to tackle MDR pathogens of animals and animal products because, over time, bacteria have evolved and produced new β-lactamases that are not inhibited by the old BLIs, leading to outbreaks of bacterial pathogens in animals that harbor these β-lactamases, raising concern for human health. Significant developments in the search of mechanisms of action of efflux pumps and their inhibitors will be a new area of research that might open up a new window to solve the emerging resistance in pathogens from veterinary sources. Also, the combinational therapy of EPIs and BLIs with antibiotics will be a promising intervention in the upcoming years. The currently available BLIs and their BL/BLIs combinations and the limitations of the currently available BLIs emphasize the urgent need for novel broad-spectrum BLIs. Finally, in conclusion, the search for novel BLIs and EPIs may prove to be a good way to combat MDR pathogens and also reduce the transmission of these pathogens to humans.

4.7 ACKNOWLEDGMENT

Dr. Kiran Marathe is thankful to ICMR, New Delhi, for financial support in the form RA-I fellowship funded by the Indian Council of Medical Research, Govt. of India (ICMR) RA-I fellowship (File No.45/23/2022-BIO/BMS).

REFERENCES

Abdulaal, W. H., Alhakamy, N. A., Asseri, A. H., Radwan, M. F., Ibrahim, T. S., Okbazghi, S. Z. and Abdel-Halim, M. S. 2024. Redirecting pantoprazole as a metallo-beta-lactamase inhibitor in carbapenem-resistant *Klebsiella pneumoniae*. *Front. Pharmacol.* 15: 1366459.

Abraham, E. P. and Chain, E. 1988. An enzyme from bacteria able to destroy penicillin. 1940. *Rev. Infect. Dis.* 10: 677–678.

Abreu, A. C., McBain, A. J. and Simões, M. 2012. Plants as sources of new antimicrobials and resistance-modifying agents. *Nat. Prod. Rep.* 29: 1007–1021.

Agersø, Y., Andersen, V. D., Helwigh, B., Høg, B. B., Jensen, L. B., Jensen, V. F. and DANRES Study Group. 2013. *DANMAP 2012: Use of antimicrobial agents and occurrence of antimicrobial resistance in bacteria from food animals, food and humans in Denmark.* DANMAP. https://www.danmap.org/reports/2012.

Ahmed, M., Lyass, L., Markham, P. N., Taylor, S. S., Vazquez-Laslop, N. and Neyfakh, A. A. 1995. Two highly similar multidrug transporters of Bacillus subtilis whose expression is differentially regulated. *J. Bact.* 177: 3904–3910.

Akilandeswari, K. and Ruckmani, K. 2016. Synergistic antibacterial effect of apigenin with Î²-lactam antibiotics and modulation of bacterial resistance by a possible membrane effect against methicillin resistant *Staphylococcus aureus*. *Cell. Mol. Biol.* 62: 74–82.

Ambler, R. P. 1980. The structure of β-lactamases. *Philos. Trans. R. Soc. B, Biol. Sci* 289: 321–331.

Aspevall, O., Dobric, S., Jagdmann, J., Nilsson, O. and Pringle, M. 2022. *Swedres-Svarm 2022:* Sales of antibiotics and occurrence of resistance in Sweden. Public Health Agency of Sweden; National Veterinary Institute. https://www.sva.se/media/s3ggt1ny/swedres-svarm-2022-edit-230808.pdf.

Avci, F. G., Atas, B., Aksoy, C. S., Kurpejovic, E., Toplan, G. G., Gurer, C. and Akbulut, B. S. 2019. Repurposing bioactive aporphine alkaloids as efflux pump inhibitors. *Fitoterapia*. 139: 104371.

Bassetti, M., Vena, A., Sepulcri, C., Giacobbe, D. R. and Peghin, M. 2020. Treatment of bloodstream infections due to Gram-negative bacteria with difficult-to-treat resistance. *Antibiotics*. 9: 632.

Bhardwaj A. K. and Mohanty P. 2012. Bacterial efflux pumps involved in multidrug resistance and their inhibitors: Rejuvinating the antimicrobial chemotherapy. *Recent Pat. Antiinfect. Drug. Discov.* 7: 73–89.

Bhowmick, T. and Weinstein, M. P. 2020. Microbiology of meropenem-vaborbactam: A novel carbapenem beta-lactamase inhibitor combination for carbapenem-resistant *Enterobacterales* infections. *Infect. Dis. Ther.* 9: 757–767.

Biedenbach, D. J., Kazmierczak, K., Bouchillon, S. K., Sahm D. F. and Bradford, P. A. 2015. In vitro activity of aztreonam-avibactam against a global collection of Gram-negative pathogens from 2012 and 2013. *Antimicrob. Agen. Chemother.* 59: 4239–4248.

Blair, J. M., Webber, M. A., Baylay, A. J., Ogbolu D. O. and L. J. Piddock. 2015. Molecular mechanisms of antibiotic resistance. *Nat. Rev. Microbiol.* 13: 42–51.

Blomquist, K. C. and Nix, D. E. 2021. A critical evaluation of newer β-lactam antibiotics for treatment of *Pseudomonas aeruginosa* infections. *Ann. Pharmacother.* 55: 1010–1024.

Bohnert, J. A. and Kern, W. V. 2005. Selected arylpiperazines are capable of reversing multidrug resistance in *Escherichia coli* overexpressing RND efflux pumps. *Antimicrob Agents Chemother.* 49: 849–852.

Bojar, B., Sheridan, J., Beattie, R., Cahak, C., Liedhegner, E., Munoz-Price, L. S. and Skwor, T. 2021. Antibiotic resistance patterns of *Escherichia coli* isolates from the clinic through the wastewater pathway. *Int. J. Hyg. Environ. Health.* 238: 113863.

Bolte, J., Zhang, Y., Wente, N., Mahmmod, Y. S., Svennesen, L. and Krömker, V. 2020. Comparison of phenotypic and genotypic antimicrobial resistance patterns associated with *Staphylococcus aureus* mastitis in German and Danish dairy cows. *J. Dairy Sci.* 103: 3554–3564.

Breijyeh, Z., Jubeh, B. and Karaman, R. 2020. Resistance of Gram-negative bacteria to current antibacterial agents and approaches to resolve it. *Molecules*. 25: 1340.

Bush, K. and Bradford, P. A. 2016. β-Lactams and β-lactamase inhibitors: An overview. *Cold Spring Harb. Perspect. Med.* 6: a025247.

Bush, K. and Jacoby, G. A. 2010. Updated functional classification of β-lactamases. *Antimicrob. Agents Chemother.* 54: 969–976.

Bush, K., Jacoby, G. A. and Medeiros, A. A. 1995. A functional classification scheme for beta-lactamases and its correlation with molecular structure. *Antimicrob. Agents Chemother.* 39: 1211–1233.

Carmeli, Y., Armstrong, J., Laud, P. J., Newell, P., Stone, G., Wardman, A. and Gasink, L. B. 2016. Ceftazidime-avibactam or best available therapy in patients with ceftazidime-resistant Enterobacteriaceae and Pseudomonas aeruginosa complicated urinary tract infections or complicated intra-abdominal infections (REPRISE): A randomised, pathogen-directed, phase 3 study. *Lancet Infect. Dis.* 16: 661–673.

Castillo-Polo, J. A., Hernández-García, M., Maruri-Aransolo, A., de la Vega, C., Ruiz-Garbajosa, P. and Cantón, R. 2024. Evolution of ceftazidime-avibactam and cefiderocol resistance in ST131-H30R1-*Escherichia coli* isolates with KPC-3 mutants and application of FTIR biotyping. *Microbiol. Spectr.* 12: e02776–23.

Chan, B. C., Ip, M., Gong, H., Lui, S. L., See, R. H., Jolivalt, C., . . . and Lau, C. B. 2013. Synergistic effects of diosmetin with erythromycin against ABC transporter over-expressed methicillin-resistant *Staphylococcus aureus* (MRSA) RN4220/pUL5054 and inhibition of MRSA pyruvate kinase. *Phytomedicine*. 20: 611–614.

Chan, B. C., Ip, M., Lau, C. B., Lui, S. L., Jolivalt, C., Ganem-Elbaz, C., . . . and Leung, P. C. 2011. Synergistic effects of baicalein with ciprofloxacin against NorA over-expressed methicillin-resistant *Staphylococcus aureus* (MRSA) and inhibition of MRSA pyruvate kinase. *J. Ethnopharmacol*. 137: 767–773.

Crandon, J. L. and Nicolau, D. P. 2013. Human simulated studies of aztreonam and aztreonam-avibactam to evaluate activity against challenging Gram-negative organisms, including metallo-β-lactamase producers. *Antimicrob. Agents Chemother.* 57: 3299–3306.

Daigle, D., Burns, C. and Pevear, D. 2018. Kinetic mechanism and parameters of inhibition of serine KPC-2, CTX-M-15, P99 AmpC and metallo VIM-2 by the broad-spectrum β-lactamase inhibitor VNRX–5133. *European Congress of Clin. Microbiol. Infect Dis.* O0606.

De Oliveira, D. M., Forde, B. M., Kidd, T. J., Harris, P. N., Schembri, M. A., Beatson, S. A., . . . and Walker, M. J. 2020. Antimicrobial resistance in ESKAPE pathogens. *Clin. Microbiol. Rev.* 33: e00181–19.

De Sousa, J. N., Barreto, J. D. A. F., de Alcântara Oliveira, F. A., Ferreira, J. H. L., Arcanjo, D. D. R., Nogueira, C. E. S., . . .and Barreto, H. M. 2024. ADMET study and inhibition of *Staphylococcus aureus* efflux pumps by a synthetic p-aminochalcone. *Results Chem.* 7: 101449.

Diallo, O. O., Baron, S. A., Abat, C., Colson, P., Chaudet, H. and Rolain, J. M. 2020. Antibiotic resistance surveillance systems: A review. *J. Glob. Antimicrob. Resist.* 23: 430–438.

Dou, Q., Zou, M., Li, J., Wang, H., Hu, Y. and Liu, W. 2017. AdeABC efflux pump and resistance of Acinetobacter baumannii against carbapenem. *Zhong nan da xue xue bao. Yi xue ban Journal of Central South University. Med. Sci.* 42: 426–433.

Duarte, A. S. R., Attauabi, M., Sandberg, M., Lindegaard, M., Sönksen, U. W. and Andersen, V. D. 2023. *DANMAP 2022: Use of antimicrobial agents and occurrence of antimicrobial resistance in bacteria from food animals, food and humans in Denmark*. DANMAP. https://www.danmap.org/reports/2022.

Eckmann, C. and Solomkin, J. 2015. Ceftolozane/tazobactam for the treatment of complicated intra-abdominal infections. *Expert Opin. Pharmacother.* 16: 271–280.

European Medicines Committee. 2020. *Categorisation of antibiotics used in animals promotes responsible use to protect public and animal health*. European Medicines Agency: Amsterdam, The Netherlands. https://www.ema.europa.eu/en/documents/report/infographic-categorisation-antibiotics-use-animals-prudent-and-responsible-use_en.pdf.

Falagas, M. E. and Karageorgopoulos, D. E. 2008. Pandrug resistance (PDR), extensive drug resistance (XDR), and multidrug resistance (MDR) among Gram-negative *bacilli:* Need for international harmonization in terminology. *Clin. Infect. Dise.* 46: 1121–1122.

Falgenhauer, L., Ghosh, H., Guerra, B., Yao, Y., Fritzenwanker, M., Fischer, J. . . . and Chakraborty, T. 2017. Comparative genome analysis of IncHI2 VIM-1 carbapenemase-encoding plasmids of *Escherichia coli* and *Salmonella enterica* isolated from a livestock farm in Germany. *Veter. Microbial.* 200: 114–117.

Fischer, J., Rodríguez, I., Schmoger, S., Friese, A., Roesler, U., Helmuth, R. and Guerra, B. 2012. *Escherichia coli* producing VIM-1 carbapenemase isolated on a pig farm. *J. Antimicrob. Chemother.* 67: 1793–1795.

Fischer, J., Rodríguez, I., Schmoger, S., Friese, A., Roesler, U., Helmuth, R. and Guerra, B. 2013. *Salmonella enterica* subsp. enterica producing VIM-1 carbapenemase isolated from livestock farms. *J. Antimicrob. Chemother.* 68: 478–480.

Fratoni, A. J., Berry, A. V., Liu, X., Chen, X., Wu, Y., Nicolau, D. P. and Abdelraouf, K. 2023. Imipenem/funobactam (formerly XNW4107) in vivo pharmacodynamics against serine carbapenemase-producing Gram-negative bacteria: A novel modelling approach for time-dependent killing. *J. Antimicrob. Chemother.* 78: 2343–2353.

Fuda, C. C. S., Fisher, J. F. and Mobashery, S. 2005. β-Lactam resistance in *Staphylococcus aureus:* The adaptive resistance of a plastic genome. *Cell. Mol. Life Sci.* 62: 2617–2633.

Gagetti, P., Bonofiglio, L., García Gabarrot, G., Kaufman, S., Mollerach, M., Vigliarolo, L. . . . and Lopardo, H. A. 2019. Resistencia a los β-lactámicos en enterococos. *Rev. Argent. Microbiol.* 51: 179–183.

Galarce, N., Arriagada, G., Sánchez, F., Venegas, V., Cornejo, J. and Lapierre, L. 2021. Antimicrobial use in companion animals: Assessing veterinarians' prescription patterns through the first national survey in chile. *Animals.* 11: 348. https://doi.org/10.3390/ani11020348

Gauba, A. and Rahman, K. M. 2023. Evaluation of antibiotic resistance mechanisms in Gram-negative bacteria. *Antibiotics.* 12: 1590.

González-Bello, C. 2017. Antibiotic adjuvants–A strategy to unlock bacterial resistance to antibiotics. *Bioorg. Med. Chem. Lett.* 27: 4221–4228.

González-Torralba, A., Oteo, J., Asenjo, A., Bautista, V., Fuentes, E. and Alós, J. I. 2016. Survey of carbapenemase-producing *Enterobacteriaceae* in companion dogs in Madrid, Spain. *Antimicro. Agents Chemother.* 60: 2499–2501.

Grinius, L., Dreguniene, G., Goldberg, E. B., Liao, C. H. and Projan, S. J. 1992. A staphylococcal multidrug resistance gene product is a member of a new protein family. *Plasmid.* 27: 119–129.

Gupta, V. K., Tiwari, N., Gupta, P., Verma, S., Pal, A., Srivastava, S. K. and Darokar, M. P. 2016. A clerodane diterpene from *Polyalthia longifolia* as a modifying agent of the resistance of methicillin resistant *Staphylococcus aureus*. *Phytomedicine*, 23: 654–661.

Hayashi, Y., Roberts, J. A., Paterson, D. L. and Lipman, J. 2010. Pharmacokinetic evaluation of piperacillin-tazobactam. *Expert Opin. Drug Metab. Toxicol.* 6: 1017–1031.

He, T., Wang, Y., Sun, L., Pang, M., Zhang, L. and Wang, R. 2016. Occurrence and characterization of bla NDM-5-positive *Klebsiella pneumoniae* isolates from dairy cows in Jiangsu, China. *J. Antimicrob. Chemother.* 72: 90–94.

Hellman, J., Olsson-Liljequist, B., Bengtsson, B. and Greko, C. 2013. *Use of antimicrobials and occurrence of antimicrobial resistance in Sweden*. Solna/Uppsala: Swedish Institute for Communicable Disease Control and National Veterinary Institute. https://www.folkhalsomyndigheten.se/contentassets/93c7cf9c91ae41b79bac167bba2d31db/swedres-svarm-2012.pdf.

Heo, Y. A. 2021. Imipenem/cilastatin/relebactam: A review in Gram-negative bacterial infections. *Drugs.* 81: 377–388.

Holler, J. G., Christensen, S. B., Slotved, H. C., Rasmussen, H. B., Gúzman, A., Olsen, C. E. . . . and Mølgaard, P. 2012. Novel inhibitory activity of the *Staphylococcus aureus* NorA efflux pump by a kaempferol rhamnoside isolated from Persea lingue Nees. *J. Antimicrob. Chemother.* 67: 1138–1144.

Hou, L., Ye, M., Wang, X., Zhu, Y., Sun, X., Gu, R. and Fang, B. 2024. Synergism with shikimic acid restores β-lactam antibiotic activity against methicillin-resistant *Staphylococcus aureus*. *Molecules.* 29: 1528.

Impey, R. E., Hawkins, D. A., Sutton, J. M. and Soares da Costa, T. P. 2020. Overcoming intrinsic and acquired resistance mechanisms associated with the cell wall of Gram-negative bacteria. *Antibiotics.* 9: 623.

Joshi, P., Singh, S., Wani, A., Sharma, S., Jain, S. K., Singh, B. and Vishwakarma, R. A. 2014. Osthol and curcumin as inhibitors of human Pgp and multidrug efflux pumps of

Staphylococcus aureus: Reversing the resistance against frontline antibacterial drugs. *Med. Chem. Commun.* 5: 1540–1547.

Jung, H. R., Lee, Y. J., Hong, S., Yoon, S., Lim, S. K. and Lee, Y. J. 2023. Current status of β-lactam antibiotic use and characterization of β-lactam-resistant *Escherichia coli* from commercial farms by integrated broiler chicken operations in Korea. *Poult. Sci.* 102: 103091.

Kakarla, P., Floyd, J., Mukherjee, M., Devireddy, A. R., Inupakutika, M. A., Ranweera, I. . . . and Varela, M. F. 2017. Inhibition of the multidrug efflux pump LmrS from *Staphylococcus aureus* by cumin spice *Cuminum cyminum*. *Arch. Microbiol.* 199: 465–474.

Kalia, N. P., Mahajan, P., Mehra, R., Nargotra, A., Sharma, J. P., Koul, S. and Khan, I. A. 2012. Capsaicin, a novel inhibitor of the NorA efflux pump, reduces the intracellular invasion of *Staphylococcus aureus*. *J. Antimicrob. Chemother.* 67: 2401–2408.

Keane, O. M., Budd, K. E., Flynn, J. and McCoy, F. 2013. Pathogen profile of clinical mastitis in Irish milk-recording herds reveals a complex aetiology. *Vet. Rec.* 173: 17–17.

Keserű, J. S., Szabó, I., Gál, Z., Massidda, O., Mingoia, M., Kaszanyitzky, É. and Biró, S. (2011). Identification of β-lactamases in human and bovine isolates of *Staphylococcus aureus* strains having borderline resistance to penicillinase-resistant penicillins (PRPs) with proteomic methods. *Vet. Microbiol.* 147: 96–102.

Köck, R., Daniels-Haardt, I., Becker, K., Mellmann, A., Friedrich, A. W., Mevius, D. . . . and Jurke, A. 2018. Carbapenem-resistant *Enterobacteriaceae* in wildlife, food-producing, and companion animals: A systematic review. *Clin. Microbiol. Infect.* 24: 1241–1250.

Konaté, K., Mavoungou, J. F., Lepengué, A. N., Aworet-Samseny, R. R., Hilou, A., Souza, A. and M'Batchi, B. 2012. Antibacterial activity against β-lactamase producing Methicillin and Ampicillin-resistants *Staphylococcus aureus*: Fractional Inhibitory Concentration Index (FICI) determination. *Ann. Clin. Microbiol. Antimicrob.* 11: 1–12.

Kourtesi, C., Ball, A. R., Huang, Y. Y., Jachak, S. M., Vera, D. M. A., Khondkar, P. . . . and Tegos, G. P. 2013. Suppl 1: Microbial efflux systems and inhibitors: Approaches to drug discovery and the challenge of clinical implementation. *Open Microbiol. J.* 7: 34.

Kumar, A., Khan, I. A., Koul, S., Koul, J. L., Taneja, S. C., Ali, I. and Qazi, G. N. 2008. Novel structural analogues of piperine as inhibitors of the NorA efflux pump of *Staphylococcus aureus*. *J. Antimicrob. Chemother.* 61: 1270–1276.

Lee, M. D., Galazzo, J. L., Staley, A. L., Lee, J. C., Warren, M. S., Fuernkranz, H. . . . and Miller, G. H. 2001. Microbial fermentation-derived inhibitors of efflux-pump-mediated drug resistance. *IL Farmaco*. 56: 81–83.

Li, B., Yao, Q., Pan, X. C., Wang, N., Zhang, R., Li, J. and Zhou, H. 2011. Artesunate enhances the antibacterial effect of β-lactam antibiotics against *Escherichia coli* by increasing antibiotic accumulation via inhibition of the multidrug efflux pump system AcrAB-TolC. *J. Antimicrob. Chemother.* 66: 769–777.

Li, X. Z., Mehrotra, M., Ghimire, S. and Adewoye, L. 2007. β-lactam resistance and β-lactamases in bacteria of animal origin. *Vet. Microbiol.* 121: 197–214.

Li, X. Z., Plésiat, P. and Nikaido, H. 2015. The challenge of efflux-mediated antibiotic resistance in Gram-negative bacteria. *Clin. Microbiol. Rev.* 28: 337–418.

Li, Y., Yan, M., Xue, F., Zhong, W., Liu, X., Chen, X. and Lv, Y. 2022. In vitro and in vivo activities of a novel β-lactamase inhibitor combination imipenem/XNW4107 against recent clinical Gram-negative bacilli from China. *J. Glob. Antimicrob. Res.* 31: 1–9.

Liu, Z., Wang, Y., Walsh, T. R., Liu, D., Shen, Z., Zhang, R. . . . and Shen, J. 2017. Plasmid-mediated novel bla NDM-17 gene encoding a carbapenemase with enhanced activity in a sequence type 48 *Escherichia coli* strain. *Antimicrob. Agents Chemother.* 61: e02233–16. https://doi.org/10.1128/aac.02233-16.

Livermore, D. M. 1995. Beta-lactamases in laboratory and clinical resistance. *Clin. Microbiol. Rev.* 8: 557–584.

Livermore, D. M., Canton, R., Gniadkowski, M., Nordmann, P., Rossolini, G. M., Arlet, G. . . . and Woodford, N. (2007). CTX-M: Changing the face of ESBLs in Europe. *J. Antimicrob. Chemother.* 59: 165–174.

Longshaw, C., Manissero, D., Tsuji, M., Echols, R. and Yamano, Y. 2020. In vitro activity of the siderophore cephalosporin, cefiderocol, against molecularly characterized, carbapenem-non-susceptible Gram-negative bacteria from Europe. *JAC Antimicrob. Resist.* 2: dlaa060.

Mahmood, H., Jamshidi, S., Mark Sutton, J. and M Rahman, K. 2016. Current advances in developing inhibitors of bacterial multidrug efflux pumps. *Curr. Medicin. Chem.* 23: 1062–1081.

Malik, S., Christensen, H., Peng, H. and Barton, M. D. 2007. Presence and diversity of the β-lactamase gene in cat and dog staphylococci. *Vet. Microbiol.* 123: 162–168.

MARAN.2022. *Monitoring of antimicrobial resistance and antibiotic usage in animals in Netherland in 2021.* https://swab.nl/en/exec/file/download/197.

Matsumura, Y., Peirano, G., Motyl, M. R., Adams, M. D., Chen, L., Kreiswirth, B. . . . and Pitout, J. D. 2017. Global molecular epidemiology of IMP-producing *Enterobacteriaceae*. Antimicrob. *Agents Chemother.* 61: e02729–16. https://doi.org/10.1128/aac.02729-16.

Melo, L. C., Boisson, M. N., Saras, E., Médaille, C., Boulouis, H. J., Madec, J. Y. and Haenni, M. 2017. OXA-48-producing ST372 Escherichia coli in a French dog. *J. Antimicrob. Chemother.* 72: 1256–1258.

Mollenkopf, D. F., Stull, J. W., Mathys, D. A., Bowman, A. S., Feicht, S. M., Grooters, S. V. . . . and Wittum, T. E. 2017. Carbapenemase-producing Enterobacteriaceae recovered from the environment of a swine farrow-to-finish operation in the United States. *Antimicrob. Agents Chemother.* 61: e01298–16. https://doi.org/10.1128/aac.01298-16.

Morris, D. O., Rook, K. A., Shofer, F. S. and Rankin, S. C. 2006. Screening of *Staphylococcus aureus, Staphylococcus intermedius* and *Staphylococcus schleiferi* isolates obtained from small companion animals for antimicrobial resistance: A retrospective review of 749 isolates (2003–04). *Vet. Dermatol.* 17: 332–337.

Mueller, M.T.C.R. 2023. *Escherichia coli.* In StatPearls; StatPearls Publishing: Treasure Island, FL, USA,

Murray, B. E. 1992. Beta-lactamase-producing enterococci. *Antimicrob. Agents Chemother.* 36: 2355–2359.

Murray, C. J., Ikuta, K. S., Sharara, F., Swetschinski, L., Aguilar, G. R., Gray, A. . . . and Tasak, N. 2022. Global burden of bacterial antimicrobial resistance in 2019: A systematic analysis. *The Lancet.* 399: 629–655.

Mushtaq, S., Garello, P., Vickers, A., Woodford, N. and Livermore, D. M. 2021. Cefepime/tazobactam compared with other tazobactam combinations against problem Gram-negative bacteria. *Int. J. Antimicrob. Agents.* 57: 106318.

Naranjo-Lucena, A. and Slowey, R. 2023. Invited review: Antimicrobial resistance in bovine mastitis pathogens: A review of genetic determinants and prevalence of resistance in European countries. *J. Dairy Sci.* 106: 1–23.

Nelson, M. L. and Levy, S. B. 1999. Reversal of tetracycline resistance mediated by different bacterial tetracycline resistance determinants by an inhibitor of the Tet (B) antiport protein. *Antimicrob. Agents Chemother.* 43: 1719–1724.

Neyfakh, A. A., Bidnenko, V. E. and Chen, L. B. 1991. Efflux-mediated multidrug resistance in Bacillus subtilis: Similarities and dissimilarities with the mammalian system. *Proc. Natl. Acad. Sci. U S A.* 88: 4781–4785.

Nobrega, D. B., De Buck, J. and Barkema, H. W. 2018. Antimicrobial resistance in non-aureus staphylococci isolated from milk is associated with systemic but not intramammary administration of antimicrobials in dairy cattle. *J. Dairy Sci.* 101: 7425–7436.

Nocera, F. P., Ambrosio, M., Fiorito, F., Cortese, L. and De Martino, L. 2021. On Gram-positive- and Gram-negative-bacteria-associated canine and feline skin infections: A 4-year retrospective study of the University Veterinary Microbiology Diagnostic Laboratory of Naples, Italy. *Animals.* 11: 1603.

NORM/NORM-VET. 2012. *Usage of antimicrobial agents and occurrence of antimicrobial resistance in Norway.* Tromsø/Oslo 2013. ISSN: 1502–2307 (print)/1890–9965 (electronic). https://www.vetinst.no/en/surveillance-programmes/norm-norm-vet-report.

NORM/NORM-VET. 2022. *Usage of antimicrobial agents and occurrence of antimicrobial resistance in Norway.* Tromsø/Oslo 2023. ISSN: 1502–2307 (print)/1890–9965 (electronic). https://www.vetinst.no/en/surveillance-programmes/norm-norm-vet-report.

O'Donnell, J., Tanudra, A., Chen, A., Miller, A. A., McLeod, S. M. and Tommasi, R. 2024. In vitro pharmacokinetics/pharmacodynamics of the β-lactamase inhibitor, durlobactam, in combination with sulbactam against *Acinetobacter baumannii*-calcoaceticus complex. *Antimicrob. Agents Chemother.* 68: e00312–e00323.

Oliveira-Tintino, C. D. D. M., Tintino, S. R., Justino de Araújo, A. C., dos Santos Barbosa, C. R., Ramos Freitas, P., Araújo Neto, J. B. D. and Coutinho, H. D. M. 2023. Efflux pump (QacA, QacB, and QacC) and β-lactamase inhibitors? An evaluation of 1, 8-naphthyridines against *Staphylococcus aureus* strains. *Molecules.* 28: 1819.

Olsen, J. E., Christensen, H. and Aarestrup, F. M. 2006. Diversity and evolution of blaZ from Staphylococcus aureus and coagulase-negative *Staphylococci. J. Antimicrob. Chemother.* 57: 450–460.

Oluwatuyi, M., Kaatz, G. W. and Gibbons, S. 2004. Antibacterial and resistance modifying activity of *Rosmarinus officinalis. Phytochemistry.* 65: 3249–3254.

Omollo, C., Singh, V., Kigondu, E., Wasuna, A., Agarwal, P., Moosa, A. . . . and Warner, D. F. 2021. Developing synergistic drug combinations to restore antibiotic sensitivity in drug-resistant Mycobacterium tuberculosis. *Antimicrob. Agents Chemother.* 65: e02554–20. https://doi.org/10.1128/aac.02554-20.

Palma, E., Tilocca, B. and Roncada, P. 2020. Antimicrobial resistance in veterinary medicine: An overview. *Int. J. Mol. Sci.* 21: 1914.

Papp-Wallace, K. M. 2019. The latest advances in β-lactam/β-lactamase inhibitor combinations for the treatment of Gram-negative bacterial infections. *Expert Opin Pharmacother.* 20: 21692184.

Papp-Wallace, K. M., Bethel C. R, Calllon J, Barnes MD., Potel G., Bajaksouzian S., Rutter J. D, Reghal A., Shapiro S., Taracila M. A., Jacobs M. R., Bonomo RA., Jacqueline C. Beyond Piperacillin-Tazobactam: Cefepime and AAI101 as a Potent β-Lactam-β-Lactamase Inhibitor Combination. Antimicrob Agents Chemother. 2019 Apr 25; 63(5):e00105-19. doi: 10.1128/AAC.00105-19. PMID: 30858223; PMCID: PMC6496078.

Papp-Wallace, K. M., McLeod, S. M. and Miller, A. A. 2023. Durlobactam, a broad-Spectrum serine β-lactamase inhibitor, restores Sulbactam activity against Acinetobacter species. *Clin. Infect. Dis.*76: S194–S201.

Parvaiz, N., Ahmad, F., Yu, W., MacKerell Jr, A. D. and Azam, S. S. 2021. Discovery of beta-lactamase CMY-10 inhibitors for combination therapy against multi-drug resistant Enterobacteriaceae. *PLoS One*, 16: e0244967.

Patterson, J. E., Wanger, A., Zscheck, K. K., Zervos, M. J. and Murray, B. E. 1990. Molecular epidemiology of beta-lactamase-producing enterococci. Antimicrob. *Agents Chemother.* 34: 302–305.

Pereira da Cruz, R., Sampaio de Freitas, T., do Socorro Costa, M., Lucas dos Santos, A. T., Ferreira Campina, F., Pereira, R. L. S. . . . and Bezerra Morais-Braga, M. F. 2020. Effect of α-bisabolol and its β-cyclodextrin complex as TetK and NorA efflux pump inhibitors in *Staphylococcus aureus* strains. *Antibiotics.* 9: 28.

Piddock, L. J. 2006. Clinically relevant chromosomally encoded multidrug resistance efflux pumps in bacteria. *Clin. Microbiol. Rev.* 19: 382–402.

Piddock, L. J., Garvey, M. I., Rahman, M. M. and Gibbons, S. 2010. Natural and synthetic compounds such as trimethoprim behave as inhibitors of efflux in Gram-negative bacteria. *J. Antimicrob. Chemother.* 65: 1215–1223.

Pleydell, E. J., Souphavanh, K., Hill, K. E., French, N. P. and Prattley, D. J. 2012. Descriptive epidemiological study of the use of antimicrobial drugs by companion animal veterinarians in New Zealand. *N. Z. Vet. J.* 60: 115–122.

Poirel, L., Berçot, B., Millemann, Y., Bonnin, R. A., Pannaux, G. and Nordmann, P. 2012. Carbapenemase-producing Acinetobacter spp. in cattle, France. *Emerg. Infect. Dis.* 18: 523.

Pomba-Féria, C. and Caniça, M. 2003. A novel sequence framework (bla TEM–1G) encoding the parental TEM-1 beta-lactamase. *FEMS Microbiol. Lett.* 220: 177–180.

Prestinaci, F., Pezzotti, P. and Pantosti, A. 2015. Antimicrobial resistance: A global multifaceted phenomenon. *Pathog. Glob. Health.* 109: 309–318.

Pulingam, T., Parumasivam, T., Gazzali, A. M., Sulaiman, A. M., Chee, J. Y., Lakshmanan, M. . . . and Sudesh, K. 2022. Antimicrobial resistance: Prevalence, economic burden, mechanisms of resistance and strategies to overcome. *Eur. J. Pharm. Sci.* 170: 106103.

Qian, M., Tang, S., Wu, C., Wang, Y., He, T., Chen, T. and Xiao, X. 2015. Synergy between baicalein and penicillins against penicillinase-producing *Staphylococcus aureus. Int. J. Med. Microbiol.* 305: 501–504.

Ramírez-Castillo, F. Y., Guerrero-Barrera, A. L. and Avelar-González, F. J. 2023. An overview of carbapenem-resistant organisms from food-producing animals, seafood, aquaculture, companion animals, and wildlife. *Front. Vet. Sci.*10: 1158588.

Reading, C. and Cole, M. 1977. Clavulanic acid: A beta-lactamase-inhibiting beta-lactam from *Streptomyces clavuligerus. Antimicrob. Agents Chemother.* 11: 852–857.

Reygaert, W. C. 2018. An overview of the antimicrobial resistance mechanisms of bacteria. *AIMS Microbiol.* 4: 482.

Roschanski, N., Friese, A., von Salviati-Claudius, C., Hering, J., Kaesbohrer, A., Kreienbrock, L. and Roesler, U. 2017. Prevalence of carbapenemase producing Enterobacteriaceae isolated from German pig-fattening farms during the years 2011–2013. *Vet. Microbiol.* 200: 124–129.

Roy, S. K., Kumari, N., Pahwa, S., Agrahari, U. C., Bhutani, K. K., Jachak, S. M. and Nandanwar, H. 2013. NorA efflux pump inhibitory activity of coumarins from *Mesua ferrea. Fitoterapia.* 90: 140–150.

Ruiz, J. 2019. Transferable mechanisms of quinolone resistance from 1998 onward. *Clin. Microbiol. Rev.* 32: e00007–19. https://doi.org/10.1128/cmr.00007-19.

Samantaray, S., Kumar, D., Bohra, G. K., Meena, D. S., Agarwal, A., Chaudhary, G. R. and Garg, M. K. 2023. Effect of Ceftriaxone+ sulbactam+ disodium EDTA Combination in Treatment of Complicated Urinary Tract Infections Caused by Metallo Beta-lactamase Producing *Enterobacterales:* An Open Label Randomized Controlled Trial. *Open Forum Infect. Dis.* 10: ofad500. 451.

Sharma, A., Gupta, V. K. and Pathania, R. 2019. Efflux pump inhibitors for bacterial pathogens: From bench to bedside. *Indain J. Med. Res.* 149: 129–145.

Shiota, S., Shimizu, M., Mizusima, T., Ito, H., Hatano, T., Yoshida, T. and Tsuchiya, T. 2000. Restoration of effectiveness of β-lactams on methicillin-resistant *Staphylococcus aureus* by tellimagrandin I from rose red. *FEMS Microbiol. Lett.* 185: 135–138.

Shiota, S., Shimizu, M., Sugiyama, J. I., Morita, Y., Mizushima, T. and Tsuchiya, T. 2004. Mechanisms of action of corilagin and tellimagrandin I that remarkably potentiate the activity of β-lactams against methicillin-resistant *Staphylococcus aureus. Microbiol. Immunol.* 48: 67–73.

Silva, N., Carvalho, I., Currie, C., Sousa, M., Igrejas, G. and Poeta, P. 2019. Extended-spectrum-β-lactamase and carbapenemase-producing enterobacteriaceae in food-producing animals in Europe: An impact on public health?. *Antibiotic Drug Resistance*. 261–273.

Siriyong, T., Chusri, S., Srimanote, P., Tipmanee, V. and Voravuthikunchai, S. P. 2016. *Holarrhena antidysenterica* extract and its steroidal alkaloid, conessine, as resistance-modifying agents against extensively drug-resistant *Acinetobacter baumannii. Microb. Drug Resist.* 22: 273–282.

Siriyong, T., Srimanote, P., Chusri, S., Yingyongnarongkul, B. E., Suaisom, C., Tipmanee, V. and Voravuthikunchai, S. P. 2017. Conessine as a novel inhibitor of multidrug efflux pump systems in *Pseudomonas aeruginosa. BMC Complement. Altern. Med.* 17: 1–7.

Smith, E. C., Kaatz, G. W., Seo, S. M., Wareham, N., Williamson, E. M. and Gibbons, S. 2007a. The phenolic diterpene totarol inhibits multidrug efflux pump activity in *Staphylococcus aureus. Antimicrob. Agents Chemother.* 51: 4480–4483.

Smith, E. C., Williamson, E. M., Wareham, N., Kaatz, G. W. and Gibbons, S. 2007b. Antibacterials and modulators of bacterial resistance from the immature cones of *Chamaecyparis lawsoniana. Phytochemistry.* 68: 210–217.

Stavri, M., Piddock, L. J. and Gibbons, S. 2007. Bacterial efflux pump inhibitors from natural sources. *J. Antimicrob. Chemother.* 59: 1247–1260.

Stegemann, M. R., Passmore, C. A., Sherington, J., Lindeman, C. J., Papp, G., Weigel, D. J. and Skogerboe, T. L. 2006. Antimicrobial activity and spectrum of cefovecin, a new extended-spectrum cephalosporin, against pathogens collected from dogs and cats in Europe and North America. *Antimicrob. Agents Chemother.* 50: 2286–2292.

Stolle, I., Prenger-Berninghoff, E., Stamm, I., Scheufen, S., Hassdenteufel, E., Guenther, S. . . . and Ewers, C. 2013. Emergence of OXA-48 carbapenemase-producing Escherichia coli and *Klebsiella pneumoniae* in dogs. *J. Antimicrob. Chemother.* 68: 2802–2808.

Tadesse, B. T., Ashley, E. A., Ongarello, S., Havumaki, J., Wijegoonewardena, M., González, I. J. and Dittrich, S. 2017. Antimicrobial resistance in Africa: A systematic review. *BMC Infect. Dise.* 17: 1–17.

Tang, M., Wei, X., Wan, X., Ding, Z., Ding, Y. and Liu, J. 2020. The role and relationship with efflux pump of biofilm formation in *Klebsiella pneumoniae. Microb. Pathogen.* 147: 104244.

Tintino, S. R., Morais-Tintino, C. D., Campina, F. F., Costa, M. D. S., Menezes, I. R., de Matos, Y. M. L. . . . and Balbino, V. Q. 2017. Tannic acid affects the phenotype of *Staphylococcus aureus* resistant to tetracycline and erythromycin by inhibition of efflux pumps. *Bioorg. Chem.* 74: 197–200.

Torres, C., Alonso, C. A., Ruiz-Ripa, L., León-Sampedro, R., Del Campo, R. and Coque, T. M. 2018. Antimicrobial resistance in *Enterococcus spp.* of animal origin. In *Antimicrobial resistance in bacteria from livestock and companion animals*, 185–227.

Van, T. T. H., Yidana, Z., Smooker, P. M. and Coloe, P. J. 2020. Antibiotic use in food animals worldwide, with a focus on Africa: Pluses and minuses. *J. Global Antimicrob. Resist.* 20: 170–177.

Van Veen, H. W., Margolles, A., Müller, M., Higgins, C. F. and Konings, W. N. 2000. The homodimeric ATP-binding cassette transporter LmrA mediates multidrug transport by an alternating two-site (two-cylinder engine) mechanism. *EMBO J.* 19: 2503–2514.

Ventola, C. L. 2015. The antibiotic resistance crisis: Part 1: Causes and threats. *Pharmacy and Therapeutics*, 40: 277.

Wang, Q., Jin, L., Sun, S., Yin, Y., Wang, R., Chen, F. . . and Wang, H. 2022. Occurrence of high levels of cefiderocol resistance in carbapenem-resistant Escherichia coli before its approval in China: A report from China CRE-network. *Microbiol. Spect.* 10: e0267021.

Wang, W., Baloch, Z., Peng, Z., Hu, Y., Xu, J., Fanning, S. and Li, F. 2017b. Genomic characterization of a large plasmid containing a bla NDM-1 gene carried on *Salmonella enterica* serovar Indiana C629 isolate from China. *BMC Infect. Dis.* 17: 1–8.
Wang, Y., Wang, J., Wang, R. and Cai, Y. 2020. Resistance to ceftazidime-avibactam and underlying mechanisms. *J. Global Antimicrob. Resist.* 22: 18–27.
Wang, Y., Zhang, R., Li, J., Wu, Z., Yin, W., Schwarz, S. . . . and Shen, J. 2017a. Comprehensive resistome analysis reveals the prevalence of NDM and MCR-1 in Chinese poultry production. *Nat. Microbiol.* 2: 1–7.
Watts, J. L. and Salmon, S. A. 1997. Activity of selected antimicrobial agents against strains of *Staphylococcus aureus* isolated from bovine intra mammary infections that produce β-lactamase. *J. Dairy Sci.* 80: 788–791.
Webber, M. A. and Piddock, L. J. 2003. The importance of efflux pumps in bacterial antibiotic resistance. *J. Antimicrob. Chemother.* 51: 9–11.
Woerde, D. J., Reagan, K. L., Byrne, B. A., Weimer, B. C., Epstein, S. E., Schlesener, C. . . . and Sykes, J. E. 2023. Characteristics of extended-spectrum β-lactamase producing enterobacterales isolated from dogs and cats, 2011–2021. *Vet. Sci.* 10: 178.
Wong, M. H. Y., Chan, E. W. C. and Chen, S. 2015. Evolution and dissemination of OqxAB-like efflux pumps, an emerging quinolone resistance determinant among members of Enterobacteriaceae. *Antimicrob. Agents Chemothera.* 59: 3290–3297.
World Organisation for Animal Health, 2023. *Annual report on antimicrobial agents intended for use in animals* (7th ed.). World Organisation for Animal Health. https://www.woah.org/app/uploads/2023/05/a-seventh-annual-report-amu-final.pdf.
World Organization for Animal Health. 2019. *OIE list of antimicrobial agents of veterinary importance*. World Organization for Animal Health. https://www.oie.int/fileadmin/Home/eng/Our_scientific_expertise/docs/pdf/AMR/A_OIE_List_antimicrobials_July2019.pdf.
Yang, L., Lu, S., Belardinelli, J., Huc-Claustre, E., Jones, V., Jackson, M. and Zgurskaya, H. I. 2014. RND transporters protect *Corynebacterium glutamicum* from antibiotics by assembling the outer membrane. *Microbiol. Open.* 3: 484–496.
Yang, X., Ye, W., Qi, Y., Ying, Y. and Xia, Z. 2021. Overcoming multidrug resistance in bacteria through antibiotics delivery in surface-engineered nano-cargos: Recent developments for future nano-antibiotics. *Front. Bioeng. Biotechnol.* 9: 696514.
Zasowski, E. J., Rybak, J. M. and Rybak, M. J. 2015. The β-lactams strike back: Ceftazidime-avibactam. *Pharmacotherapy: The Journal of Human Pharmacology and Drug Therapy.* 35: 755–770.
Zhang, J., Sun, Y., Wang, Y., Lu, M., He, J., Liu, J., . . . and Sun, X. 2014. Non-antibiotic agent ginsenoside 20 (S)-Rh2 enhanced the antibacterial effects of ciprofloxacin in vitro and in vivo as a potential NorA inhibitor. *Eur. J. Pharmacol.* 740: 277–284.
Zhang, S., Liao, X., Ding, T. and Ahn, J. 2024b. Role of β-lactamase inhibitors as potentiators in antimicrobial chemotherapy targeting Gram-negative bacteria. *Antibiotics.* 13: 260.
Zhang, Y., Chen, C., Cheng, B., Gao, L., Qin, C., Zhang, L. . . . and Wan, Y. 2022. Discovery of quercetin and its analogs as potent OXA-48 beta-lactamase inhibitors. *Front. Pharmacol.* 13: 926104
Zhang, Z., Xu, Q., Wang, Y., Qu, S., Tan, J., Tang, Y. . . . and Zheng, X. 2024a. Exploiting the synergistic antibacterial activity of shikimic acid and ceftiofur against methicillin-resistant *Staphylococcus aureus. World. J. Microbiol. Biotechnol.* 40: 78
Zhou, X., Jia, F., Liu, X., Yang, J., Zhang, Y. and Wang, Y. 2011. In vitro synergistic interaction of 5-O-methylglovanon and ampicillin against ampicillin resistant *Staphylococcus aureus* and *Staphylococcus epidermidis* isolates. *Arch. Pharmacal. Res.* 34: 1751–1757.

5 An Open Traditional Poultry Production
Toward Healthier Animals

Amro Abd Al Fattah Amara and Arti Gupta

5.1 INTRODUCTION

The origin of the domestic chicken is believed to be the wild red jungle fowl of India. It belongs to the species *Gallus gallus* (Marchewka et al. 2020). They were domesticated nearly 4,000 years ago. They grow historically beside sheep to satisfy the meat and eggs demand (Poultry 2013a). The name *poultry* (1300–1400; Language: Old French; Origin: pouleterie, from polet 'young chicken') represents birds that can grow in houses or close to houses. Poultry includes chickens, turkeys, ducks, quails, pigeons, and geese (Figure 5.1). They are maintained for their eggs, meat, and feathers. Meat from the described birds is also referred to as poultry. Guinea fowl, squabs (a young, newly hatched or unfledged pigeon), and quail are chiefly of indigenous interest. Quail was recorded in a tomb in Egypt (3000 BC). Broiler is the flesh of a small young chicken. Its weight typically does not exceed 2 1/2 lb., suitable for roasting (Alnahhas et al. 2016; Arrazola and Torrey 2021; Poultry 2013b). *Broiler* as a noun is used as an alternative name for chicken. Modern high-volume controlled poultry farms started to flourish in Great Britain around 1920 and after World War II in the United States. Mature hens and younger pullets (females) are grown for meat and/or for egg production (Hedlund and Jensen 2022). Some breeds and varieties are environmentally adapted and become compatible with the farmers' resources (Pessato, McKechnie, and Mariette 2022). They support the farmers' life and satisfy their demands for eggs, meat, and feathers (Bamidele et al. 2020). Mature males are named cocks, or roosters (Attia et al. 2019), and were originally used for breeding purposes and for sport. The meat production usually is a derivative of egg production. The farmers used to exclude hens (Bai et al. 2021) that could no longer produce enough eggs. They were killed/culled and sold for meat (or used locally). During the mid-20th century, poultry meat production became a self-standing industry worldwide. Different studies were conducted to adjust the best food formula and production conditions (Nawaz et al. 2021; Poultry 2013b). Domestic ducks (female) or drake (male) belong to the subfamily Anatinae (of the waterfowl), family Anatidae; such as Muscovy duck (*Cairina moschata*) (Adeola et al. 2022), which was domesticated in Colombia and Peru by the pre-Columbian Indians. Another example is the wild mallard (*Anas platyrhyncos*), domesticated in China about 2,000 years ago.

 DOI: 10.1201/9781003269847-5

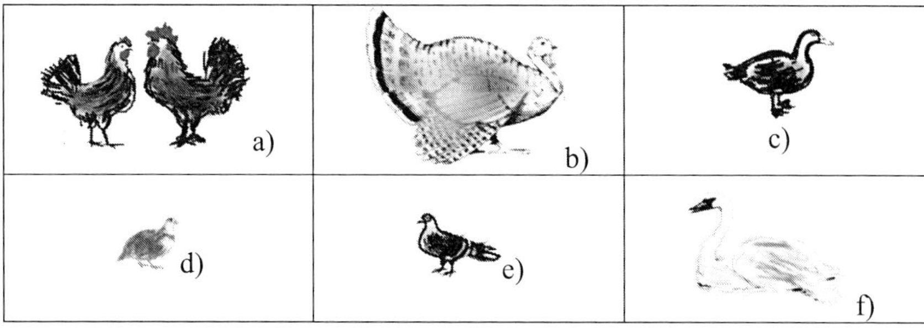

FIGURE 5.1 The common method of growing poultry in Egyptian villages: (a) chickens, (b) turkeys, (c) duck, (d) quail, (e) pigeon, and (f) geese. (Prepared in this study.)

Duck is usually classified as meat producing, egg producing, and ornamental (Sultana et al. 2016; Poultry 2013b). Turkeys apportioned to either family Phasianidae or family Meleagrididae in order Galliformes are native to North America (Poultry 2013b). The Aztec and Zuni Indians domesticated them for food and sacrifice and used their feathers for decoration. Turkeys started to be raised for meat on a wide scale after World War II. Male turkeys reach up to 12 kg in about 26 weeks. Hen turkeys mature earlier and rarely exceed 9 to 10 kg. Different genetic changes lead to different phenotypes of turkeys (Derks et al. 2018; Poultry 2013b). A goose is a member of the family Anatidae (Parada et al. 2012). Modern breeds are mostly descended from the greylag (*Anser anser*) (Bao et al. 2022), a wild goose of northern Eurasia. Unlike its monogamous wild cousin, the domestic goose is polygamous (Gumułka, Avital-Cohen, and Rozenboim 2021; Poultry 2013b). Squabs are immature pigeons. A squab is a member of the family Columbidae. Squab production is carried on locally but is rare in most countries with established poultry industries (Poultry 2013a; Jiang et al. 2019; Xu et al. 2020; Ye et al. 2016; Poultry 2013b; Yin et al. 2022). The United States successfully developed the poultry production sector (Huang and Huang 2011; Kennedy et al. 2015). This sector has concentrated nutrition, improved breeding stock, smart management, and planned disease control. A modern broiler chick gains over 43 times its first weight in an eight-week period. The feed requirement for each dozen eggs is only slightly more than half as high as it was in the early 1900s. Annual egg production for each hen has increased from 104 to 244/year since 1910 (Poultry 2013b). The ideal controlled environment should avoid crowding, chilling, frightening, and overheating the hen. Some processes are well controlled and mechanized, such as feeding, watering, egg gathering, and cleaning operations. The most chicks hatched each year are used for broiler production and the remainder for egg production. *Pullet* is flesh of a medium-sized young chicken suitable for frying and *pullet* is also used as a name for a young hen usually less than a year old (Du et al. 2022). Studies have described tools, lifestyles, foods, different formulas, farming mechanisms, and other suggestions to raise chicken weight. Raising the meat/kilo food, the number of produced eggs/hen/year, egg weight and size, are the most important ways to recover investment and gain profits (Iannotti et al. 2020). After years of trials, some fixed chicken farming style processes were established and became dominant (1-day chicks, 45 days growing and feeding, 1 or more days for selling: the total number of days is usually 47). This style is based mainly on purchasing chicks at one day of age (usually) (Oliveira Gda et al. 2021). Based on the chick's type and the producer's recommendations, the overall process could be changed from one farm to another (Grabež et al. 2022). Studies were conducted to understand the best conditions for growing chicks (Bona et al. 2018; Li et al. 2020). Some traditional recommendations are used to support the modern production process (Rossi et al. 2022). These recommendations were extracted from the experience of farmers from previous generations who are interested in growing chicks and chickens. For example, they avoided supplying chicks with water in the first days (in case of using the chicks to be grown via the traditional slow farming style [TSFS]) (personal communication). The chicks then were isolated from any source of contamination in a large, cleaned, land-layered, closed space with warm lighting that enables access to food and drink sources. In the modern study, the process is different due to giving chicks antibiotics in drinking water and applying a scheduled vaccines program. In the modern growing style, the birds are grown under a nearly continuous source of light. If chickens must sleep, they will sleep in light. The birds eat and drink excessively for nearly 40 to 45 days. After that, they are ready to reach the market as living birds or be sold as killed/culled and packed, cold, or frozen.

5.2 THE PROBLEMS AND THE BIRDS' CONDITIONS

It is a general remark by anyone who is in close connection with modern poultry farms that those chickens are subjected to a nearly continuous source of electric light, feed all the time, grow fast, are given a remarkable amount of antibiotics and vaccines, have little space in which to move, have weak legs, have weak immune systems, are nervous but with weak responses, and their sizes are incompatible with their ages. In general, they are forced to eat and drink continuously in protective mode by using vaccines and antibiotics. They have no welfare except that they are given food in excess amounts. A normal balanced life will give normal growth with normal activity and *vice versa* (Amara 2013). It is important to highlight that alternatives to antibiotics can be used safely (Amara 2023; Amara et al. 2024; Amara, Salem-Bekhit, and Alanazi 2013b; Amara 2022; El-Baky and Al Fattah Amara 2023; Sharaf et al. 2013; Sheweita et al. 2022).

5.2.1 Environmental Impacts

The side effects of poultry production on the environment are important. They could negatively affect the production process as well as many other associated components. The crowded fast farming style (CFFS) for poultry production is a source of different pollutants that reach our bodies and the environment. Types of pollutants could include the following:

1. Oxygen-demanding substances could reduce oxygen from the environment, such as in the aquatic environment causing fish death, and might cause a reduction in total biological biodiversity.
2. Ammonia, upon reaching water streams, could kill fish and reduce or change ecosystem biodiversity (Adler et al. 2021; Linhares et al. 2022). Ammonia could mutagenize living cells, particularly microbial cells.

Volatilized ammonia is able to pollute the air causing acidification and damage to vegetation and sensitive ecosystem components.
3. Solid wastes from the poultry industry, particularly feathers, could cause serious environmental pollution.
4. Nutrients as wastes, specifically those that contain nitrogen and phosphorus, might cause additional problems, such as eutrophication, and enhance blooms of toxic algae and other toxic/pathogenic microorganisms.
5. Besides enhancing the growth of pathogens in the environment, the poultry industrial wastes have their own pathogenic microbes, including antibiotic-resistant microbes that contaminate the environment.
6. Heavy metals that contaminated food used (e.g., copper, arsenic, selenium, and zinc) in poultry production (including sanitizing and disinfecting agents) can harm human health and cause ecological risks. The main point is that if heavy elements reach the environment, they should be removed. Other methods such as precipitation are not effective. Heavy metals are not degradable, and upon reaching the biological creatures' organs and cells, they accumulate in their bodies, particularly in the liver (such as in fish).
7. Antibiotics are usually used excessively in poultry production and reach the environmental ecosystem, including our bodies. They have long-term ecosystem effects. However, the major problem they cause is the elevation of antibiotic-resistant microbes (Hussein and Amara 2006; Mansaray et al. 2022; Mensah et al. 2022; Nguyen et al. 2021; Shakir et al. 2021).
8. Hormones that are used extensively in the fast crowded CFFS, like antibiotics, also have long-term ecosystem effects (Uyanga et al. 2022).
9. The emission of different odors and other airborne discharges are numerous, such as hydrogen sulfide, other bad odor compounds, fine particles and dusts, and components that contribute to global warming (carbon dioxide and nitrous oxide) (Zhan et al. 2022).

In contrast, the TSFS enables fast drying of bird excretion before distributing it regularly in fields. The open style provides better benefit from the energy of the sun, good aeration, less disease distribution, better building of the birds' immune systems, better adaptation to the environment, less food consumption, self-dependence, better welfare, natural selection for the best male able to breed (by fighting), and the like. In fact, it contains all the positive factors that are lost in the controlled directed farming style. One additional point is (in the open style) that birds did not receive antibiotics or vaccines (or minimally received), which put consumers in better health and encouraged their consumers to purchase their product. In contrast, eating some products of the poultry produced in CFFS, like liver, heart, and the like, is not recommended.

5.2.2 Welfare of Animals

5.2.2.1 Cages and Captivity

Wild mammals, in captivity practice, cause considerable environmental disorders. For example, birds face stressful conditions of controlled diet, climate, social factors, reduction in free available movement space, and irritation (Rodrigues da Costa and Diana

2022). Captivity is believed to negatively affect immunologic responses (Crawley et al. 2021). In one type of production, hens are housed in wire cages with two or three hens in each cage (Baxter and Bédécarrats 2019). Cages are found to increase production, lower mortality, decrease cannibalism, lower feeding requirements, lower losses in food, decrease diseases and parasite transmission, improve culling, and reduce space and labor requirements. However, it is a painful way for the animals to live and puts great stress on them. Cages direct the birds to eat all the time. Other parameters were also studied in combination with the effect of cages on hens (Baxter and Bédécarrats 2019; Morrissey et al. 2016; Shi et al. 2019; Yan et al. 2021). The most interesting points for the producer are establishing better control, decreasing food loss, decreasing hen movement, and enabling better reorganization, treatments, breeding, and the like. Collectively, cages limit the birds' welfare (Khanal et al. 2020).

5.2.2.2 About One Thousand Hours of Eating, Drinking, and Digesting

There is no known animal on the earth, including birds, that spends its life eating and drinking only. Normally, animals eat and drink when they are hungry. They live like us and need what we need. They need to reproduce, to sleep enough, to enjoy the weather, to enjoy the environment, to move in an accepted free area, and the like. In reality, there was no shortage in poultry production over the ages, except during and after the industrial revolution, when cities became crowded by workers and the villages lost a great part of their manpower due to their workers' movements to cities. To cover the demand of such crowded cities, the poultry CFFS was invented, and interest in the TSFS decreased. As a solution, the modern production structure was applied on living candidates, poultry. For example, poultry production has fixed start and end points, the process is a subject of continuous optimization, the process neglects some side effects on the birds, aiming to fulfill the main industrial goal which is usually concern with optimizing the profits and decreasing the overall costs using minimum utilities and the like. In the viewpoint of an industrial specialist, bulk poultry production achieved tremendous success. All success points were accomplished perfectly. However, and forever, the product is a living candidate. When subjected to stress, the hormonal systems of living creatures begin to produce adrenaline and other hormones. The birds' biological systems over time will be affected, the immune system will be compromised, and the animal will be susceptible to infections and the like. Those described factors and others obligate producers to rethink and to reevaluate production conditions and take steps to save human health as well as improve the animals' welfare. Being not long into the journey is something to be grateful for. The journey of fast-growing chicken production begins after the first day of hatching. The process starts using one-day-old chicks. In addition, in most cases, their lives at the farm start from their second day of life and typically last 45 days (46 total days' age = 1104 hours). During this time, they are growing fast, they are eating all day and night, their digestive system never stops digesting food, their weak immune systems are nearly replaced by antibiotics, and any other methods that encourage weight gain are welcome! Modern chicken production gives heavier but less vital birds than slow-growing chickens produces. Modern chicken production raises chickens by gathering them into sheds where their wastes produce ammonia-laden air and wet litter. They are surrounded by their own waste. These conditions create wounds followed by infection with pathogens like *Salmonella* and *Campylobacter* (Suresh et al. 2020). To overcome these unhealthy conditions, to treat the birds or to protect them, they are subjected to excessive use of antibiotics. Poultry production is one of the major sources of antibiotic-resistant microbes. There are a number of researchers conducting research that compares CFFS and TSFS (Ben-Gigi et al. 2021). In general, many find it favorable to return to the TSFS for the best results for both animals and humans.

5.2.2.3 Poultry Diseases

Poultry is susceptible to several diseases, including common diseases such as fowl typhoid (Chaudhari et al. 2012), pullorum (Chen, Ishfaq, and Wang 2022), fowl cholera (Mbuthia et al. 2001), chronic respiratory disease (Fahey, Crawley, and Dunlop 1953; Kamaruzaman et al. 2021), infectious sinusitis, infectious coryza (Nsengimana et al. 2022), avian infectious hepatitis (Makbul, Nassal, and Böttcher 2020), infectious synovitis, bluecomb, Newcastle disease, fowlpox (Bertino et al. 2013), avian leukosis complex (Qu et al. 2022), coccidiosis (van Eerden et al. 2022), blackhead (Tyzzer 1919), infectious laryngotracheitis (Yegoraw et al. 2021), infectious bronchitis, and erysipelas (Imaeda et al. 2018). Parasitic diseases of poultry, including hexamitiasis of turkeys, are caused by roundworms, tapeworms, lice, and mites. Understanding the conditions that might lead to acquisition of any of these diseases and infections is important for better protection.

5.2.2.4 Antibiotics and Antibiotic Resistance

Antibiotics have helped the poultry industry to grow by preventing negative impacts of bacterial diseases (Bermudez 2003). In dense poultry farming, antibiotics such as tetracycline, tylosin, bacitracin, virginiamycin, salinomycin, and bambermycin are often used (Shakir et al. 2021). In the United States, tetracyclines make up more than two-thirds of antimicrobials administered (Gonzalez and Angeles Hernandez 2017), while in the European Union, they represent only 37% (Carvalho and Santos 2016). In 2015, sales of veterinary antimicrobial agents were 8,361 tonnes in the European Union according to report from the European Surveillance of Veterinary Antimicrobial Consumption (ESVAC 2017). This figure is deliberate without considering the impact of the growth promoters in animal production (Kummerer 2009). Antibiotics used as growth factors are not permitted in ESVAC contributing countries (ESVAC 2017). In 2014, 81% of antimicrobial compounds applied in Canada on broiler farms were for prevention purposes. In feed, 84% of these antimicrobials were used to prevent necrotic enteritis caused by *Clostridium perfringens* and coccidiosis (CSCRA 2016). Antibiotic resistance is one of the major problems that face the CFFS. The problem comes from those associated pathogenic strains that could infect humans and birds. Pathogenic microbes become resistant to antibiotics such as *Salmonella enterica* serovar Typhimurium and other strains (DiMarzio et al. 2013), *Salmonella* Typhimurium DT104 (Hawkins et al. 2019), *Escherichia coli* (Halfaoui, Menoueri, and Bendali 2017; Al-Marri et al. 2021), *E. coli* O157:H7 (Gaougaou et al. 2018), *Campylobacter* (Luangtongkum et al. 2009; Audu et al. 2022; Gharbi et al. 2022), which were reported in different countries (Kaviani Rad et al. 2022). Nowadays, most antibiotics are ineffective due to antibiotic resistance (Montoro-Dasi et al. 2020). Meanwhile, science offers alternatives to antibiotic treatment, such as treatment with bacteriophages (Lorenzo-Rebenaque et al. 2022), vaccines (Gupta et al. 2014; Johnson and Heneine 2001; Stachyra et al. 2016; Acevedo-Villanueva et al. 2021), active peptides with antimicrobial activity (Zaky, Shim, and Abd El-Aty 2021), enzymes such as lysozymes that are proven to be efficient in the growing of some animals (Abu Hafsa et al. 2022), and the like. Additionally, one promising method showing increasing interest for evacuating microbes and controlling pathogens is the sponge-like protocol, and its concept could be used to control microbes (Amara 2015, 2017; Amara, Salem-Bekhit, and Alanazi 2013a, 2015). H_2O_2 was recommended by El-Baky and Amara (2014) to control Newcastle virus as well as other viruses (particularly in the water) (El-Baky and Amara 2014).

5.2.2.5 Vaccines in High-Density Farming

Vaccines are more costly and time consuming than antibiotics. Meanwhile, due to the increase in antibiotic resistance, some bacterial pathogens were prepared as vaccines. Examples of those vaccines (and vaccine trials) include Newcastle disease (Tatár-Kis et al. 2020), genotype VII Newcastle disease virus (Wang et al. 2021), *Salmonella*-vectored vaccine delivering three *Clostridium perfringens* antigens to protect poultry against necrotic enteritis (Wilde et al. 2019), *Campylobacter jejuni* vaccine (Zeng, Xu, and Lin 2010), airborne Gumboro vaccine virus (Zhao et al. 2012), live attenuated *Salmonella* Typhimurium vaccine (van der Wolf et al. 2021), and many others. In dense production, high-density nurturing led to transmission of infection to each animal. Vaccines still one of the major protectors for the birds that grown in the CFFS (Hughey, Wilger, and Johnson 1989). Vaccination policies and protocols vary from country to country, company to company, and farm to farm. Various factors can influence choice of vaccination time/method/source. They include the birds' species, nature of the growing place, workforce, type of product (egg/meat), and production cycles. Other factors include health status, innate immunity, and vaccine costs (Rauw et al. 2009). Livestock vaccination against specific diseases is compulsory in some countries (Belgium, Netherlands, and Germany). In France, only long-lived poultry (lying and breeding) are vaccinated (Rauw et al. 2009).

5.2.2.6 Feeding

Poultry feeding is an optimized science that ensures the most benefit from the feeding process for body growth and fat production. Using well-balanced, high-quality protein sources supports the growth of muscle, organ, skin, and feather growth. The essential minerals are important for the production of bones and eggs: 3% to 4% of live birds and 10% of eggs are composed of minerals. The most important minerals are calcium, sodium, phosphorus, chlorine, iron, copper, potassium, sulfur, manganese, cobalt, magnesium, and zinc. Vitamins are important for the health of the birds and are required for their normal growth.

5.2.2.7 Light

The source, type, and intensity of the light are effective factors in the physiological and behavioral aspects of broilers. Light is needed for the birds' vision and color discrimination. The birds' body temperature, feeding rate, metabolism, and digestion are directly or indirectly affected by light. Light also affects hormone amounts and hence affects growth and maturation. In addition, some other physical activity can be affected by intensity, color, and duration of light that increase physical activity, affecting the development of broilers. Broiler producers must consider light management activities (Pandey 2019). Young chicks prefer brighter light. Meanwhile reports mentioned that their growth and development are improved in blue or green light over red or white light (Pandey 2019). Lighting length and duration are dependent on chickens' age. Broilers need to be provided four hours/day for sleep, but they may require more hours at points during the growing period. In contrast, darkness is important for the health and welfare of the birds, as their bodies' essential hormones, such as melatonin, do not secrete in the presence of continuous light. Daily dark periods are needed to ascertain traditional patterns of essential hormone secretion. Birds grown in imbalanced light/dark days have health issues. Meanwhile, light in crowded poultry farming is essential to avoid death, spiking mortality, and leg problems. Nerveless, continuous light disrupts diurnal behavior. Green light accelerates muscle growth, and blue light stimulates growth (Pandey 2019).

5.3 EARLY RESPONSE AGAINST THE MODERN POULTRY CHICKEN

When the chicken produced in CFFS start to reach Egyptians' villages, farmers (already having open spaces for chickens) took some steps to ensure the quality of the new type of white meat. They did not consume the meat directly, instead they brought these big (lazy) white chickens to their houses and put them in a separate clean open place. Then they let them grow like their own hens. They subjected them to free feeding with ground seeds/grains (that they harvested themselves), for two to three weeks to 40 days. They mimicked their natural environment and the folk beliefs (also religious) that bad effects from food are gone after 40 days. This period was enough (as they thought) to satisfy their own beliefs and morals. Such practice could be improved scientifically as a downstream (in-house or in the market) control process. Other controlling steps could be applied to validate the birds' health. For example, blood samples from the end product can be investigated before the birds reach consumers. In fact, an eye examination of an expert might be enough to distinguish between healthy and less healthy/thick birds. In time, such practices that were common among Egyptians were dropped after the acceptance of the new white chicken products.

5.4 EGYPTIAN TRADITIONAL SLOW FARMING STYLE FOR GROWING POULTRY

In Egypt, the TSFS is already running from unknown times. For countries that have not practiced this poultry growing style, they might start to use the proposed model in this chapter or even the Egyptians' TSFS original model. This will not break down running a CFFS but will enable slower advances toward better production that satisfies demand, supports the health of the public, and satisfies different consumers. There is a rising belief that slow-growing styles improve poultry welfare. The proposed model can guarantee one big chicken each day after the first 90 days and five big chickens to sell each month. Or one can sell them all (35 big chickens each month). In a country that has 25 million farmers' houses, practicing this style in their own houses will support the net national income. If it is practiced by most farmers, the net product for one year will be 124×10^6 chickens per month. If 50% of this net product is reached in market; that will give 62.5×10^6 chickens that are grown in the TSFS. They are using mainly farmer food rest, suitable agriculture by-products, and seed/grains that are locally produced.

5.4.1 Forty-Year-Old Successful Open Slow-Growing Poultry Farming

It is better to describe a real experience. For that, a style of growing nearly all types of poultry is described. This example takes place in a farm region, in a village named Bardala, near the city of Kafr Al Dawar in the Behaira governorates. In the Amara family house, there are nearly all types of poultry grown with each other but in different numbers. They have four turkeys, five to six ducks, ten chickens and two cocks (Figure 5.2), five geese, in addition to three pigeons' houses: one are in the upper place of the house for domesticated pigeons (big variety) and one in a separate house for wild pigeons (small variety) and one in a small partition in the first floor. There are also some rabbits (around five to eight, including one male). This small collection satisfies the Amara family for generations. It was a perfect system. They were grown in a house built in the border of the family agriculture land, considered to be totally self-dependent from most types of food and drinks. Most of the old Egyptian houses in the village were self-dependent too.

FIGURE 5.2 The original traditional Egyptians' slow model for poultry production. The Egyptian classical model: one (or two) cocks and ten chickens. (Prepared in this study.)

5.4.2 The Egyptian Old Model Resists Changes (A Family's Experiences)

It might be interesting to highlight one significant point that keeping this number of birds (as discussed earlier) was ideal for a family whose members are working on agricultural lands. In addition, some of their members have other duties, such as educational, government serving, etc. During the last 50 years, there were some trials to increase the number of the birds. Such trials were successful for a short time but did not continue over time. Usually, the number of different birds and domestic animals returned to original numbers as listed earlier. One story tells a pioneering trial done by the head of the Amara family during which he developed something new where he grew a flock of turkey with about 25 members and employed one farmer to care for them and take them to the agricultural land to feed freely in open air. He tried to grow turkey like sheep by letting them feed from the land directly. Why not? They would do that in the wild habitats. In fact, he succeeded. Nerveless, after a time, the Egyptians' TSFS original model for poultry production proved itself as the most adapted one, at least in their case and environment. It demonstrated continuity, stability, and sustainability. After time passed, the number of turkey returned to four to five turkey. That happened because of the cost of labor, the need for human control, and other factors. This example highlights the effectiveness of introducing a model that needs minimal care, management, and understanding of the key factors that conserve the success of such a model through the generations. Concerning the bulk production, the farmer's family style of production is different in some important ways due to many facts, such as the following: (1) They have another important duty (e.g., the agricultural activities). (2) They need, in most cases, to save their resources for their own use (seeds, grains, and the like). (3) They have a relaxed lifestyle (not in a hurry to gain money) and need a production model that has a relaxed controlling system with minimum daily responsibility. (4) They give some of their responsibility to their children. Light duties that could be managed are suitable for children. (5) They accept the model that can be easily processed. Some of them are interested in satisfying their own demand and not interested in gaining extra profits. Meanwhile, they might accept the idea of gaining some extra profit if their relaxed mode of life did not change (or minimally changed). (6) In the open-style system, birds grow nearly free (no crowding). (7) In ideal conditions, birds will consume the rest of the human food (after each meal). Minimum support might be required for their feeding in special weather or cases from the harvested seeds and grains.

The explanation for the success of the original traditional Egyptian TSFS model for poultry production could be summarized in the following points:

1. It is flexible and light. For example, in the days where all the family work on gathering crops, all birds can get self-care or need minimum care.
2. The different types of poultry are healthy, semi-wild, well adapted to the environment, and derived from local verities that could have competely different conditions and adapt themselves.
3. They are in the correct place, the farmer house, where there are nearly no food problems. Foods could be obtained from the rest of the family meals,

seeds, grains, crops, vegetables, and the like. And the houses usually contain different types of seeds that can be used at any time.
4. Growing birds in the open style is an open training course for any one of the family members interested in gaining such experiences. Additionally, birds are a source of happiness for the children.
5. Before plastic products reach the countryside, one could say that the house output from wastes was 0.
6. The members of the family growing birds are healthy, while birds supply them with an essential protein source (e.g., meat and eggs) from different varieties.
7. Their by-products (faces) contain no antibiotics, no drugs, and are taken directly to a special sunny place where they become drier and more aged to be used at the correct time as fertilizer.
8. Birds were not a burden on the family concerning space, feed, care, and the like.
9. They are part of the house security, where there is a response to any sound.
10. They wake up early (cocks sound).
11. For some families, they are essential to the total family income.

From all of those points, one should observe that at that time, there were no refrigerators or electricity in Egyptian houses. Everything ran smoothly and naturally. Thus, there was no need to increase the number of birds particularly if they were grown to satisfy family demand and not for sale. By observing other houses in the farming region and following some stories, various numbers exist in different periods.

5.4.3 Building a Model for the Slow-Open Style of Poultry Farming

5.4.3.1 Specifying the Aim of the Model

The main criterion of the proposed model is to be accepted by farmers (the main preferable place is the farmland). But it should have some additional criteria, such as (1) mobility, (2) sustainability, (3) safety, (4) productivity, (5) manageability, (6) cost-effectiveness (not being risky in case birds are lost; farmers should be able to replace any losses), (7) flexibility, and (8) the possibility of enlarging and scaling up.

5.4.3.1.1 The Running Models

There are some models for growing fowl and chickens, such as bulk broiler production, old Egyptian farmer slow style for growing fowl; fowl production in cages or closed rooms; natural feeding style using open/close area; food produced by companies for growing fowl in open/close areas and the like. They could be classified based on three main criteria: (1) TSFS or CFFS –growing; (2) the type of food, either completely crud or natural food produced and supplied after industrial processes; and (3) open or closed growing style. In some cases, mixed styles are used. For example, one style that was studied was a slow-growing open style, but the fowl were given commercially produced food that contained additives that promoted weight gain. This style can be found in places where their owners did not have their own resources from the cultivated seeds. In fact, the most native and close to wildlife is the TSFS that is processed by Egyptian farmers in relaxed modes (e.g., ten chickens and two cocks). Global demand has increased, and there is a need to get more benefit from the different styles (Santos et al. 2022). Flexible new models like the traditional Egyptian model can be designed and built with more productivity and minimum cost risk (in the form of losing birds). It should introduce a fixable model that can be managed by any member of the family, does not put a load on the family, and avoids negative points in both the TSFS and CFFS models. Some other major points should be included in the proposed models, such as the need for (1) naturally grown healthy birds; (2) birds that have adapted to the local environment and can resist diseases and microbes (to avoid using antimicrobial drugs); (3) an appropriate number of birds so that the flock can be portable, simple to handle, fast to replace, and easy to control; (4) maintenance of acceptable poultry welfare (e.g., natural day/light exposure); (5) healthcare for the poultry; (6) a market to sell the products; (7) a supplier to supply the chicks; and (8) housing that should include both open and closed spaces to enable chicken to enjoying the natural environment and have stable daily lives.

5.4.3.2 The Chosen Template – The Egyptians' Traditional Slow Farming Style Model for Poultry Production

Models from all over the world, were analyed in the literature. Based on observations of the old small Egyptian style for growing different poultry and the other models that are currently being used, the Egyptians' original TSFS model for poultry production is selected to be the one that may be used to build a bigger model.

5.4.3.3 Description of the Egyptians' Original Traditional Slow Farming Style Model for Chicken Production

The proposed TSFS for poultry farming is extracted from the traditional style of poultry production in the Egyptian village. This style could be described as a style that did not put any excess costs on the farmer and gave the birds free space to live naturally with no or minimal care from the farmers. From an economic viewpoint, this style satisfies the demands of one family for eggs and meat. Chicken losses will not break down the farmer's budget. The farmers are always able to replace the lost birds with other ones. It is a model established over ages. It exists from a collection of different poultry birds (as earlier). In most cases, it contains eight to ten chickens and one or two cocks. The chickens and the cocks have a small place to sleep (or are left free). They are left to select a suitable place to sleep in their free living area. Some times they sleep like flying birds on any wire or linear stand. They wake up early and start to collect their food from the surrounding area. If they were not fed manually, nothing will happen, because they will look for food around their place. Alternatively, they will find their way to outside the house and will eat from the surrounding farming area or from the garden. Additionally, they are always fed food leftovers (of any meal), like the rest of the rice, bread, uncooked parts of vegetables, fruits (e.g., the fruit peels), and the like. Additionally, they are supplied with suitable agriculture by-products or with ground seeds and grains. It is important to highlight that eating the leftovers of recently cooked seeds and grains that are freshly harvested will save the birds from the toxins that could be found in preserved grains. For example, processed chickens' food (from seeds and grains and other constituents) is usually subjected to a long step of storing, movement, industrial processes, restoring, and the like. The long processing time of the birds' food might expose the food to fungal contamination, which is risky in most cases. Consuming human leftovers is the best way to get benefits from the meal by-products. In the old Egyptian farming style, there is nothing left, and they have 0% waste (before the use of plastic). Traditional Egyptian farmers were self-dependent and well organized.

5.5 OTHER FACTORS THAT COULD AFFECT THE PROCESS

Other factors include the following: (1) house space and its free surrounding area; (2) available space for bird farming; (3) available people to care for the birds; (4) amount of food that can be apportioned to the birds, although as discussed above, birds in most Egyptian farmers' houses depend on meal leftovers and the grains and seeds from the cultivated surrounding lands and are able to access surrounding open areas for self-feeding and (5) ways to substitute any lost, slotted or dead birds, which might depend on the presence of the local market near the farmer's house.

5.6 BIRD WELFARE IN ORIGINAL TRADITIONAL EGYPTIANS' SLOW FARMING MODEL FOR POULTRY PRODUCTION

Some traditional songs describe how the farmers and their birds and animals are happy. In the farmer's house, the birds move freely. Only doors prevent them from entering special areas. In fact, they never pass certain lines. They can jump to the other houses around and return at any time. They can go outside the house to feed themselves from the food here and there and return, without any control from the farmers. The birds in some places communicate with the house members. Everything is calm and systematic. There is some sort of understanding between the birds and the farmers, especially those who are in daily contact with them. For example, a female duck gives signals to lay eggs. One could observe that Egyptians who live near the running channels take their ducks in the early morning to the channel and leave them swimming and eating. The ducks did not leave certain areas even though the water is not static. At the end of the day, the one who cares for them (usually women) calls them with one word (Asal, Asal . . .). The ducks leave the channel in regular order in the form of a line and come back home. This is a cost-free type of duck farming practiced by Egyptians over the years. Other types of practice include turkey immunization. Egyptians follow an unusual practice to immunize and grow newly hatched young turkeys. Young turkeys do not survive in the Egyptian environment unless they are fed just after their hatching on a mixture of egg, cheese, and the *Allium porrum,* a traditional folk mixture. Other sick birds are given fresh-cut pieces of onion. Some put cut onion pieces in the poultry's drinking water. They invented that to protect the birds. There is a general old belief in Egypt that the onion is a strong health supporter for humans and animals, including birds. Additionally, there is a general belief that onion and garlic can protect against epidemics. And the common words are "the onion keep your heart strong and the blood clean". Egypt has some extra opportunities to support any kind of open poultry growing model or animal growing model. Egypt is a land with moderate weather, flat land, and dominant farming regions. Thc climate is acceptable for birds throughout the year, as temperatures rarely go below 10°C or higher than 34°C (in the delta region), and there are no winds or unusual weather conditions (except for a few days/years). For that, farmers have minimal requirements to install a place to start poultry farming. Some are growing them in open places. Their food is derived from the surrounding resources. If they made a house for birds, it cost nothing and was built from the resources around the house. For example, mud is mixed with seed covers from rice, formed to a suitable size, left to dry, and then put side by side with others. Cotton tree woody stems are used to cover the top of the cage. Usually, the house is kept without a door and closed instead by any suitable portable barrier. A small cage could take 4 hours to build. The portability of traditional Egyptian poultry farming for the chicken allows for a suitable source of eggs and meat. For these reasons, this method has been chosen to be the template to build a bigger model that could adapt some important characteristics of the Egyptian model to be used everywhere. A personal experience could support this choice. Fifteen chicks (one day old) can be grown in the first suitable carton. They are grown, and 13 of them are chocks and 2 are chickens. They are growing with nearly no effort. After the first week passes, they are left free and fed with the food available for other existing birds. They use no medicine and do not take any kind of care, they are just observed as they grow. However, after 40 days, they become young cocks and chickens from different varieties and start to leave the houses to go to the surrounding neighborhood houses. That was really the style of growing chicken by a child in the Amara family house. At 60 days, they all were culled and consumed by members of the family. There is no single case of death or illness. It might be interesting here to report that the chicken grown using the Egyptians' original TSFS model for poultry production did not receive any kind of treatment. Generally, the only problem they faced was one disease called by Egyptians "Ferra" (meaning "so fast") because this disease comes on suddenly in a particular time of the year and kills all chickens. The scientific name of this disease is Newcastle virus disease. What the Egyptians did if that happened was tell each other, and the only practice (at that time 1981 to the present) is to isolate the chickens in a closed room or cull them all before being infected. Culling 10 to 12 chickens did not cause economic problems because they were consumed by members of the house. After the disappearance of the "Ferra" from the area, the farmer bought another group either as chicks or as mature chickens and cocks. During a period of nearly 30 years, "Ferra" attacked chickens about six times.

5.7 PROPOSED MODEL AND PROPOSED NAME: 110 CHICKEN (CAPACITY)/90 DAYS (OUTPUT/MONTH)

The proposed model is described in its simplest form so it is easy to understand. The model has two mobile wings and one fixed part. The first mobile wing is the input of the production process, where each month 35 chicks (one day old) enter the farming house. The one-day-old chicks will be left to grow until they are 31 days old (1 day [original age] + 30 farming days) in place (a) (Figure 5.3), then will be moved to

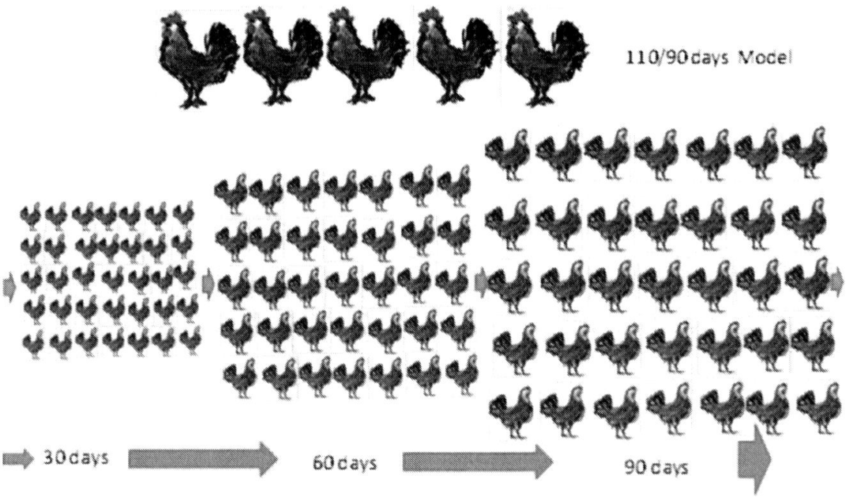

FIGURE 5.3 Chicks and chickens at 30 days, 60 days, and 90 days. The proposed model guarantees at least one big chicken every day or 35 chickens each month, (Prepared in this study.)

another place (b), and another 35 one-day-old chicks replace them in place (a). Then both chicks in place (a) and chickens in place (b) are left for another 30 days, and then each move one more step. Birds in place (a) move to (b) and another group of 35 one-day-old chicks enter place (a), and birds in place (b) move to place (c). All are left another 30 days. Their ages at the end of 90 days will be as follows: for place (a), 31 days; for place (b), 61 days; and for place (c), 91 days. Alternatively, three groups of one-day-old, 31-day-old, and 61-day-old birds can be bought from the market to fill in places (a), (b), and (c) at the same time. After (or during) reaching 91 days in place (c), there are two pathways (concerning chickens in place [c]). The first is that the farmer can consume one chicken daily. Farmers can manage this step directly from place (c) or can use an alternative place. The second is that farmers can stay until the end of the 30 days and sell them all. This model enables at least one big chicken every day or 35 chickens every month (Figure 5.4). The farmer can change the plan based on his demand and his resources. The chicken will be able to enjoy normal day/light and will have a better life. Eggs could be produced at place (c). Or additional places can be used to grow the best chickens that are able to produce eggs. Five cocks are found in place (c) following the original Egyptians' TSFS model for poultry production. It is recommended to select cocks growing with the other chickens. That keeps them calmer when together. One could reduce the number of cocks by selecting the strongest ones. The cocks play an organizing role and are part of their welfare (Figure 5.5). And, they can be used for any further expansion, fertilized eggs, self-breeding, research, behavior study, and so on. For some reason, Egyptian farmers grow chickens in the presence of cocks. Even sellers in the market selling live chickens have one or two cocks from local areas (usually not for sale). They are usually used to keep some sort of control of the chickens.

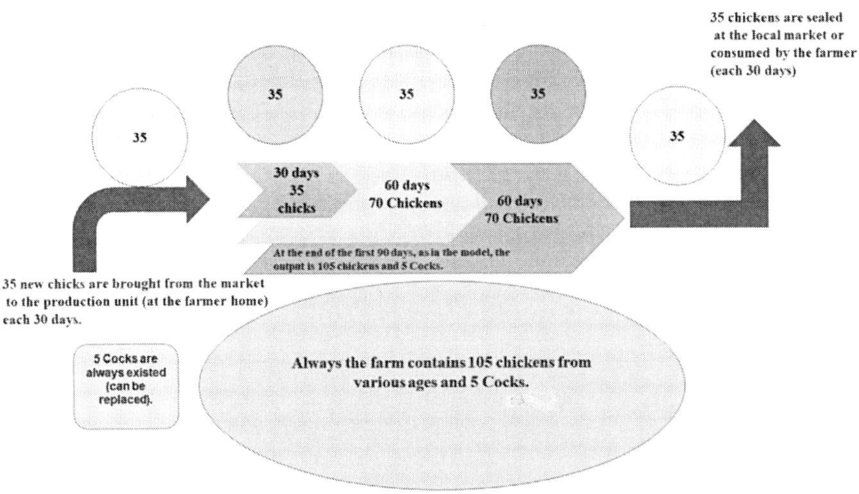

FIGURE 5.4 Schematic illustration of 90 days and the three stages of the proposed model for growing chickens. (Prepared in this study.)

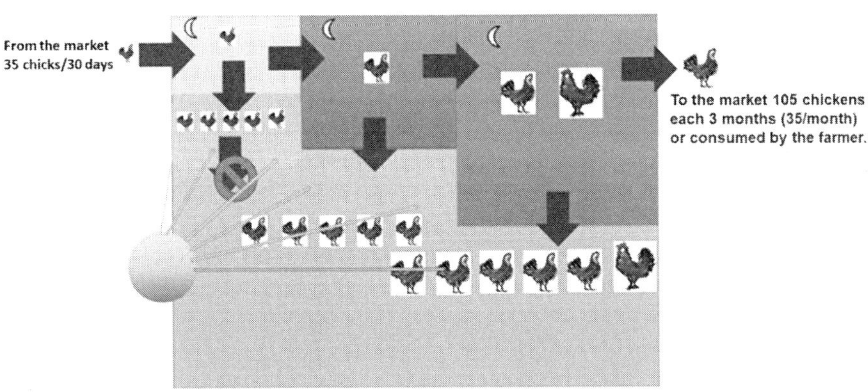

FIGURE 5.5 Schematic illustration of the three stages of the proposed model for growing showing the welfare of the birds. (Prepared in this study.)

5.8 CONCLUSIONS

Some old models for growing poultry used over the ages have proven successful due to their simplicity, sustainability, flexibility, and manageability. One simple model is the original Egyptians' TSFS model for poultry production. There is quite enough published data comparing CFFS and TSFS. This research is mostly concerned with the birds' welfare, food, antibiotics, antibiotic resistance, growing conditions, impact on the environment, and human health. It might be interesting to enlarge the old Egyptian model to be more reproducible but without changing its main criteria, such as the open style, the bird's welfare, the minimal use of antibiotics and vaccines, and the minimal need for care and food waste. The proposed model in this study is flexible with minimal cost needed to start up or to recover any losses. If it can be applied in a wider scale, it will be able to support the economy of any government. Those who are interested in enlarging it are still able to do so.

REFERENCES

Abu Hafsa, S. H., A. E. M. Mahmoud, A. M. A. Fayed, and A. A. S. Abdel-Azeem. 2022. "The Effect of Exogenous Lysozyme Supplementation on Growth Performance, Caecal Fermentation and Microbiota, and Blood Constituents in Growing Rabbits." *Animals (Basel)* 12 (7). https://doi.org/10.3390/ani12070899.

Acevedo-Villanueva, K. Y., G. O. Akerele, W. G. Al Hakeem, S. Renu, R. Shanmugasundaram, and R. K. Selvaraj. 2021. "A Novel Approach against *Salmonella*: A Review of Polymeric Nanoparticle Vaccines for Broilers and Layers." *Vaccines (Basel)* 9 (9):1041. https://doi.org/10.3390/vaccines9091041.

Adeola, A. C., F. E. Sola-Ojo, Y. A. Opeyemi, A. O. Oguntunji, L. M. Nneji, M. K. Ewuola, S. F. Bello, W. A. Olaniyi, A. T. Adesoji, A. P. Karuno, O. J. Sanke, and E. L. Daniel. 2022. "Genetic Diversity and Population Structure of Muscovy Duck (Cairina Moschata) from Nigeria." *PeerJ* 10. https://doi.org/10.7717/peerj.13236.

Adler, C., A. J. Schmithausen, M. Trimborn, S. Heitmann, B. Spindler, I. Tiemann, N. Kemper, and W. Büscher. 2021. "Effects of a Partially Perforated Flooring System on Ammonia Emissions in Broiler Housing—Conflict of Objectives between Animal Welfare and Environment?" *Animals (Basel)* 11 (3):707. https://doi.org/10.3390/ani11030707.

Al-Marri, T., A. Al-Marri, R. Al-Zanbaqi, A. A. Ajmi, and M. Fayez. 2021. "Multidrug Resistance, Biofilm Formation, and Virulence Genes of *Escherichia coli* from Backyard Poultry Farms." *Vet World* 14 (11):2869–2877. https://doi.org/10.14202/vetworld.2021.2869–2877.

Alnahhas, N., C. Berri, M. Chabault, P. Chartrin, M. Boulay, M. C. Bourin, and E. Le Bihan-Duval. 2016. "Genetic Parameters of White Striping in Relation to Body Weight, Carcass Composition, and Meat Quality Traits in Two Broiler Lines Divergently Selected for the Ultimate pH of the Pectoralis Major Muscle." *BMC Genet* 17:61. https://doi.org/10.1186/s12863-016-0369-2.

Amara, A. A. . 2013. "The Inevitability of Balanced Lives: Genes and, Foods in Action and, Interactions." *IIOAB J* 4 (2):1–27.

Amara, A. A. 2015. *Kostenlose viral Ghosts, Bacterial Ghosts, Microbial Ghosts and More*. Vol. Book. Germany: Schüling Verlage.

Amara, A. A. 2017. "Cell Evacology: Could It Be a New Science." *Cell Mol Med* 3 (2):9.

Amara, A. A. 2023. "Fermented Food Microbes and Their Preservative Byproducts.' In Edited by A. Gupta and R. Prasad, *Antimicrobials in Food Science and Technology*: CRC Press.

Amara, A. A. , N. Abd El-Baky, M. M. Salem-Bekhit, Khairy M. A. Zoheir, and M. M. M. Ahmed. 2024. "Antibiotic Resistance from Theory to Facts." In Edited by A. Gupta, R. Prasad, A. Gupta, R. Prasad, *Antimicrobial Resistance in Agriculture and Its Consequences*: CRC, Press.

Amara, A. A., M. M. Salem-Bekhit, and F. K. Alanazi. 2013a. "Sponge-Like: A New Protocol for Preparing Bacterial Ghosts." *TSWJ* 2013:545741.

Amara, A. A., M. M. Salem-Bekhit, and F. K. Alanazi. 2013b. "Sponge-Like: A New Protocol for Preparing Bacterial Ghosts." *Sci World J* 2013. https://doi.org/10.1155/2013/545741.

Amara, A. A., M. M. Salem-Bekhit, and F. K. Alanazi. 2015. "Plackett-Burman Randomization Method for Bacterial Ghosts Preparation form E. Coli JM109." *Saudi Pharm J* 22 (3):273–279.

Amara, Amro Abd Al Fattah. 2022. "The Role of Divalent Cations in Antibiotic Sensitivity." In Edited by V. K. Gupta, S. D. Sarker, M. Sharma, M. E. Pirovani, Z. Usmani, and C. Jayabaskaran, *Biomolecules from Natural Sources*: Wiley.

Arrazola, A., and S. Torrey. 2021. "Welfare and Performance of Slower Growing Broiler Breeders During Rearing." *Poult Sci* 100 (11):101434. https://doi.org/10.1016/j.psj.2021.101434.

Attia, Y. A., A. S. El-Naggar, B. M. Abou-Shehema, and A. A. Abdella. 2019. "Effect of Supplementation with Trimethylglycine (Betaine) and/or Vitamins on Semen Quality, Fertility, Antioxidant Status, DNA Repair and Welfare of Roosters Exposed to Chronic Heat Stress." *Animals (Basel)* 9 (8):547. https://doi.org/10.3390/ani9080547.

Audu, B. J., S. Norval, L. Bruno, R. Meenakshi, M. Marion, and K. J. Forbes. 2022. "Genomic Diversity and Antimicrobial Resistance of Campylobacter spp. from Humans and Livestock in Nigeria." *J Biomed Sci* 29. https://doi.org/10.1186/s12929-022-00786-2.

Bai, Y., R. Wang, Y. Yang, R. Li, and X. Wu. 2021. "Folic Acid Absorption Characteristics and Effect on Cecal Microbiota of Laying Hens." *Front Vet Sci* 8:720851. https://doi.org/10.3389/fvets.2021.720851.

Bamidele, O., E. B. Sonaiya, O. A. Adebambo, and T. Dessie. 2020. "On-Station Performance Evaluation of Improved Tropically Adapted Chicken Breeds for Smallholder Poultry Production Systems in Nigeria." *Trop Anim Health Prod* 52 (4):1541–1548. https://doi.org/10.1007/s11250-019-02158-9.

Bao, Q., Y. Yao, K. Weng, M. Zheng, Y. Zhang, Y. Zhang, G. Chen, and Q. Xu. 2022. "Research Note: Comparison on Laying Behavior and Clutch Traits Among Zhedong White Geese (Anser cygnoides), Sichuan White Geese (Anser cygnoides), and Hungarian Geese (Anser anser)." *Poult Sci* 101 (2). https://doi.org/10.1016/j.psj.2021.101594.

Baxter, M., and G. Y. Bédécarrats. 2019. "Evaluation of the Impact of Light Source on Reproductive Parameters in Laying Hens Housed in Individual Cages." *J Poult Sci* 56 (2):148–158. https://doi.org/10.2141/jpsa.0180054.

Ben-Gigi, R., A. Haron, D. Shinder, M. Ruzal, and S. Druyan. 2021. "Differential Physiological Response of Slow- and Fast-Growing Broiler Lines to Hypoxic Conditions During Chorioallantoic Membrane Development." *Poult Sci* 100 (2):1192–1204. https://doi.org/10.1016/j.psj.2020.10.068.

Bermudez, A. J. 2003. *Principles of disease prevention: Diagnosis and control*, 3e60. Ames, IA, USA: Iowas State University Press.

Bertino, P., M. Panigada, E. Soprana, V. Bianchi, S. Bertilaccio, F. Sanvito, A. H. Rose, H. Yang, G. Gaudino, P. R. Hoffmann, A. Siccardi, and M. Carbone. 2013. "Fowlpox-Based Survivin Vaccination for Malignant Mesothelioma Therapy." *Int J Cancer* 133 (3):612–623. https://doi.org/10.1002/ijc.28048.

Bona, L., N. van Staaveren, B. B. Pokharel, M. van Krimpen, and A. Harlander-Matauschek. 2018. "The Effect of Low Protein Energy-Rich Diets on Plasma Hepatic Markers, Hepatic Damage, and Discrimination Reversal Learning in Young Female Chicks." *Front Vet Sci* 5:107. https://doi.org/10.3389/fvets.2018.00107.

Carvalho, I. T., and L. Santos. 2016. "Antibiotics in the Aquatic Environments: A Review of the European Scenario." *Environ Int* 94:738–757.

Chaudhari, A. A., C. V. Jawale, S. W. Kim, and J. H. Lee. 2012. "Construction of a Salmonella Gallinarum ghost as a Novel Inactivated Vaccine Candidate and Its Protective Efficacy Against Fowl Typhoid in Chickens." *Vet Res* 43 (1):44. https://doi.org/10.1186/1297-9716-43-44.

Chen, X., M. Ishfaq, and J. Wang. 2022. "Effects of Lactobacillus Salivarius Supplementation on the Growth Performance, Liver Function, Meat Quality, Immune Responses and Salmonella Pullorum Infection Resistance of Broilers Challenged with Aflatoxin B1." *Poult Sci* 101 (3). https://doi.org/10.1016/j.psj.2021.101651.

Crawley, J. A. H., O. Liehrmann, D. J. Franco dos Santos, J. Brown, U. K. Nyein, H. H. Aung, W. Htut, Z. M. Oo, M. W. Seltmann, J. L. Webb, M. Lahdenperä, and V. Lummaa. 2021. "Influence of Handler Relationships and Experience on Health Parameters, Glucocorticoid Responses and Behaviour of Semi-Captive Asian Elephants." *Conserv Physiol* 9 (1):coaa116. https://doi.org/10.1093/conphys/coaa116.

CSCRA. 2016. *Systeme Canadien de Surveillance de la Resistance aux Antimicrobiense Rapport De.* Gouvernement du Canada.

Derks, M. F. L., J. M. Herrero-Medrano, Rpma Crooijmans, A. Vereijken, J. A. Long, H. J. Megens, and M. A. M. Groenen. 2018. "Early and Late Feathering in Turkey and Chicken: Same Gene but Different Mutations." *Genet Sel Evol* 50 (1):7. https://doi.org/10.1186/s12711-018-0380-3.

DiMarzio, M., N. Shariat, S. Kariyawasam, R. Barrangou, and E. G. Dudley. 2013. "Antibiotic Resistance in Salmonella enterica Serovar Typhimurium Associates with CRISPR Sequence Type." *Antimicrob Agents Chemother* 57 (9):4282–4289. https://doi.org/10.1128/aac.00913–13.

Du, X., P. Qin, Y. Liu, F. K. Amevor, G. Shu, D. Li, and X. Zhao. 2022. "Effects of Key Farm Management Practices on Pullets Welfare—A Review." *Animals (Basel)* 12 (6). https://doi.org/10.3390/ani12060729.

El-Baky, A. N., and A. A. Amara. 2014. "Newcastle Disease Virus (LaSota strain) as a Model for Virus Ghosts Preparation Using H2O2 Bio-Critical Concentration." *Int Sci Investig J* 3 (5):38–50.

El-Baky, N. A., and A. A. Al Fattah Amara. 2023. "Natural Antimicrobial Therapeutic Peptides: Milk Lactoferricin and *Spirulina platensis* Peptides." In edited by A. Gupta and R. Prasad, *Antimicrobials in Pharmaceutical and Medicinal Research*, 19–48. CRC Press.

ESVAC. 2017. *Sales of Veterinary Antimicrobial agents in 30 european countries in 2015. Trends from 2010 to 2015.* Seventh Esvac Report:Ema - 184855 – 2017.

Fahey, J. E., J. F. Crawley, and W. R. Dunlop. 1953. "Studies on Chronic Respiratory Disease of Chickens; Observations on Outbreaks in Canada in a Two Year Period 1." *Can J Comp Med Vet Sci* 17 (7):294–298.

Gaougaou, G., Y. Ben-Fadhel, E. Déziel, and M. Lacroix. 2018. "Effect of β-lactam antibiotic resistance gene expression on the radio-resistance profile of *E. coli* O157:H7." *Heliyon* 4 (12):e00999. https://doi.org/10.1016/j.heliyon.2018.e00999.

Gharbi, M., A. Béjaoui, C. Ben Hamda, N. Alaya, S. Hamrouni, G. Bessoussa, A. Ghram, and A. Maaroufi. 2022. "Campylobacter spp. in Eggs and Laying Hens in the North-East of Tunisia: High Prevalence and Multidrug-Resistance Phenotypes." *Vet Sci* 9 (3). https://doi.org/10.3390/vetsci9030108.

Gonzalez, R. M., and J. C. Angeles Hernandez. 2017. "Antibiotic and Synthetic Growth Promoters in Animal Diets: Review of Impact and Analytical Methods." *Food Contr* 72:255–267.

Grabež, V., B. Egelandsdal, A. Cruz, E. Hallenstvedt, L. T. Mydland, O. Alvseike, K. Kåsin, L. Ruud, V. Karlsen, and M. Øverland. 2022. "Understanding Metabolic Phenomena Accompanying High Levels of Yeast in Broiler Chicken Diets and Resulting Carcass Weight and Meat Quality Changes." *Poult Sci* 101 (5). https://doi.org/10.1016/j.psj.2022.101749.

Gumułka, M., N. Avital-Cohen, and I. Rozenboim. 2021. "Determination of Annual Plasma Hormone Levels Associated with Reproduction in Long-Day Breeding Domestic Geese." *Animals (Basel)* 11 (8):2363. https://doi.org/10.3390/ani11082363.

Gupta, S. K., P. Bajwa, R. Deb, M. M. Chellappa, and S. Dey. 2014. "Flagellin A Toll-Like Receptor 5 Agonist as an Adjuvant in Chicken Vaccines." *Clin Vaccine Immunol* 21 (3):261–270. https://doi.org/10.1128/cvi.00669–13.

Halfaoui, Z., N. M. Menoueri, and L. M. Bendali. 2017. "Serogrouping and Antibiotic Resistance of *Escherichia coli* Isolated from Broiler Chicken With Colibacillosis in Center of Algeria." *Vet World* 10 (7):830–835. https://doi.org/10.14202/vetworld.2017.830–835.

Hawkins, J. L., J. Uknalis, T. P. Oscar, J. G. Schwarz, B. Vimini, and S. Parveen. 2019. "The Effect of Previous Life Cycle Phase on the Growth Kinetics, Morphology, and Antibiotic Resistance of *Salmonella Typhimurium* DT104 in Brain Heart Infusion and Ground Chicken Extract." *Front Microbiol* 10:1043. https://doi.org/10.3389/fmicb.2019.01043.

Hedlund, L., and P. Jensen. 2022. "Effects of Stress During Commercial Hatching on Growth, Egg Production and Feather Pecking in Laying Hens." *PLoS One* 17 (1). https://doi.org/10.1371/journal.pone.0262307.

Huang, K. S., and S. W. Huang. 2011. "Changes in the Nutrient Content of American Diets." *Health Econ Rev* 1:19. https://doi.org/10.1186/2191-1991-1-19.

Hughey, V. L., P. A. Wilger, and E. A. Johnson. 1989. "Antibacterial Activity of Hen Egg White Lysozyme Against Listeria Monocytogenes Scott A in Foods." *Appl Environ Microbiol* 55 (3):631–638.

Hussein, Z. M., and A. A. Amara. 2006. "Case-by-case Study Using Antibiotic-EDTA Combination to Control *Pseudomonas aeruginosa*." *Pak J Pharm Sci* 19 (3):236–243.

Iannotti, L. L., M. Chapnick, J. Nicholas, C. A. Gallegos-Riofrio, P. Moreno, K. Douglas, D. Habif, Y. Cui, C. Stewart, C. K. Lutter, and W. F. Waters. 2020. "Egg Intervention Effect on Linear Growth no Longer Present After Two Years." *Matern Child Nutr* 16 (2):e12925. https://doi.org/10.1111/mcn.12925.

Imaeda, N., A. Ando, M. Takasu, T. Matsubara, N. Nishii, S. Takashima, A. Shigenari, T. Shiina, and H. Kitagawa. 2018. "Influence of Swine Leukocyte Antigen Haplotype on Serum Antibody Titers Against Swine Erysipelas Vaccine and Reproductive and Meat Production Traits of SLA-Defined Selectively Bred Duroc Pigs." *J Vet Med Sci* 80 (11):1662–1668. https://doi.org/10.1292/jvms.18–0027.

Jiang, S. G., N. X. Pan, M. J. Chen, X. Q. Wang, H. C. Yan, and C. Q. Gao. 2019. "Effects of Dietary Supplementation with dl-Methionine and dl-Methionyl-dl-Methionine in Breeding Pigeons on the Carcass Characteristics, Meat Quality and Antioxidant Activity of Squabs." *Antioxidants (Basel)* 8 (10):435. https://doi.org/10.3390/antiox8100435.

Johnson, J. A., and W. Heneine. 2001. "Characterization of Endogenous Avian Leukosis Viruses in Chicken Embryonic Fibroblast Substrates Used in Production of Measles and Mumps Vaccines." *J Virol* 75 (8):3605–3612. https://doi.org/10.1128/jvi.75.8.3605–3612.2001.

Kamaruzaman, I. N. A., K. Y. Ng, R. H. Hamdan, N. Shaharulnizim, Cwscw Zalati, M. Mohamed, M. L. Nordin, Nzim Rajdi, L. Abu-Bakar, and M. F. H. Reduan. 2021. "Complex Chronic Respiratory Disease Concurrent With Coccidiosis in Broiler Chickens in Malaysia: A Case Report." *J Adv Vet Anim Res* 8 (4):576–580. https://doi.org/10.5455/javar.2021.h547.

Kaviani Rad, A., S. K. Balasundram, S. Azizi, Y. Afsharyzad, M. Zarei, H. Etesami, and R. R. Shamshiri. 2022. "An Overview of Antibiotic Resistance and Abiotic Stresses Affecting Antimicrobial Resistance in Agricultural Soils." *Int J Environ Res Public Health* 19 (8). https://doi.org/10.3390/ijerph19084666.

Kennedy, D. A., J. R. Dunn, P. A. Dunn, and A. F. Read. 2015. "An Observational Study of the Temporal and Spatial Patterns of Marek's-Disease-Associated Leukosis Condemnation of Young Chickens in the United States of America." *Prev Vet Med* 120 (0):328–335. https://doi.org/10.1016/j.prevetmed.2015.04.013.

Khanal, T., G. Y. Bédécarrats, T. M. Widowski, and E. G. Kiarie. 2020. "Rearing Cage Type and Dietary Limestone Particle Size: II, Effects on Egg Production, Eggshell, and Bone

Quality in Lohmann Selected Leghorn-Lite Hens." *Poult Sci* 99 (11):5763–5770. https://doi.org/10.1016/j.psj.2020.06.079.

Kummerer, K. 2009. "Antibiotics in the Aquatic Environmentea Reviewe Part I." *Chemosphere*: 417–434.

Li, C., S. Schallier, C. Lamberigts, J. Lesuisse, N. Everaert, W. Merckx, and J. Buyse. 2020. "Management Factors Resulting in a Severe Reduction in Feed Intake–Induced Spiking Mortality Syndrome in Young Broiler Chicks." *Poult Sci* 99 (10):4939–4946. https://doi.org/10.1016/j.psj.2020.06.032.

Linhares, F. G., K. A. A. Torres-Cordido, M. S. Sthel, M. G. da Silva, and L. Mota. 2022. "Monitoring of Ammonia Concentrations from Coir-Husk Litter of Brazilian Poultry House Using Diode Laser Photoacoustic Spectroscopy." *Environ Monit Assess* 194 (6). https://doi.org/10.1007/s10661-022-10070-3.

Lorenzo-Rebenaque, L., D. J. Malik, P. Catalá-Gregori, C. Marin, and S. Sevilla-Navarro. 2022. "Gastrointestinal Dynamics of Non-Encapsulated and Microencapsulated Salmonella Bacteriophages in Broiler Production." *Animals (Basel)* 12 (2). https://doi.org/10.3390/ani12020144.

Luangtongkum, T., B. Jeon, J. Han, P. Plummer, C. M. Logue, and Q. Zhang. 2009. "Antibiotic Resistance in Campylobacter: Emergence, Transmission And Persistence." *Future Microbiol* 4 (2):189–200. https://doi.org/10.2217/17460913.4.2.189.

Makbul, C., M. Nassal, and B. Böttcher. 2020. "Slowly Folding Surface Extension in the Prototypic Avian Hepatitis B Virus Capsid Governs Stability." *eLife* 9:e57277. https://doi.org/10.7554/eLife.57277.

Mansaray, A. H. D., D. P. Y. Yankson, R. A. B. Johnson, F. L. Moses, J. S. Kanu, I. F. Kamara, R. Zachariah, A. M. V. Kumar, and K. Selvaraj. 2022. "Bacterial Isolates and Antibiotic Resistance of *Escherichia coli* Isolated from Fresh Poultry Excreta Used for Vegetable Farming in Freetown, Sierra Leone." *Int J Environ Res Public Health* 19 (9). https://doi.org/10.3390/ijerph19095405.

Marchewka, J., P. Sztandarski, Ż. Zdanowska-Sąsiadek, K. Damaziak, F. Wojciechowski, A. B. Riber, and S. Gunnarsson. 2020. "Associations Between Welfare and Ranging Profile in Free-Range Commercial and Heritage Meat-Purpose Chickens (*Gallus gallus* domesticus)." *Poult Sci* 99 (9):4141–4152. https://doi.org/10.1016/j.psj.2020.05.044.

Mbuthia, P. G., H. Christensen, M. Boye, K. M. D. Petersen, M. Bisgaard, P. N. Nyaga, and J. E. Olsen. 2001. "Specific Detection of Pasteurella multocida in Chickens with Fowl Cholera and in Pig Lung Tissues Using Fluorescent rRNA In Situ Hybridization." *J Clin Microbiol* 39 (7):2627–2633. https://doi.org/10.1128/jcm.39.7.2627–2633.2001.

Mensah, G. I., V. Y. Adjei, E. K. Vicar, P. S. Atsu, D. L. Blavo, S. A. M. Johnson, and K. K. Addo. 2022. "Safety of Retailed Poultry: Analysis of Antibiotic Resistance in *Escherichia coli* From Raw Chicken and Poultry Fecal Matter From Selected Farms and Retail Outlets in Accra, Ghana." *Microbiol Insights* 15. https://doi.org/10.1177/11786361221093278.

Montoro-Dasi, L., A. Villagra, S. Sevilla-Navarro, M. Pérez-Gracia, S. Vega, and C. Marin. 2020. "The Dynamic of Antibiotic Resistance in Commensal *Escherichia coli* Throughout the Growing Period in Broiler Chickens: Fast-Growing vs. Slow-Growing Breeds." *Poult Sci* 99 (3):1591–1597. https://doi.org/10.1016/j.psj.2019.10.080.

Morrissey, K. L. H., S. Brocklehurst, L. Baker, T. M. Widowski, and V. Sandilands. 2016. "Can Non-Beak Treated Hens be Kept in Commercial Furnished Cages? Exploring the Effects of Strain and Extra Environmental Enrichment on Behaviour, Feather Cover, and Mortality." *Animals (Basel)* 6 (3):17. https://doi.org/10.3390/ani6030017.

Nawaz, A. H., K. Amoah, Q. Y. Leng, J. H. Zheng, W. L. Zhang, and L. Zhang. 2021. "Poultry Response to Heat Stress: Its Physiological, Metabolic, and Genetic Implications on Meat

Production and Quality Including Strategies to Improve Broiler Production in a Warming World." *Front Vet Sci* 8:699081. https://doi.org/10.3389/fvets.2021.699081.

Nguyen, T. K., L. T. Nguyen, T. T. H. Chau, T. T. Nguyen, B. N. Tran, T. Taniguchi, H. Hayashidani, and K. T. L. Ly. 2021. "Prevalence and Antibiotic Resistance of *Salmonella* Isolated from Poultry and Its Environment in the Mekong Delta, Vietnam." *Vet World* 14 (12):3216–3223. https://doi.org/10.14202/vetworld.2021.3216-3223.

Nsengimana, O., G. Habarugira, L. Ojok, D. Ruhagazi, A. Kayitare, and A. Shyaka. 2022. "Infectious Coryza in a Grey Crowned Crane (Balearica regulorum) Recovered from Captivity." *Vet Med Sci* 8 (2):822–826. https://doi.org/10.1002/vms3.766.

Oliveira Gda, S., S. T. Nascimento, V. M. dos Santos, and B. S. Lima Dallago. 2021. "Spraying Hatching Eggs with Clove Essential Oil Does Not Compromise the Quality of Embryos and One-Day-Old Chicks or Broiler Performance." *Animals (Basel)* 11 (7):2045. https://doi.org/10.3390/ani11072045.

Pandey, U. 2019. "Effect of Lighting in Broiler Production." *Acta Sci Agric* 3 (6):114–116.

Parada, R., J. Książkiewicz, M. Kawka, and K. Jaszczak. 2012. "Studies on Resources of Genetic Diversity in Conservative Flocks of Geese Using Microsatellite DNA Polymorphic Markers." *Mol Biol Rep* 39 (5):5291–5297. https://doi.org/10.1007/s11033-011-1327-8.

Pessato, A., A. E. McKechnie, and M. M. Mariette. 2022. "A Prenatal Acoustic Signal of Heat Affects Thermoregulation Capacities at Adulthood in an Arid-Adapted Bird." *Sci Rep* 12. https://doi.org/10.1038/s41598-022-09761-1.

Poultry. 2013a. *Encyclop}dia Britannica. Encyclop}dia Britannica Student and Home Edition*. Chicago: Encyclop{\ae}dia Britannica.

Poultry. 2013b. *Encyclopedia Britannica. Encyclopedia Britannica Student and Home Edition*. Chicago: Encyclopedia Britannica.

Qu, X., X. Li, Z. Li, M. Liao, and M. Dai. 2022. "Chicken Peripheral Blood Mononuclear Cells Response to Avian Leukosis Virus Subgroup J Infection Assessed by Single-Cell RNA Sequencing." *Front Microbiol* 13. https://doi.org/10.3389/fmicb.2022.800618.

Rauw, F., Y. Gardin, B. T. van den, and B. Lambrecht. 2009. "La Vaccination Contre la Maladie de Newcastle Chez le Poulet (Gallus gallus)." *Societe et Environnement* 13:587.

Rodrigues da Costa, M., and A. Diana. 2022. "A Systematic Review on the Link Between Animal Welfare and Antimicrobial Use in Captive Animals." *Animals (Basel)* 12 (8). https://doi.org/10.3390/ani12081025.

Rossi, L., S. Berni Canani, L. Censi, L. Gennaro, C. Leclercq, U. Scognamiglio, S. Sette, and A. Ghiselli. 2022. "The 2018 Revision of Italian Dietary Guidelines: Development Process, Novelties, Main Recommendations, and Policy Implications." *Front Nutr* 9. https://doi.org/10.3389/fnut.2022.861526.

Santos, M. N., T. M. Widowski, E. G. Kiarie, M. T. Guerin, A. M. Edwards, and S. Torrey. 2022. "In pursuit of a Better Broiler: Tibial Morphology, Breaking Strength, and Ash Content in Conventional and Slower-Growing Strains of Broiler Chickens." *Poult Sci* 101 (4). https://doi.org/10.1016/j.psj.2022.101755.

Shakir, Z. M., A. O. Alhatami, Y. Ismail Khudhair, and H. Muhsen Abdulwahab. 2021. "Antibiotic Resistance Profile and Multiple Antibiotic Resistance Index of *Campylobacter* Species Isolated from Poultry." *Arch Razi Inst* 76 (6):1677–1686. https://doi.org/10.22092/ari.2021.356400.1837.

Sharaf, M., A. Amara, A. Aboul-Enein, S. Helmi, A. Ballot, and P. Schnitzler. 2013. "Antiherpetic Efficacy of Aqueous Extracts of the Cyanobacterium *Arthrospira fusiformis* from Chad." *Pharmazie* 68 (5):376–380. https://doi.org/10.1691/ph.2013.2807.

Sheweita, S. A., A. A. Amara, H. Gamal, A. A. Ghazy, A. Hussein, and M. Bahey-el-din. 2022. "Bacterial Ghosts of Pseudomonas aeruginosa as a Promising Candidate Vaccine and Its Application in Diabetic Rats." *Vaccines* 10 (6). https://doi.org/10.3390/vaccines10060910.

Shi, H., B. Li, Q. Tong, W. Zheng, D. Zeng, and G. Feng. 2019. "Effects of LED Light Color and Intensity on Feather Pecking and Fear Responses of Layer Breeders in Natural Mating Colony Cages." *Animals (Basel)* 9 (10):814. https://doi.org/10.3390/ani9100814.

Stachyra, A., P. Redkiewicz, P. Kosson, A. Protasiuk, A. Góra-Sochacka, G. Kudla, and A. Sirko. 2016. "Codon Optimization of Antigen Coding Sequences Improves the Immune Potential of DNA Vaccines Against Avian Influenza Virus H5N1 in Mice and Chickens." *Virol J* 13 (1):143. https://doi.org/10.1186/s12985-016-0599-y.

Sultana, H., D. W. Seo, M. S. A. Bhuiyan, N. R. Choi, M. R. Hoque, K. N. Heo, and J. H. Lee. 2016. "Genetic Diversity and Phylogenetic Analysis of South-East Asian Duck Populations Based on the mtDNA D-loop Sequences." *Asian-Australas J Anim Sci* 29 (12):1688–1695. https://doi.org/10.5713/ajas.15.1017.

Suresh, G., H. Pakdel, T. Rouissi, S. K. Brar, M. Diarra, and C. Roy. 2020. "Evaluation of Pyroligneous Acid as a Therapeutic Agent Against Salmonella in a Simulated Gastrointestinal Tract of Poultry." *Braz J Microbiol* 51 (3):1309–1316. https://doi.org/10.1007/s42770-020-00294-1.

Tatár-Kis, T., E. A. Fischer, C. Cazaban, E. Walkó-Kovács, Z. G. Homonnay, F. C. Velkers, V. Palya, and J. A. Stegeman. 2020. "A Herpesvirus of Turkey-Based Vector Vaccine Reduces Transmission of Newcastle Disease Virus in Commercial Broiler Chickens with Maternally Derived Antibodies." *Vaccines (Basel)* 8 (4):614. https://doi.org/10.3390/vaccines8040614.

Tyzzer, E. E. 1919. "Developmental Phases of the Protozoon of "Blackhead" in Turkeys " *J Med Res* 40 (1):1–30.3.

Uyanga, V. A., Q. Xin, M. Sun, J. Zhao, X. Wang, H. Jiao, O. M. Onagbesan, and H. Lin. 2022. "Research Note: Effects of Dietary L-Arginine on the Production Performance and Gene Expression of Reproductive Hormones in Laying Hens Fed Low Crude Protein Diets." *Poult Sci* 101 (5). https://doi.org/10.1016/j.psj.2022.101816.

van der Wolf, P., M. Meijerink, E. Libbrecht, G. Tacken, E. Gijsen, K. Lillie-Jaschniski, and V. Schüller. 2021. "*Salmonella Typhimurium* Environmental Reduction in a Farrow-to-Finish Pig Herd Using a Live Attenuated Salmonella Typhimurium Vaccine." *Porcine Health Manag* 7:43. https://doi.org/10.1186/s40813-021-00222-1.

van Eerden, E., R. R. Santos, F. Molist, M. Dardi, L. A. Pantoja-Millas, J. Molist-Badiola, M. Baratelli, and M. Pages. 2022. "Efficacy of an Attenuated Vaccine Against Avian Coccidiosis in Combination With Feed Additives Based on Organic Acids and Essential Oils on Production Performance and Intestinal Lesions in Broilers Experimentally Challenged With Necrotic Enteritis." *Poult Sci* 101 (6). https://doi.org/10.1016/j.psj.2022.101848.

Wang, Z., X. Zhao, Y. Wang, C. Sun, M. Sun, X. Gao, F. Jia, C. Shan, G. Yang, J. Wang, H. Huang, C. Shi, W. Yang, A. Qian, C. Wang, and Y. Jiang. 2021. "*In vivo* Production of HN Protein Increases the Protection Rates of a Minicircle DNA Vaccine against Genotype VII Newcastle Disease Virus." *Vaccines (Basel)* 9 (7):723. https://doi.org/10.3390/vaccines9070723.

Wilde, S., Y. Jiang, A. M. Tafoya, J. Horsman, M. Yousif, L. A. Vazquez, and K. L. Roland. 2019. "Salmonella-Vectored Vaccine Delivering Three Clostridium Perfringens Antigens Protects Poultry Against Necrotic Enteritis." *PLoS One* 14 (2):e0197721. https://doi.org/10.1371/journal.pone.0197721.

Xu, Q., H. Li, W. Zhou, X. Zou, and X. Dong. 2020. "Age-Related Changes in Serum Lipid Levels, Hepatic Morphology, Antioxidant Status, Lipid Metabolism Related Gene Expression and Enzyme Activities of Domestic Pigeon Squabs (Columba livia)." *Animals (Basel)* 10 (7):1121. https://doi.org/10.3390/ani10071121.

Yan, W., Q. Zhou, Z. Yuan, L. Fu, C. Wen, N. Yang, and C. Sun. 2021. "Impact of the Gut Microecology on Campylobacter Presence Revealed by Comparisons of the Gut Microbiota from Chickens Raised on Litter or in Individual Cages." *BMC Microbiol* 21 (1):290. https://doi.org/10.1186/s12866-021-02353-5.

Ye, M., B. Zhou, S. Wei, M. Ding, X. Lu, X. Shi, J. Ding, S. Yang, and W. Wei. 2016. "Transcriptomic Analysis Identifies Candidate Genes Related to Intramuscular Fat Deposition and Fatty Acid Composition in the Breast Muscle of Squabs (Columba)." *G3 (Bethesda)* 6 (7):2081–2090. https://doi.org/10.1534/g3.116.029793.

Yegoraw, A. A., A. M. Assen, P. F. Gerber, and S. W. Walkden-Brown. 2021. "Transmission of Infectious Laryngotracheitis Virus Vaccine and Field Strains: The Role of Degree of Contact and Transmission by Whole Blood, Plasma and Poultry Dust." *Vet Res* 52 (1):91. https://doi.org/10.1186/s13567-021-00959-1.

Yin, Z., W. Zhou, H. Mao, X. Dong, X. Huang, H. Zhang, and H. Liu. 2022. "Identification of Genes Related to Squab Muscle Growth and Lipid Metabolism from Transcriptome Profiles of Breast Muscle and Liver in Domestic Pigeon (Columba livia)." *Animals (Basel)* 12 (9). https://doi.org/10.3390/ani12091061.

Zaky, A. A., J. H. Shim, and A. M. Abd El-Aty. 2021. "A Review on Extraction, Characterization, and Applications of Bioactive Peptides From Pressed Black Cumin Seed Cake." *Front Nutr* 8:743909. https://doi.org/10.3389/fnut.2021.743909.

Zeng, X., F. Xu, and J. Lin. 2010. "Development and Evaluation of CmeC Subunit Vaccine against Campylobacter jejuni." *J Vaccines Vaccin* 1 (3):112. https://doi.org/10.4172/2157–7560.1000112.

Zhan, F., L. Sun, G. Zhao, M. Li, and C. Zhu. 2022. "Multiple Technologies Combined to Analyze the Changes of Odor and Taste in Daokou Braised Chicken during Processing." *Foods* 11 (7). https://doi.org/10.3390/foods11070963.

Zhao, Y., A. J. A. Aarnink, R. Dijkman, T. Fabri, M. C. M. de Jong, and P. W. G. Groot Koerkamp. 2012. "Effects of Temperature, Relative Humidity, Absolute Humidity, and Evaporation Potential on Survival of Airborne Gumboro Vaccine Virus." *Appl Environ Microbiol* 78 (4):1048–1054. https://doi.org/10.1128/aem.06477–11.

6 Perspectives of Antimicrobials

Poultry and Fiber Production

K.R. Preethy, M. Chamundeeswari, K. I. Leroy, and S. Sharmila

6.1 INTRODUCTION

6.1.1 Antimicrobials

An antimicrobial substance is used to destruct or inhibit the growth of microorganisms. Some of the classifications under antimicrobial substances include antibiotic, antiviral, antimycotic, sporicide, germicide, antiparasitic, and antifungal agents. Antimicrobial agents can also be classified based on their function. Substances that inhibit the growth of bacteria reversibly are known as bacteriostatic. Similarly, the agents that have deadly irreversible action on bacteria are known as bactericidal. Uses of antimicrobial agents include treatment of diseases, disinfection of skin, sterilization of ancillary equipment, and so on. Examples of antimicrobial agents include aminoglycosides, alcohols, penicillins, quinolones, macrolides, etc. Some antimicrobials are effective against many microorganisms (broad spectrum), whereas others are effective against either molds or yeasts or only bacteria (narrow spectrum) [1,2].

6.1.2 Antimicrobials in Poultry

Spoilage of poultry meat is mainly due to sulfide-like odors caused by bacteria, and the bacteria responsible for such spoilage include *Pseudomonas* spp., *Proteus* spp., *Citrobacter* spp., etc. *Campylobacter* and *Salmonella* are the main microbial species that spoil poultry meat. These are present at high concentrations in the gastrointestinal tracts of birds, but after contamination of poultry meat, even at low-level concentration these microbes must be analyzed [3].

6.1.3 Poultry Pathogens

In many nations, the poultry business has grown to be a significant economic sector. One of the main challenges limiting the progress of this industry is the effect of diseases on the production of poultry. The poultry sector is being compelled to create cutting-edge antimicrobial techniques as a result of the increasing prevalence

DOI: 10.1201/9781003269847-6

of bacterial diseases and the rise of drug-resistant pathogens. Several organizations are actively looking into the possibility of employing bacteriocins to prevent the bacterial infections that harm the chicken industry and result in significant economic losses. Two significant pathogens in the poultry business, *C. perfringens* and *E. coli*, have been the subjects of the majority of studies on this topic. The bacterium *C. perfringens*, a Gram-positive spore-forming anaerobic, colonizes the chickens' intestines and is the cause of necrotic enteritis. The primary disease of concern for poultry producers globally is *C. perfringens* because of production losses and animal death. Antimicrobial substances that might hinder or kill this bacterial species may help prevent potential food safety issues because it is also a foodborne disease linked to chicken [4].

Six *Enterococcus faecalis* strains that inhibit *C. perfringens* were found in pig feces after they searched for bacteria with antagonistic activity against *C. perfringens*. They proposed that these bacteria or their antibacterial components may be helpful substitutes for antibiotics in the chicken industry, despite the fact that they did not purify the bacteriocin-like inhibitory chemicals. It has been noted that a number of *Bacillus* species, including *Bacillus cereus, Bacillus subtilis, Bacillus pumilus,* and *Bacillus licheniformis,* exert antagonistic effects on *C. perfringens* via producing bacteriocins. A recent study by Udompijitkul et al. showed that nisin exerts strong antimicrobial activity against *C. perfringens* in vitro. In addition, Ben Lagha et al. and Vet Res Jozefak et al. reported that dietary nisin significantly and dose-dependently increases feed conversion by and the growth performance of broiler chickens. Like salinomycin, a widely used ionophore coccidiostat, nisin exerts a modulating effect on the microbiota of the gastrointestinal tract by decreasing counts of Bacteroides and Enterobacteriaceae. It thus represents an effective dietary supplement for broiler chickens. Similar effects were reported for other bacteriocins, including diversion produced by *Carnobacterium divergens* and albusin B produced by *Ruminococcus albus* [4].

Gram-negative *E. coli* is a typical component of the gastrointestinal microflora in chicken. However, some strains have the potential to cause serious conditions such as chicken colibacillosis. When several avian pathogenic *E. coli* enter the circulation via the respiratory or digestive tract, systemic infections can result, especially if the animal is under stress or has a weak immune system. Two isolates (*Lactobacillus fermentum* and *Lactobacillus rhamnosus*) of lactic acid bacteria from chicken intestine samples were discovered to have the ability to inhibit the growth of *E. coli* in vitro. These isolates may create bacteriocins because of the nature of their inhibitory compounds. In an experimental *E. coli* infection of broiler chickens, Ogunbanwo et al. looked into the potential therapeutic efficacy of bacteriocin and bacteriocin-producing *Lactobacillus plantarum* strain [4]. Refer Table 6.1 to know about the comparison of bacterial counts at spoilage times after storage.

6.1.4 Types of Antimicrobials in Poultry

The types of antimicrobials used in poultry are β-lactam, polypeptides, aminoglycosides, macrolides, florfenicol, tetracyclines, sulfonamides, quinolones, and fluoroquinolones. These examples are discussed in detail in the following topics.

Maximum residue limit (MRL) is defined as the highest concentration of a metabolite that results by using medicinal products that may be legally permitted or acceptable in or on a food. Refer Table 6.2 for MRL for antimicrobials in poultry products.

6.1.4.1 Tetracycline

Tetracycline (Figure 6.1) is also a broad-spectrum bacteriostatic agent that works by inhibiting protein synthesis. Tetracyclines have activity against Gram-positive

TABLE 6.1
Comparison of Bacterial Counts at Spoilage Times After Storage [5]

Comparison of bacterial counts at spoilage time after storage under air or under two different modified atmosphere packaging (MAPs).

Bacterial Counts (log CFU/g) When Spoilage Was Reached

	Air	Air	30% CO_2 – 70% N_2	30% CO_2 – 70% N_2	70% CO_2 – 30% N_2
Lactic acid bacteria	2.91	3.2	6.88	4.1	6.89
Pseudomonas	6.28	6.2	6.83	6.1	6.71
B. thermosphacta	6.9	ND_3	6.39	ND	7.21
Enterobacteriaceae	6.15	5.8	6.42	5.8	6.71

TABLE 6.2
Maximum Residue Limits (MRLs) for Antimicrobials in Poultry Products [5]

Substance	Matrix	MRL, mg/kg
Tetracycline	Muscle	100
	Liver	300
	Kidney	600
	Eggs	200
Streptomycin	Muscle	600
	Fat	600
	Liver	600
	Kidney	1,000
Tilmicosin	Muscle	150
	Liver	2,400
	Kidney	600
	Skin and fat	250
Florfenicol	Muscle	100
	Skin and fat	20
	Liver	2,500
	Kidney	750
Tiamulin	Muscle	100
	Liver	1,000
	Skin and fat	100
	Eggs	1,000

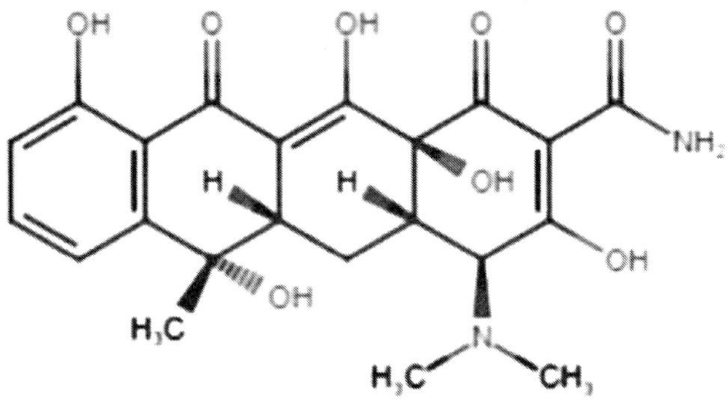

FIGURE 6.1 Structure of tetracycline.

and Gram-negative bacteria, including some anaerobes. Tetracycline's activity range affects specific bacteria such as *E. coli*, *Klebsiella* spp., *Pasteurella* spp., and *Streptococcus* spp. Tetracycline is given by means of diet. It can be added to the flock's drinking water or feed [6].

Storage/Stability: Tetracycline oral tablets and capsules should be stored in tight, light-resistant containers at room temperature (15°C to 30°C).
Dosage: 40 to 200 mg/L drinking water or 100 to 600 mg/kg feed

6.1.4.2 Streptomycin

Streptomycin is active against mycobacteria, leptospira, *Francisella tularensis, Yersinia pestis,* but only some mycoplasma, Gram-negative organisms, and *Staphylococcus* spp.

Absorption: Streptomycin is rapidly and well absorbed in intramuscular (IM) and subcutaneous (SC) regions of administration. When administered orally, it is poorly absorbed by adult birds.
Dosage: 25 to 50 mg/kg PO, IM q12–24h [7].

6.1.4.3 Tilmicosin

Tilmicosin is a semisynthetic macrolide antibiotic that was synthesized from tylosin. It has an antibacterial spectrum with enhanced activity against *Pasteurella* spp. It has also shown effectiveness against *Mycoplasma* spp.

Storage/Stability: Stored at or below room temperature. The drug should not be stored under direct sunlight.
Limitations: The drug must be handled with extreme caution, as it is potentially fatal to humans [8].

6.1.4.4 Florfenicol

Florfenicol (Figure 6.2) is a broad-spectrum antibiotic similar to chloramphenicol, which helps in inhibition of protein synthesis. Florfenicol is active against

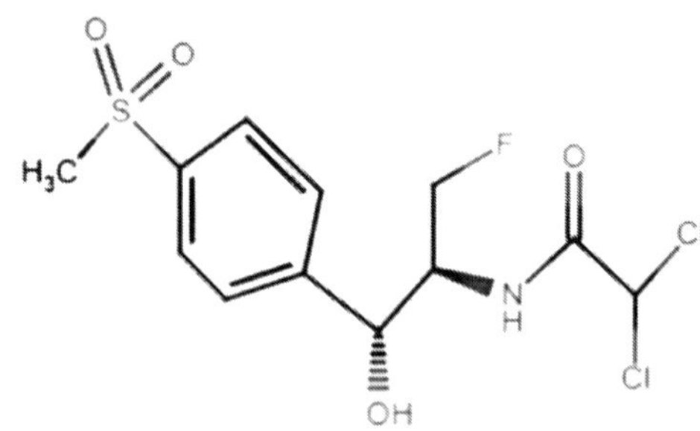

FIGURE 6.2 Structure of florfenicol.

both Gram-negative and Gram-positive organisms such as *E. coli, Salmonella typhimurium*, and *Staphylococcus aureus* [9].

Uses: Helps in treatment of respiratory problems.
Limitations: Florfenicol causes kidney damage to broilers.
Dosage: 15 to 30 mg/kg, given PO or IM.

6.2 ENHANCEMENT OF POULTRY PRODUCTION

Poultry chicken's growth was steadily affected by administering a small dose of the drug streptomycin. Therefore, it paved way for the usage of these drugs, such as chlortetracycline, doxycycline, and sulfonamides, which helped growth promotion in chicken. These drugs were approved as growth promoters for animals in 1951 by the US Food and Drug Administration. In the early 1990s, more than 50% of poultry were given doses of antimicrobial drugs. they were first administered at subtherapeutic levels for improving animal production in the United States during the 1990s. Supplement of antimicrobials (tiamulin, nosiheptide, salinomycin, and tylosin) could also enhance the quality of meat in poultry animals [10].

6.3 USE OF ANTIBIOTICS IN POULTRY

Antibiotics in poultry are generally administered for the treatment and prevention of diseases and as growth promoters. Use of antimicrobials for the prevention of diseases is allowed in all large poultry producing nations across the globe. The diseases (colibacillosis, necrotic enteritis, etc.) generally caused by *Salmonella, E. coli,* or *Clostridium* spp. in intestinal regions were treated by using antimicrobial agents. These infections were a major concern to poultry producing sectors by badly affecting the economy of those nations. Depending on the country's economy, the usage of different antibiotics varies with respect to development of industry, animal husbandry, and the type of species. The volumes of antibiotic administered vary depending on the stage of poultry production and the percentage at which it gets affected by the disease. Generally, there are different methods of tracking antibiotic usage: one common method is to follow antibiotic sales, finding out the designated antibiotic for the particular animal species, and to identify the usage of antibiotics in farm animals. A total of 106 metric tonnes of active antibiotics were sold for use in chicken in 2016. In 2016, tetracyclines, penicillins, polymyxins, and then sulfonamides and trimethoprim were primarily used to treat poultry herds. In major nations involved in a large amount of poultry production, currently there are no quantitative monitoring statistics on antibiotics use in broilers. Active antibiotic compounds that may be used in feed, water, or parenterally were examined in national listings of all pharmaceuticals that are permitted for use in poultry in the United States, Brazil, China, Poland, United Kingdom, Germany, France, and Spain [3]. Due to the lack of effective alternatives for the treatment of bacterial infections, the World Health Organization classifies fluoroquinolones, third- and fourth-generation cephalosporins, macrolides, glycopeptides, and polymyxins as "highest priority critically important" antibiotics for human medicine. These antibiotics were proven to be effective against severe human infections. Enrofloxacin use in poultry was prohibited in the United States by the FDA in 2005.

The use of antibiotics along with stringent biosecurity and hygiene standards over the past 50 years have helped the poultry industry to expand by preventing the negative impacts of many avian diseases. Vaccination can also be used as an additional measure when biosecurity remains efficient. The immune system is protected by a vaccination from specific pathogens, such as viruses or bacteria, to which it may be subjected to in the future. From nation to nation and farm to farm, there are differences in vaccination practices and vaccine types. The species, area, number of workers, kind of production, and production cycle are just a few of the parameters that can affect the method of choice for vaccination. The general health of the poultry, maternal immunity, and the cost of the vaccine all have a role in the process of immunization. Many nations, including Belgium, the Netherlands, and Germany, require that livestock be vaccinated against specific diseases (such as Newcastle disease), while others, like France, only long-lived fowl (laying and breeding) are immunized [11].

Antibiotics are ineffective against infections caused by viruses and fungi. They exclusively deal with infectious conditions where bacterial causal agents are present. In general, antibiotics are employed in veterinary or human medicine, fish farming, animal feed, phytosanitary treatments, and as a preventive or curative measure. The chemical characteristics, method of action, and bacterial species that an antibiotic affects are used to categorize them. Antibiotics that are bacteriostatic instead of bacteriocidal weaken bacteria by preventing them from proliferating and making it easier for the immune system to phagocytose them. Animal resistance increases, lowering the mortality rate. Tetracycline, bacitracin, tylosin, salinomycin, virginiamycin, and bambermycin are only a few examples of the antibiotics that are frequently used in industrial chicken raising, particularly in North America [11].

6.4 IMPACTS

6.4.1 Impact on Chicken Growth, Digestive Tract, and Immune System

Antibiotics are used in the poultry business to promote meat production through better feed conversion, accelerated growth, and illness control. In poultry production, antibiotics can be used successfully at subtherapeutic doses to promote growth and protect the health of birds by altering the immune status of broiler chickens. This is mostly because gut flora has been altered, and gastrointestinal infections have been controlled. Although the exact process is yet unknown, antibiotics are expected to work by altering the diversity and relative abundance of bacteria in the intestine to create the best microbiota possible for growth (Dibner and Richards, 2005). For instance, studies using meta-genome sequencing methods have shown that meals containing salinomycin (60 ppm) have an effect on the microbiome dynamics in chicken ceca (Fung et al., 2013). Similar to this, it has been discovered that the proximal ileum of the broiler duodenal loop has more *Lactobacillus* species when virginiamycin (100 ppm) is used as a growth promoter. This demonstrates that virginiamycin modifies the microbiota in chicken guts. Additionally, populations of the bacteriostatic *Lactobacillus* spp. are dramatically reduced in the ileum of hens getting tylosin-containing feed compared to birds receiving no tylosin. Other researchers have shown that the number of *Lactobacillus* species decreases after the administration of antibiotics. For a reminder, *Lactobacillus* are the main commensal bacteria responsible for bile hydrolase salt synthesis. The reduction in the *Lactobacillus* population in animals receiving antibiotics likely decreases the intestinal activity of the bile hydrolase salts, increasing the relative abundance of conjugated bile salts, promoting lipid metabolism and energy harvesting, causing an increase in animal weight gain (Lin et al., 2013). A change in the gut microbiota of chickens can affect their immunity and their health. However, alterations in the intestinal microbiota of chickens can be influenced by numerous causes. These elements include the environment in which animals live, their exposure to pathogens, their diet, and the presence of antibiotics in their food (Lee et al., 2012) [11].

6.4.2 Impact on Meat Quality and Eggs

Foodborne diarrheal illnesses in humans are primarily brought on by *Campylobacter*. When a person has a damaged immune system, if and older adult of a very small child, a campylobacter infection can be serious or even fatal. *E. coli* bacteria are widespread and have the potential to be disease-causing. *E. coli* O157:H7 is the most prevalent strain of *E. coli* infection that sickens individuals. One of the most prevalent and pervasive foodborne infections in the world is salmonellosis. Mild gastroenteritis is frequently brought on by *Salmonella* infections. These bacterial infections affect individuals in tens of millions of cases each year. Non-typhoidal *Salmonella* serovars, Campylobacter *spp.* are responsible for 1.4 million infections, 325,000 hospitalizations, 5,000 fatalities, and 5,000 cases. And, 270,000 cases and nearly 2.4 million cases, respectively, are brought on by pathogenic *E. coli*, including *E. coli* O157:H7. Eggs are usually linked to the spread of *Salmonella* (Singh et al., 2010). The growth of bacteria in the intestines is the primary cause of this contamination. Foods of animal origin are secondarily contaminated by resistant germs along the manufacturing line. According to retail samples had a greater incidence of multidrug-resistant *Salmonella* than samples from slaughterhouses [12].

6.5 ANTIBIOTIC AND BACTERIAL RESISTANCE

According to scientific research, treating animals with antibiotics may cause the animals to develop bacterial resistance. The capacity of germs to multiply when exposed to an antibiotic that typically suppresses or kills microorganisms of the same species is known as antibiotic resistance. Resistance is acquired through gene acquisition or mutation in mobile genetic elements like transposons, integrons, plasmids, or phages. In its digestive tract and in its flesh, chickens have a high number of Enterobacteriaceae that are tetracycline resistant. Several researchers have focused on bacterial resistance to antibiotics. More than 43% of the *Salmonella enterica* isolates in one investigation on samples taken from chicken farms in British Columbia (Canada) were concurrently resistant to ampicillin, amoxicillin-clavulanic acid, ceftiofur, cefoxitin, and ceftriaxone, according to emphasize the occurrence of many *Salmonella* strains that are isolated from broiler farms and are both single- and multidrug-resistant to antibiotics. Additionally, broiler-isolated *Enterococci, Mycoplasma gallisepticum*, and *Salmonella* spp. have all been found to exhibit antibiotic resistance. Chicken meat contains several resistant and multidrug-resistant isolates, according to a German study. According to another Italian study, 86% of *S. enterica* isolates from chicken carcasses were tetracycline resistant, and 30% of isolates exhibited multi-pharmacological phenotypic resistance to ampicillin, sulfamethoxazole, and tetracycline. According to research by, 78% of Ecuador's production birds had tetracycline resistance (broilers and laying hens). Sulfisoxazole and trimethoprim-sulfamethoxazole resistance affected more than half of the isolates (69% and 63%, respectively). Antibiotic resistance in bacteria is a problem for public health. For instance, poultry meat may contribute to human infections in Canada also discovered that *S. enterica* isolates from egg and chicken carcasses were resistant to quinolones, sulfisoxazole, streptomycin, penicillins, and sulfisoxazole. *S. enterica* isolates exhibited resistance to at least 21 of the authors' prescribed medications. The majority of isolates had genes known to be linked to human infections, including the *spv* operon, SPI-1, and SPI–2. The safety of people is at danger because of this. The overuse of specific antibiotics such as penicillins, tetracyclines, macrolides, and aminoglycosides is mostly to blame for this condition. Antibiotic abuse and the selection pressure that has resulted from it have reduced therapeutic efficacy and increased populations of bacteria that are resistant to antibiotics. Despite the cessation of antibiotic use, antibiotic resistance may still advance over time. Indeed, it has been demonstrated that *E. coli* strains resistant to trimethoprim and streptomycin can survive for a number of weeks in a chicken farm without the use of the aforementioned antibiotics. On the other side, organic farms have lower levels of antibiotic resistance. Therefore, it is essential to identify the precise sources and ecology of resistant bacteria in order to devise methods to prevent their spread [13].

6.6 PERSPECTIVES

Alternatives to antibiotics should have the following qualities: (1) be nontoxic or have no negative effects on animals, (2) be simple to remove from the body or able to leave only short-term residues, (3) do not cause bacterial resistance, (4) are stable in feed and animal gastrointestinal tracts, (5) can be easily decomposed and have no negative effects on the environment, (6) do not affect palatability, and (7) do not disrupt the normal intestinal flora of animals. In actuality, there are currently no antibiotic alternatives that satisfy all of the aforementioned criteria. When it comes to the efficiency of preventing sickness and promoting growth, antibiotic alternatives still fall well short of antibiotics. Only a few bacterial infectious diseases may currently be controlled by vaccines, and antibacterial vaccines are often used to prevent bacterial infections. Animal health is mostly preserved via immunomodulators and feed enzymes, which do not directly harm or inhibit bacteria. Currently, bacteriophages are exclusively utilized in food, and their safety is still debatable. Plant extracts and probiotics have complicated chemical makeups and poor stability, which can have a variety of effects and safety issues [6,14].

Antimicrobial peptides can treat bacterial infections, but their high price and limited antibacterial spectrum prevent them from being widely used. They can also cause bacterial resistance. However, compared to all other antibiotic alternatives, antibiotics have a higher ability to directly inhibit or kill germs. Additionally, antibiotics are produced using a single, relatively pure active ingredient, and good manufacturing practices ensure high levels of stability, consistency, and quality. Humans have not yet discovered a method that is more clinically effective than choosing the right medications to treat the targeted bacterium. The main justification behind the European Union's restriction on low-dose antibiotics as feed additives is antimicrobial resistance. Penicillium, the bacterium that makes penicillin, has actually coexisted with other bacteria for tens of thousands of years. Although antibiotic alternatives have not yet been linked to bacterial resistance, with so many of them in use, germs may eventually change to develop resistance. Drug pressure is proportionate to the likelihood of drug resistance developing and spreading. The irregular and unlawful use of antibiotic substitutes may have the same detrimental effects as the conventional usage of antibiotics. Meanwhile, we must remember that "prevention is better than cure." Because of unfavorable farming conditions and high disease incidence in many developing nations, antibiotics continue to be a useful tool in the prevention and management of animal diseases [15]. Without a significant change in the production environment, the prevalence of bacterial illnesses in the target animals will probably rise as the amount of "old" antibiotics used in feed decreases. This might result in more advanced antibiotics being used therapeutically, which might have unexpected consequences that present new problems for public health. Furthermore, there is no scientific data that distinguish between the use of antibiotics for treatment and prevention in terms of the emergence of resistance. Political and social pressure may cause the implementation of such a policy, but bacteria may not always "listen" to the policymakers, thus it is important to weigh the benefits and risks before doing so. In conclusion, to ensure the long-term sustainable development of animal husbandry, judicious use of antibiotics and ongoing research of antibiotic substitutes are required. For responsible use of antibiotics and the creation of regulations and policies governing their use, we must firmly specify the target animals, the length of the treatment, and the withdrawal time. In order to keep antibiotic resistance and food-chain residues within predetermined safe levels, we must also strengthen the oversight and enforcement of legislation [16].

6.7 ALTERNATIVES

6.7.1 Herbal Alternatives

Spices and herbal extracts are important for enhancing the health and production of bird species. Plant extracts or other active ingredients in bird food may have advantages such as enhancing the synthesis of endogenous digestive enzymes, enhancing appetite and feed intake, boosting immunity, and having antiviral, antibacterial, anthelminthic, and antioxidant effects. The physiological and chemical process of the bird's gut may be impacted by isoprene derivatives, flavonoids, glucosinolates, and other herbal metabolites. There may also be intermediate nutrient metabolism in addition to the stabilizing effect on the intestinal microbiota. Numerous infections pose a hazard to the production of poultry and other animals, resulting in significant monetary losses. The use of herbal feed additives is expanding in the chicken industry due to restrictions on the use of some antibiotics, risky side effects, and cost-effectiveness. There are many uses for herbs, spices, and their botanical extracts. They encourage feed intake, stimulate the immune system, and function as antioxidants in addition to having antibacterial, coccidiostat, and anthelmintic qualities. The most popular herbal feed additions used in chicken feed include nutmeg, cinnamon, cloves, cardamom, coriander, cumin, anise, celery, parsley, fenugreek, capsicum, pepper, horseradish, mustard, ginger, garlic, onion, rosemary, thyme, mint, and turmeric [17].

6.7.2 Essential Oils

The plant essence, benzene, or terpene derivatives derived through water and/or steam distillation are referred to as essential oils. The antibacterial, antiviral, antioxidant, and antiparasitic properties of essential oils make them beneficial. The effectiveness of feed additives has been investigated for many plant sources of essential oils. The addition of essential oil improved the performance of the birds and the qualities of their carcasses. Essential oils have the ability to lower pathogen levels in the stomach, including *Salmonella* and Gram-negative *E. coli* [17].

6.8 ANTIMICROBIALS IN FIBER PRODUCTION

World textile consumption has been increasing for decades. From 1950 to 2008, per capita consumption increased from 3.7 to 10.4 kg, and continued development pushed the demand for natural fibers, including cotton and wool, to 25.4 million tons in 2014. Due to their large surface area and ability to retain moisture, textiles are known to encourage the growth of microorganisms such as bacteria and fungi.

These microbes are found almost everywhere and can multiply rapidly depending on humidity, nutrients, and temperature levels. Some bacterial populations can double every 20 to 30 minutes under ideal conditions (36°C to 40°C, pH 5 to 9). Microbial growth on fabrics causes many undesirable effects, not only on the fabric itself but also on the user. These effects include the development of unpleasant odors, reduced mechanical strength, staining and discoloration, and increased potential for user contamination. Increased public awareness of pathogenic effects has driven intensive research and development in recent years to minimize or even eliminate microbial growth on textiles. Demand for antimicrobial textiles has increased drastically over the last few years. Various approaches have been investigated to impart antimicrobial functionality to textiles, mainly the incorporation of antimicrobial compounds into polymer fibers that can leach out of the polymer matrix or the grafting of specific entities onto the polymer surface or the fiber surface. As for antimicrobial compounds, different types are used, such as quaternary ammonium compounds, triclosan, metal salts, polybiguanides, and even natural polymers [18–20].

6.8.1 Quaternary Ammonium Compounds

Quaternary ammonium compounds (QACs) are a subgroup of linear alkylammonium compounds consisting of a hydrophobic alkyl chain and a hydrophilic counterpart. Compounds with long alkyl chains (12 to 18 carbon atoms) are most commonly used in the textile industry, mainly for cotton, polyester, nylon, and wool. The antibacterial activity of these compounds depends on the length of the alkyl chain, the presence of perfluorinated groups, and the number of cationic ammonium groups in the molecule. These compounds are active against a wide range of microorganisms, such as Gram-positive and Gram-negative bacteria, fungi, and certain types of viruses. Despite the effectiveness of QACs, there are drawbacks. In other words, the lack of physical bonding causes it to leach out of the fiber, resulting in a rapid decrease in concentration within the fiber [18,21] (Table 6.3).

6.8.2 Triclosan

Triclosan, (2,4,4′-Trichloro-2′-hydroxydiphenyl ether) (C12H7C13O2) is an odorless synthetic chlorinated bisphenol that, unlike most cationic biocides, does not ionize in solution and improves endurance. It has a broad spectrum of activity against both Gram-negative and Gram-positive bacteria but also exhibits some antifungal and antiviral properties. This biocide works primarily by inhibiting the biosynthesis of lipids such as phospholipids, lipopolysaccharides, and lipoproteins, thereby affecting cell membrane integrity. Triclosan is mainly used in combination with polyester, nylon, polypropylene, cellulose acetate, and acrylic fibers. However, the recent widespread use of triclosan outside of healthcare is of great concern as it may create selective pressure for certain antibiotic cross-resistant strains and thereby create bacterial resistance to triclosan. In addition, the reported photochemical conversion of triclosan in aqueous solutions to 2,8-dichlorodibenzo-p-dioxin is another major concern due to its toxicity [18,22] (Table 6.3).

6.8.3 Metals and Metallic Salts

Nonessential metals, either in the free state or as compounds, can be highly toxic to most microorganisms at very low concentrations. Due to their biocidal activity, various metal, oxide, or salt compounds, mainly based on silver, have been used as antimicrobial agents for textiles for agriculture, healthcare, and other industries. Metals are often used in the form of salt-based additives due to the very expensive form of the metal. Thus, redox-active essential metals act as catalytic cofactors for various cellular enzymes by generating or catalyzing reactive oxygen species that can induce oxidative stress and damage cellular proteins, lipids, and DNA. The antimicrobial properties of silver particles have long been exploited in textiles in general, but for some biomedical textiles in particular, their broad spectrum of activity in multi-microbial colonization associated with hospital infections is of particular importance. Silver has shown bactericidal activity against a wide range of Gram-positive and Gram-negative bacteria, namely, *Pseudomonas aeruginosa, Staphylococcus aureus, Staphylococcus epidermidis, E. coli,* and *Klebsiella pneumoniae* [23] (Table 6.3).

6.8.4 Chitosan

Chitosan is a natural hydrophilic copolymer resulting from the deacetylation of chitin extracted from the exoskeletons of crustaceans such as crabs and shrimps, as well as the cell walls of some fungi. It consists of two monomeric units, D-glucosamine and *N*-acetyl-D-glucosamine, linked by a β(1–4)-glycosidic bond. This linear polysaccharide has been extensively studied in textile finishing, mainly for medical applications, for its biocompatibility, nontoxic, noncarcinogenic, and antimicrobial activity. Mainly used for finishing cotton, polyester, and wool fibers, it has an antibacterial effect against a wide variety of microorganisms, including fungi, algae, and some bacteria. Chitosan's antimicrobial mode of action is influenced by intrinsic factors and environmental conditions, namely, chitosan's molecular weight (Mw) and degree of polymerization, its degree of deacetylation, medium pH, and microbial species. Complexes based on chitosan and other more efficient biocides are being investigated to develop more efficient and durable antimicrobial treatments for textiles. The main drawbacks of chitosan as an antimicrobial agent for textiles, either in finishing or incorporation into synthetic fibers, are poor handling and dependence on pH and temperature activity. Chitosan dissolves only in acidic environments and becomes polycationic when the pH is below the pKa of the molecule (6.3 to 6.5), and thus stronger antibacterial activity has been reported in acidic conditions [18,24] (Table 6.3).

6.8.5 Natural-Based Antimicrobial Agents

Due to the emergence of antibiotic resistance in pathogenic bacteria, antimicrobial compounds extracted from herbs and plants have also been extensively investigated as alternative therapeutic strategies to combat microbial growth in textiles. Several herbal compounds with a broad spectrum of activity against various fungal and bacterial pathogens have been identified and are commercially available. A major advantage of using these natural compounds for antimicrobial purposes is that they do not have the side effects that are common with synthetic chemicals. To date, no reports of antimicrobial resistance to these natural chemicals have been published. Despite the lack of research on the mechanisms of their antimicrobial action, microbial resistance is likely prevented by their multiple mechanisms of action, thereby avoiding the selection of resistant strains. In addition, these substances can exhibit efficient antibacterial activity, are safe, are readily available, are nontoxic to the skin, and are environmentally friendly. Amino acids and peptides are also considered promising candidates for antimicrobial fiber applications. These antimicrobial peptides (AMPs) are numerous small proteins that exhibit broad antimicrobial activity against a variety of microorganisms, including both Gram-positive and Gram-negative bacteria. Different families of different AMPs have been found in invertebrates and vertebrates and are characterized by their small size (12 to 50 amino acids), arginine and lysine residues responsible for their positive charge, and their interacting parents characterized by a meridian structure film. They are usually classified according to their size, conformational structure, or primary amino acid structure. Several studies have proposed various hypotheses to explain the mechanism of action of AMP. That is, AMPs can influence several internal cellular processes, from macromolecular synthesis (i.e., RNA, DNA synthesis) to the loss of ATP from actively respiring cells. An effective AMP is L-cysteine, which is used to promote the biofunctionalization of wool and polyamides, giving these fibers a lasting antimicrobial effect [18,20,25].

6.8.6 Poly(Hexamethylene Biguanide)

Polybiguanides are polycationic amines composed of repeating cationic biguanide units separated by aliphatic chains. One of the most commonly used biguanide-based polymers is poly(hexamethylene biguanide) (PHMB) ($[C_8H_{17}N_5]n$), in which cationic biguanide groups are distributed between hydrophobic hexamethylene groups. It has a hydrophobic backbone. Both cationic and hydrophobic properties of PHMB suggest that it interacts with microbial cell membranes through electrostatic and hydrophobic interactions. These interactions with membrane phospholipids lead to cell membrane disruption and lethal leakage of cytoplasmic material, whose activity increases on a weight basis with increasing degree of polymerization. PHMB's current commercial applications (Table 6.3) are primarily in the areas of healthcare (mouthwashes and wound dressings), clothing, and household and water treatment textiles [18,26].

6.9 ENVIRONMENTAL AND HEALTH IMPACT OF ANTIMICROBIAL AGENTS

The effectiveness of a biocide outweighs the potential environmental impact and its cost of use. As an environmental benefit, antimicrobial-treated fabrics can extend the wear life of fabrics, reduce the frequency of wash cycles, lower wash temperatures, and reduce detergent consumption in each wash cycle. These active ingredients are effective at low doses and have a long-lasting function, manufacturing, the textile treatment process, and the subsequent use and disposal of the

TABLE 6.3
Action Modes of the Main Antimicrobial Agents, as well as the Main Fibers in Which They Are Used [18]

Biocide	Action Modes	Fibers
QACs	Damage cell membranes	Cotton
	Denature proteins	Polyester
	Inhibit DNA production, avoiding multiplication	Nylon
		Wool
Triclosan	Blocks lipid biosynthesis, affecting the integrity of cell membranes	Polyester
		Nylon
		Polypropylene
		Cellulose acetate
		Acrylic
Metals and metallic salts	Generate reactive oxygen species, damaging cellular proteins, lipids, and DNA	Cotton
		Wool
		Polyester
		Nylon
Chitosan	Low molecular weight: inhibits synthesis of mRNA, preventing protein synthesis	Cotton
	High molecular weight: causes leakage of intracellular substances or blocks the transport of essential solutes into the cell	Polyester
		Wool
PHMB	Interacts with membrane phospholipids, resulting in its disruption and the lethal leakage of cytoplasmic materials	Cotton
		Polyester
		Nylon

Abbreviations: PHMB, poly(hexamethylene biguanide); QAC, quaternary ammonium compound.

product [18]. When antimicrobial agents are released from fabrics, they can enter the aquatic environment if not removed during wastewater treatment. Several studies have shown effective removal of high levels of silver (including metallic Ag and Ag nanoparticles) and triclosan during wastewater treatment, i.e., 85% to 99% and 90% to 96%, respectively. For example, for QACs, cations tend to adsorb strongly to the negative surfaces of mud, soil, and sediment due to their positive charge. In addition to removal by sedimentation, biodegradation is another possible process that enables the removal of compounds in wastewater and aquatic environments. For example, QAC is commonly referred to as a "hard antimicrobial". This is because it is poorly metabolized and excreted primarily in an unmetabolized form. It is generally believed that QACs are poorly biodegradable. However, significant differences were observed depending on the substance concentration. For example, triclosan is degraded under aerobic conditions but may persist under anaerobic conditions. The photodegradation half-life of triclosan, the main degradation process, is estimated to be about 41 minutes and the half-life in seawater to be about 10 days. Triclosan degradation products, such as methyltriclosan, are potentially more toxic than the parent compound and are a major concern when using this antibacterial agent [9]. By definition, silver-based products as materials of mineral origin are not considered biodegradable. However, all forms of silver are subject to several transformation processes that can lead to their removal from the aquatic environment. Several studies have shown that nanosilver can be immobilized through the formation of stable sulfide complexes, resulting in a highly insoluble, less toxic, and bioavailable sulfide complex than dissolved silver. The formation of silver sulfide therefore significantly reduces the risk of silver-based compounds in the aquatic environment [27,28].

Furthermore, with the introduction of strict environmental legislation, companies are not only concerned with the issue of industrial waste disposal but also with the possibility of replacing traditional processes with new approaches that can offer equal or greater efficiency and reduced environmental impact. The use of plasma technology for pretreatment is increasingly replacing traditional wet chemical applications not only for economic reasons but also for environmental benefits. Dry plasma treatment does not use hazardous chemical solutions (a solvent-free process), does not produce contaminated water, and does not pose a mechanical hazard to the treated fabric. Another eco-friendly approach increasingly being used in textiles is the promotion of certain enzyme-catalyzed reactions. Enzymes most commonly used for polymer modification include, for example, glycosidases, proteases, lipases, and laccases. For example, to obtain wool fibers with antibacterial activity, proteases are used to treat the fibers with antibacterial natural dyes, and the enzymatic process has been shown to increase the amount of natural dyes consumed. Moreover, the ecological treatment of plant fibers by fungi is also an alternative to traditional chemical processes. The fungal treatment creates pits on the fiber surface, roughening the surface and increasing the interfacial adhesion between the fiber and the surrounding matrix. In addition, ultrasonic energy and ultraviolet radiation are also used as environmentally friendly techniques to facilitate textile dyeing where the dyes may have antimicrobial properties. In addition to "green" technologies, researchers are also investigating the use of "green" modified antimicrobial agents as an important resource in the development of sustainable bioactive fibers [29,30].

Antimicrobial resistance mechanisms are obtained either by acquiring genes from resistant microorganisms within the same niche (acquired resistance) or by developing new ways to prevent the entry of these agents (intrinsic resistance). For QACs, several studies have reported bacterial resistance to QACs, either inherent or acquired. Self-resistance is exhibited by Gram-negative bacteria, bacterial spores, mycobacteria, and under certain conditions, staphylococci. This is achieved by reducing cell membrane permeability by altering phospholipid synthesis or binding membrane proteins. For silver compounds, several studies have reported antimicrobial resistance to drugs, and the facile selection of silver resistance in some bacteria could be of benefit in improving surveillance of silver-resistant isolates in clinics, suggesting that better control of antibiotics would be beneficial [18,31].

The development of resistance is a normal evolutionary process of microorganisms. However, this phenomenon is greatly accelerated by the selective pressure exerted by the widespread use of antimicrobials and becomes a major global public health problem of increasing scale. Resistant strains can multiply and spread. Therefore, potential health effects must be considered in evaluating the safety of antimicrobial compounds for animals and humans. Antimicrobial mechanisms of action by diffusion or contact, concentration in fabrics, route of exposure, and frequency of use influence thc degree to which humans are likely to be exposed to fabrics. Some parameters commonly evaluated to assess the risk profile of antimicrobial agents associated with textiles are toxicity (acute and chronic), skin sensitization and irritation, and disruption of the skin ecosystem. All antibacterial agents presented in this review have no conclusions regarding skin toxicity and effects on human health. Therefore, its toxicity is a concern. According to several studies, no significant human health effects have been observed for Si-QAC (silane quaternary ammonium compounds). For example, among metal compounds, ZnPT (zinc pyrithione) has potential neurotoxicity risks. Ag ions, which act nonspecifically like other biocides, interact with skin flora, causing shedding of cell membranes from healthy bacterial cell walls and weakening the skin's protective barrier. In addition, chronic exposure to Ag can deposit Ag metal/Ag sulfide particles on the skin, causing a discoloration known as argyria and even eye discoloration known as platysma. These are not life-threatening conditions, but they are of course cosmetically undesirable [31].

Some changes need to be made to current regulations to affect antimicrobial usage patterns and the market share of specific compounds. Antimicrobial usage patterns in textiles are therefore tailored to specific applications and needs, significantly reducing antimicrobial penetration into the environment. For example, the textile industry uses the highest percentage of all triclosan used but uses QACs at the lowest percentage of all QACs used. Approximately 210 tons of triclosan used in textiles can be replaced with less than 2 tons of silver nanoparticles or 180 tons of Si-QAC to achieve similar antimicrobial efficacy with less environmental impact. Regulatory evaluations of individual antimicrobials should therefore be based on a broader view and knowledge of antimicrobials, considering the risks and benefits of their use compared to alternative antimicrobials [32].

6.10 CONCLUSION

Antimicrobials have been proven to be an effective tool to eliminate or stop the growth of microorganisms. Antimicrobial drugs are used during the production of poultry and fiber for a variety of purposes, such as tools for the prevention and treatment of diseases, as growth-promoting agents, and to boost output volume. Increased body weight and lifespan of poultry animals, disease resistance, and extended shelf life of chicken products are all benefits of employing antimicrobial medications, while the adverse effects include the emergence of microorganisms that are resistant to treatment and cause poor healing. Short-term, full-dose therapeutic usage of antibiotics has less potential for selection than long-term, low-level exposure. So, the minimum quantity of antimicrobials is always preferred by medical experts. Similarly in fiber production, antimicrobials are used to maintain the sustainability of fiber for a longer period of time as well as to eradicate any bad odor due to microbial buildup. Even though use of antimicrobials is advisable for the desired product, a control must be set up to avoid any repercussions or any reversal of damage from antimicrobials. The discovery of antimicrobial medications over the past 60 years, with over 100 substances currently on the market, has undoubtedly contributed significantly to medical science. The presence of nutritious and bioactive components leads to an improvement in intestinal health, which has a positive impact on animal growth performance. Both humans and animals produce more healthy, prolific humans and animals when these medications are used. However, it is likely that the use of these medications and the evolution of antibiotic-resistant bacteria are related, and that increased antibiotic usage is associated with a greater and quicker emergence of antimicrobial-resistant bacterial strains. However, the growing concern over antimicrobial resistance calls for responsible use, stricter regulations, and the adoption of alternative strategies, such as probiotics and improved farm management practices, to ensure long-term sustainability and public health safety. Certain food animals are developing resistance to human antibiotic classes with significant medical importance. There should be increased research and usage of antibiotic alternatives in food-animal production because the use of antibiotics in this industry is a significant contributor to antibiotic resistance [17,18,33,34].

6.11 ACKNOWLEDGMENT

We express our sincere gratitude to the Department of Biotechnology, St. Joseph's College of Engineering, for their motivation and guidance in the making of this chapter. We are deeply obliged to Dr. L. F. A. Anand Raj MSc, MTech, PhD, Professor and Head of the Department of biotechnology for his guidance and concern throughout the bachelor's program. We would like to thank our mentors, Dr. K.R. Preethy, Assistant Professor, and Dr. M. Chamundeeswari, Associate Professor, Department of Biotechnology, for their valuable support in the making of this book chapter. We also extend our gratitude to Taylor & Francis Group for the acceptance of this chapter.

REFERENCES

1) Mohammad Asif. 2017. Antimicrobial Agents. *Journal of Analytical & Pharmaceutical Research* 4(3)
2) Richard J Fair and Yitzhak Tor. 2014. Antibiotics and Bacterial Resistance in the 21st Century. *Perspectives in Medicinal Chemistry* 6: 25–64.
3) Nataliya Roth, Annemarie Kasbohrer, Sigrid Mayrhofer, Ulrike Zitz, Charles Hofacre and Konrad J Domig. 2019. The Application of Antibiotics in Broiler Production and the Resulting Antibiotic Resistance in *Escherichia coli*: A Global Overview. *Poultry Science* 98(4): 1791–1804.
4) Amel Ben Lagha, Bruno Haas, Marcelo Gottschalk and Daniel Grenier. 2017. Antimicrobial Potential of Bacteriocins in Poultry and Swine Production. *Veterinary Research* 48:22
5) Amélie Rouger, Odile Tresse and Monique Zagorec. 2017. Bacterial Contaminants of Poultry Meat: Sources, Species, and Dynamics. *Microorganisms* 5(3):50
6) Poultry DVM – Tetracycline. http://www.poultrydvm.com/drugs/tetracycline
7) Poultry DVM – Streptomycin. http://www.poultrydvm.com/drugs/streptomycin
8) Poultry DVM – Tilmicosin. http://www.poultrydvm.com/drugs/tilmicosin
9) Poultry DVM – Florfenicol. Florfenicol for Chickens and Ducks (poultrydvm.com)
10) Haihong Hao, Guyue Cheng, Zahid Iqbal, Hafiz I. Hussain, Lingli Huang and Menghong Dai. 2014. Benefits and Risks of Antimicrobial Use in Food-Producing Animals. *Frontiers in Microbiology* 5: 288

11) Youcef Mehdiab, Marie-Pierre Létourneau-Montminy, Marie Lou Gaucher, Younes Chorfi, Gayatri Suresh et al. 2018. Use of Antibiotics in Broiler Production: Global Impacts and Alternatives. *Animal Nutrition* 4(2): 170–178
12) J. A. Patterson and K. M. Burkholder. 2003. Application of Prebiotics and Probiotics in Poultry Production. *Poultry Science* 82: 627–631.
13) Anil Kumar Garg and Ranjan Kumar Mohanta. 2012. Antibiotics Use In Animal Husbandry Sector: What Has To Be Done?. *Agriculture Today* 15 https://www.researchgate.net/publication/275019219 Antibiotics use in animal husbandry sector what has to be done
14) Thi Thu HaoVan, Zuwera Yidana, Peter M. Smooker and Peter J. Coloe. 2020. Antibiotic Use in Food Animals Worldwide, With a Focus on Africa: Pluses and Minuses. *Journal of Global Antimicrobial Resistance* 20: 170–177
15) E. R. Manson. 1968. *Antibiotics In Animal Feeds.* A.V.A Conference Paper
16) Heather K. Allen, Uri Y. Levine, Torey Looft, Meggan Bandrick and Thomas A. Casey. 2013. Treatment, Promotion, Commotion: Antibiotic Alternatives in Food-Producing Animals. *Trends in Microbiology* 21(3): 114–119
17) Mohamed E. Abd El-Hack, Mohamed T. El-Saadony, Heba M. Salem, Amira M. El-Tahan *et al.* 2022. Alternatives to Antibiotics for Organic Poultry Production: Types, Modes of Action and Impacts on Bird's Health and Production. *Poultry Science* 101(4)
18) Diana Santos Morais, Rui Miranda Guedes and Maria Ascensao Lopes. 2016. Antimicrobial Approaches for Textiles: From Research to Market. *Materials (Basel)* 9(6): 498.
19) L. Windler, M. Height and B Nowack. 2013. Comparative Evaluation of Antimicrobials for Textile Applications. *Environment International* 53: 62–73.
20) M. Shahid and F. Mohammad. 2013. Perspectives for Natural Product Based Agents Derived from Industrial Plants in Textile Applications—A Review. *Journal of Cleaner Production* 57: 2–18.
21) Y. A. Son and G. Sun. 2003. Durable Antimicrobial Nylon 66 Fabrics: Ionic Interactions With Quaternary Ammonium Salts. *Journal of Applied Polymer Science* 90: 2194–2199.
22) Mehmet Orhana, Dilek Kut and Cem Gunesoglu. 2007. Use of Triclosan as Antibacterial Agent in Textiles. *Indian Journal of Fibre & Textile Research* 32: 114–117
23) M. Ghoranneviss and S. Shahidi. 2013. Effect of Various Metallic Salts on Antibacterial Activity and Physical Properties of Cotton Fabrics. *Journal of Industrial Textiles* 42: 193–203.
24) X. Wang, Y. Du, L. Fan, H. Liu and Y. Hu. 2005. Chitosan-Metal Complexes as Antimicrobial Agent: Synthesis, Characterization and Structure-Activity Study. *Polymer Bulletin* 55: 105–113.
25) Vinay G. Nadiger and Sanjeev R. Shukla. 2015. Antimicrobial Activity of Silk Treated With Aloe-Vera. *Fibers and Polymers* 16: 1012–1019
26) A. Varesano, C. Vineis, A. Aluigi and F. Rombaldoni. 2011. Antimicrobial Polymers for Textile Products. *Science Against Microbial Pathogens: Communicating Current Research and Technological Advance* 3: 99–110
27) Vinay G. Nadiger and Sanjeev R. Shukla. 2016. Antibacterial Properties of Silk Fabric Treated with Silver Nanoparticles. *The Journal of The Textile Institute* 107(12): 1543–1553
28) S. M. Gawish, H. M. Mashaly, A. M. Ramadan and R. Farouk. 2017. Effect of Mordant on UV Protection and Antimicrobial Activity of Cotton, Wool, Silk and Nylon Fabrics Dyed with Some Natural Dyes. *Journal of Nanomedicine & Nanotechnology* 8: 421
29) Slater K. 2003. *Environmental Impact of Textiles: Production, Processes and Protection.* Vol. 27. Woodhead Publishing, Cambridge, UK.

30) Kan C.-W. 2014. *A Novel Green Treatment for Textiles: Plasma Treatment as a Sustainable Technology.* CRC Press, Boca Raton, FL, USA.
31) S. Kalia, K. Thakur, A. Celli, M. A. Kiechel and C. L. Schauer. 2013; Surface Modification of Plant Fibers Using Environment Friendly Methods for Their Application in Polymer Composites, Textile Industry and Antimicrobial Activities: A Review. *Journal of Environmental Chemical Engineering*. 1: 97–112.
32) H. Radhouani, N. Silva, P. Poeta, C. Torres, S. Correia and G. Igrejas. 2014. Potential Impact of Antimicrobial Resistance in Wildlife, Environment and Human Health. *Frontiers in Microbiology* 5.
33) P. Yasotha, K. M. Pachiyappan, M. U Aafrin Banu. and B. Bhanindhraa. 2019. Natural Antimicrobial Agents For Textile Materials. *World Journal of Pharmaceutical Research* 8: 167–176
34) Rifat Ullah Khan, Adia Fatima, Shabana Naz, Marco Ragni, Simona Tarricone and Vincenzo Tufarelli. 2022. Perspective, Opportunities and Challenges in Using Fennel (Foeniculum vulgare) in Poultry Health and Production as an Eco-Friendly Alternative to Antibiotics: A Review. *Antibiotics* 11(2): 278

7 Bovine Mastitis
Rethinking Antibiotics and Emerging Alternatives

**Satish V Patil, Sunil H Koli, Vikas S Patil, Jitendra D Salunkhe, Kiran R Marathe, and Bhavana V Mohite*
* Corresponding Author

7.1 INTRODUCTION

Bovine mastitis is a bacterial infectious disease that occurs in cattle worldwide. It is a major inflammatory disease in dairy cattle that causes significant economic losses in the dairy industry due to reduced milk production, altered milk quality, and increased treatment costs (Cheng and Sung 2020). Mastitis infection is a global problem, and various efforts are ongoing to fight against it, as indicated by the volume of various publications on the topic (Figure 7.1). Mastitis can be classified by two ways: based on symptoms, it is either clinical or subclinical, and based on etiology, it is categorized as either environmental or contagious (Cobirka et al. 2020). These types are occurring due to adhesion, invasion, and colonization of pathogens in the mammary gland, which leads to inflammation and damage of the mammary gland and udder tissue of dairy cattle (Abebe et al. 2016). Pathogens causing mastitis are recognized in vast numbers, i.e., ~135 microorganisms are identified in mastitis infections. However, among those, only 20 distinct microorganisms are commonly detected in mastitis infections. Mastitis pathogens include mostly all types of microbes, such as bacteria, actinomycetes, yeast, fungi, algae, and viruses. Among these, bacterial pathogens are the primary microbial cause of mastitis in dairy animals. Although the algae, fungi, and viruses are quite uncommon, there are still some reports advocating the presence of these microorganisms in mastitis cases, e.g., fungi: *Aspergillus fumigatus* and *Candida albicans*; viruses: bovine herpes viruses, BHV4, parainfluenza 3; and algae: *Prototheca* (Table 7.1) (Radostits et al. 2007; Wawron et al. 2013; Jagielski et al. 2019).

Mastitis is a multifactorial disease, and various risk factors are contributed in bovine mastitis, including host factors, breed, parity, pathogen involved, lactation stage, etc. (Radostits et al. 2007; Zygmunt et al. 2015; Sordillo 2005). For instance, in contagious mastitis, adult and lactating or dry period animals are at more risk. Sometimes early lactating and aged cows have a high risk of infection. Parity-like factors also play an important role; multiparous cows especially in old age have a high risk of infections (Abebe et al. 2016). In India, the prevalence of clinical and subclinical mastitis is quite different than in other countries. Clinical mastitis prevalence is

DOI: 10.1201/9781003269847-7

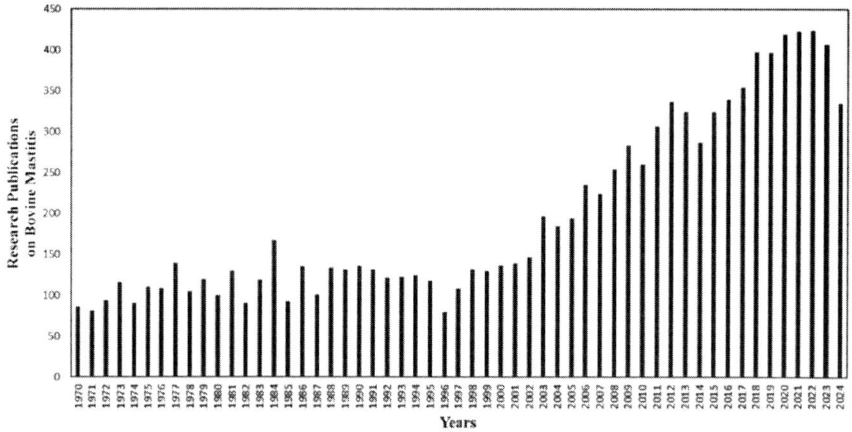

FIGURE 7.1 Research publications scenario on bovine mastitis (1974–2024).

(*Source*: **https://pubmed.ncbi.nlm.nih.gov/?term=bovine+mastitis&filter=years)**

TABLE 7.1
Various Causative Agents of Mastitis and Type of Mastitis Caused

Sr. No.	Causative Agent	Mastitis Type	References
1	*Streptococcus agalactiae*	Clinical/Subclinical	(Rosini and Margarit 2015)
2	*Enterococcus faecalis*	Clinical	(Elhadidy and Zahran 2014)
3	*Streptococcus dysgalactiae*	Clinical	(Olson et al. 2002)
4	*Streptococcus uberis*	Clinical	(Crowley et al. 2011; Krömker et al. 2014)
5	*Escherichia coli*	Clinical	(Costa et al. 2014; Milanov et al. 2015)
6	*Staphylococcus aureus*	Subclinical	(Darwish and Asfour 2013; Fabres-Klein et al. 2015)
7	Coagulase-negative Staphylococci	Subclinical	(Darwish and Asfour 2013)
8	*Corynebacterium* spp.	Clinical/subclinical	(Bitew et al. 2010; Loong et al. 2019; Kirkeby et al. 2021; Lückenet al. 2021; Yimana and Bekele 2022; Woudstra et al. 2023; Myllys et al.1998)
9	*Pseudomonas aeruginosa*	Clinical	(Mbindyo et al. 2020)
10	*Enterobacter aerogenes*	Clinical	(Gwatkin et al. 1937)
11	Tuberculous-type bacilli	Clinical/subclinical	(Jones and Ward 1989)
12	*Mycoplasma* spp.	Clinical/subclinical	(Soucie 1994)
13	*Bacillus* spp.	Clinical	(Amer et al. 2018; Etifu and Tilahun 2019; Fisseha et al. 2020; Saravanan and Palanivel 2020)

TABLE 7.1 (*Continued*)
Various Causative Agents of Mastitis and Type of Mastitis Caused

Sr. No.	Causative Agent	Mastitis Type	References
14	*Streptococcus spp., Streptococcus acidominimus*	Clinical and subclinical	(Rosell and Miller 1933; Ayers and Mudge 1922; Plastridge and Anderson 1936)
15	*Streptococcus lactis*	Subclinical	(Mathers 1916)
16	*Streptococcus viridans*	Subclinical	(Carpenter 1922)
17	*Staphylococcus* spp.	Clinical/subclinical	(Workineh et al. 2002; Nickerson et al.1995)
18	*Streptococci*	Clinical	(Savage 1920; Gwatkin et al. 1937)
19	*Prototheca zopfii*	Subclinical	(Jagielski et al. 2019; Wawron et al. 2013)
20	Bovine herpesvirus 2, vaccinia, cowpox, pseudocowpox, vesicular stomatitis, foot-and-mouth disease viruses, and bovine papillomavirus	Clinical	(Wellenberg et al. 2002)
21	*Nocardia asteroides*	Clinical	(Evans 1918; Defaala and Gharib 1958; Mtmch-Petersen 1954)
22	*Nocardia flavescens*	Clinical	(Bennur et al. 2015)
23	*Nocardia brasiliensis*	Clinical	(Ditchfield, et al. 1959)
24	*Actinomyces farcinica*	Clinical	(Gray 1935)
25	*Mycoplasma bovis*	Clinical	(Hale et al. 1962; Parker et al. 2018)
26	*Aspergillus fumigatus*	Clinical	(Radostits et al. 2007)
27	*Candida albicans*	Clinical	(Radostits et al. 2007)
28	*Trueperella pyogenes*	Clinical	(Yassin et al. 2011; Hillerton and Bramley 1989)

recorded as 4.77% to 18.74% (Sharma and Sindhu 2007; Ghose et al. 2001; Bhat et al. 2017), while subclinical prevalence contributed majorly, i.e., 32.90% to 36.27%, which leads to 70% to 80% economic loss (Rai et al. 2022). This accounted for an average loss of $147 due to mastitis infection per annum per cow and total yield loss by 70%. However, clinical mastitis infections are easily identified by visual observation for abnormalities in udder structure, swollen udder, reddening of the udder, etc. Furthermore, in the case of subclinical mastitis, the infected cattle give abnormal milk with changes in color, flakes that clot in milk, watery consistency, etc. The mastitis disease is classified into different types on the basis of the degree of inflammation.

7.2 TYPES OF MASTITIS

Mastitis infections are categorized mainly on the basis of etiology (environmental and contagious mastitis) and symptoms (clinical, subclinical, and chronic mastitis) (Cobirka et al. 2020).

7.2.1 Environmental Mastitis

Environmental mastitis is related to the infection caused due to the surrounding environment, and most of the time this environmental mastitis is also related to the hygiene of cattle. Fodder waste and dung waste management and water used to clean cattle are related to the cause of environmental mastitis. Various microorganisms have been documented as causative agents of environmental mastitis; however, *E. coli*, *Enterobacter aerogenes,* and *Klebsiella pneumonae* are the major bacterial pathogens identified in environmental mastitis. Although there are various sources of environmental mastitis, the hygiene of cattle and cleanliness in their surroundings are crucial parameters determining the degree of environmental mastitis infections; feces and urine are the major sources of mastitis caused by *E. coli* and *Enterobacter* spp. (Sheldon et al. 2002). Environmental mastitis is cause by microorganisms that infect animals during milking and after milking, because the teat canal remains exposed during the milking and also remains open for some hours after milking. These pathogens are abundant in unhygienic milking environments and enter through the teat canal. Environmental mastitis is also characterized by its symptoms, mainly inflammations of teats. Short-term infections are common consequences as the pathogens are not well adapted to grow in the mammary gland and do not have trophism toward alveolar cells.

7.2.2 Contagious Mastitis

In contagious mastitis, the pathogens generally involved are *Staphylococcus aureus, Streptococcus agalactiae, Streptococcus dysgalactiae, Corynebacterium bovis*, and *Mycoplasma* spp. (Table 7.1). These causative agents majorly infect the mammary gland through multiple routes, e.g., through previously infected cows, bedding, feces and urine, and contact with uninfected cattle (Hogan and Smith 2012). The risk of infection increases during the milking time, i.e., milking cattle have a high risk of contagious infections. In this infection, pathogens are well adapted for survival in the mammary glands and generally cause chronic infections in cattle. Although *Streptococcus agalactiae* and *Staphylococcus aureus* are the common pathogens for contagious mastitis, various microorganisms like *Pseudomonas, Mycoplasma, Serratia,* and some yeasts (Table 7.1) are also reported from a variety of reservoir organs and peripheral environments.

Besides these, mastitis is classified on the basis of symptoms in two major types: subclinical and clinical.

7.2.3 Subclinical Mastitis

Subclinical mastitis is characterized by no visual hallmark symptoms, only a decrease in milk-producing capacity and quality of milk (Batavani et al. 2007; Ogola et al. 2007). It is often difficult to identify until physical changes in the milk and udder are observed; thus, it is also known as hidden mastitis. It is the most prevalent mastitis and contributes up to 70% of losses due to infection in milk production. Subclinical mastitis is a most challenging infection. The worldwide accepted method of detection is to check the somatic cell count (SCC); if this count is in the range of 0 to 200,000 cells/ml the cow should not be infected by mastitis, but if it crosses the range above 200,000 cells/ml; then it indicates subclinical infection. It demonstrates 40 times more SCC increase than clinical mastitis (Shaheen et al. 2016; Bisutti et al. 2022).

7.2.4 Clinical Mastitis

Clinical mastitis is easier to recognize than subclinical mastitis, which is identified by various morphological and color changes in the udders with a sudden reduction in milk production. The udder changes include redness, swelling, inflammation, and induration. The change in milk quality is also one of the major symptoms – milk appears watery and bloody. Microorganisms related to these infections are *Streptococcus agalactiae, Staphylococcus aureus,* and *Streptococcus dysgalactiae* (Qadri et al. 2015). The SCC remains less than 500,000 cells/ml. The cattle always have fever, loss of appetite, and shivering (Hamadani et al. 2013). However, these symptoms vary depending on the degree of infection in cattle.

Chronic mastitis was recognized by resistant infection of the udder. This type of infection mostly remains in subclinical form. Chronic mastitis is characterized by large lumps and swelling in udders, and many times it changes to udder atrophy and fibrosis (Dogan et al. 2006). In severe cases, a total loss of milk occurs and antimicrobial therapy is ineffective (FAO 2014).

7.3 MODE OF ACTIONS OF MASTITIS-CAUSING PATHOGENS AND CONSEQUENCES

The colonization of *Staphylococcus aureus* like Gram-positive microorganisms causes the release of lipoteichoic acid, it is a phosphoribitol compound with lipids; a component of the cell walls of Gram-positive bacteria, which causes necrosis of mammary gland cells and leads to a reduction in milk production (Kerro Dego et al. 2002; Ryan and Ray 2004). The cell wall component of Gram-negative bacteria, i.e., lipopolysaccharide (LPS), is responsible for causing discomfort and pain to cattle. The LPS or endotoxin translocates from the mammary gland into blood circulation and leads to symptoms like dehydration, fever, diarrhea, and pain (Mashburn-Warren et al. 2008; Eckel and Ametaj 2016). The elevated level of LPS in a case of mastitis of cow was reported (Hakogi et al. 1989; Dosogne et al. 2002). The microorganisms related to mastitis were characterized by their potential for biofilm formation, which protects them from phagocytosis of macrophages and neutrophiles (Giovannini et al. 2017; Kerro Dego et al. 2002). The natural immune response elicited against the infections induces IL-1 and TNF-α, which are responsible for the elevated temperature, swelling, pain, and redness in teats.

7.4 DIAGNOSIS OF MASTITIS

Mastitis is a polyetiological and multifactorial disease. Age, parity, lactation stage, milk yield, and udder anatomy are all factors that contribute to a cow's susceptibility to mastitis (Zigo et al. 2021). Mastitis can be classified in two ways: based on symptoms, it is either clinical or subclinical; based on etiology, it is categorized as either environmental or contagious (Cobirka et al. 2020). Early diagnosis is critical for the efficient and effective control of mastitis. The diagnostic process typically includes a physical inspection of the udder for clinical signs, evaluation of the animal's medical history, along with the use of sensitive laboratory tests, which are crucial for early detection. Implementing this diagnostic approach, along with good management practices, can help to reduce the spread of disease (Tommasoni et al. 2023).

7.4.1 Clinical Examination

Clinical mastitis is identified based on both local and systemic signs, making it relatively easy to diagnose by a clinical examination (Gruet et al. 2001). Redness, heat, udder swelling, and the existence of blood clots, flakes, or watery milk are symptoms of mastitis infection. Upon examination, signs of inflammation and infection in the udder, including localized swelling or lumps, further confirm the diagnosis of mastitis. In contrast, subclinical mastitis, which leads to lower milk quality, production losses, and elevated SCCs, is more challenging to detect by clinical examination (Gruet et al. 2001). Despite its lack of visible symptoms, subclinical mastitis is more prevalent than clinical mastitis and is often underestimated because it does not produce obvious clinical signs.

7.4.2 Somatic Cell Count in Milk

SCC and detection of bacteria are the main tools in diagnosing mastitis. SCCs serve as an indicator of mastitis in cows, facilitating the diagnosis of subclinical mastitis in cows and herds. The number of somatic cells present in milk is a valuable marker of intramammary infection, enabling the assessment of milk quality and hygiene and ultimately aiding in treatment decisions for controlling mastitis (Sharma et al. 2011). Milk obtained from a healthy mammary gland typically contains 1×10^5 cells/ml, whereas in the case of infection, the SCC rises above 1×10^6 cells/ml (Bytyqi et al. 2010). However, various noninfectious factors can also influence SCC. These include the lactation stage, age or breed of the cow, parity, stress, season, diurnal variation, and milk transportation conditions. Understanding these factors is essential for accurately interpreting SCC results (Sharma et al. 2011).

Various automated devices are available for the rapid counting of somatic cells in milk. The commercial somatic cell counters, such as Coulter Milk Cell Counter, fluoro-opto-electronic instruments, and charge-coupled device cameras, are among the most commonly used devices for measuring SCC (Heeschen 1975; Gonzalo et al. 2003). However, in practice, the fluoro-opto-electronic instruments are primarily used over the other methods. In India, however, the direct microscopic counting method is preferred due to its low cost. Despite its affordability, this method has limitations, especially when analyzing milk with high solids content (Gonzalo et al. 2003; Sharma et al. 2011).

SCC is a key indicator for assessing milk quality and detecting mastitis. Various countries have established regulatory limits for SCC. The US Food and Drug

Administration set the maximum SCC for milk intended for internal consumption at 750,000 cells/ml. Similarly, the European Union Health Certification program has issued guidelines for companies exporting milk and milk products to the European Union, enforcing an SCC level of 400,000 cells/ml (More et al. 2013). In Canada, the limit for raw milk is set at 500,000 cells/ml. The International Dairy Federation (2013) established a cutoff limit of SCC of around 200,000 cells/ml at the cow level; exceeding this threshold suggests the presence of mastitis. According to some researchers, however, normal somatic cells can reach up to 500,000 cells/ml. However, an individual udder quarter with an SCC of 200,000 cells/ml and whole cow milk with an SCC of 400,000 cells/ml typically indicate mastitis (Sharma et al. 2011).

7.4.3 Identification of Mastitis Pathogens

Timely identification of the causative pathogens of mastitis is essential for investigating milk quality and udder health issues. Pathogen detection is primarily based on various bacterial culture methods, such as aerobic, anaerobic, and microaerophilic methods, using both selective and nonselective media, either in the lab or on the farm (Zadoks et al. 2023; Rowe et al. 2024). To isolate mastitis pathogens, a 0.01 to 0.1 ml milk sample is inoculated onto selective or nonselective media and incubated at 37°C for 24 to 48 hours. In certain cases, plates are incubated for a longer period under microaerophilic or anaerobic conditions, especially for the isolation of *Mycoplasma* spp. Once the pathogen is isolated, the isolate needs to be identified. Traditionally, this is done through phenotypic and biochemical testing. However, the reliability of these methods can vary between laboratories and studies, and they are also often time consuming.

Recently, molecular techniques including matrix-assisted laser desorption ionization-time of flight (MALDI-TOF) and quantitative real-time polymerase chain reaction (RT-PCR) have become more commonly used for identification of mastitis-causing microorganisms. These methods offer distinct advantages. MALDI-TOF allows for highly accurate identification of isolated microorganisms, even at the species level, as compared to traditional phenotypic and biochemical methods. On the other hand, RT-PCR can be applied directly to a milk sample, enabling rapid pathogen identification without the need for culturing (Loy et al. 2023).

7.4.4 Detection of Markers of Inflammation

Recent studies have explored the use of various markers to detect intramammary gland inflammation, aiding in the diagnosis of mastitis. Various genomic and proteomic approaches were used to investigate the gene and biomolecules involved and responsible for bovine mastitis (Hu et al. 2021; Algharib et al. 2024). Difference in various metabolite concentrations between healthy and affected cows have been used to confirm the presence of mastitis. Hu et al. (2021) reported variations in amino acid levels in cow's pre- and postpartum milk using gas chromatography-mass spectrometry (Hu et al. 2021). Similarly, Sundekilde et al. (2013) conducted a metabolomics study using nuclear magnetic resonance on milk samples, which showed that milk with high SCC had elevated levels of lactate, acetate, BHBA, butyrate, and isoleucine, while milk with low SCC had lower levels of lactose, hippurate, and fumarate (Tommasoni et al. 2023). In addition, certain proteins, such as cathelicidins (Addis et al. 2016) and amyloid A (Jaeger et al. 2017), have been identified as potential markers for mastitis and can be rapidly detected using enzyme-linked immunosorbent assay (ELISA). ELISA-based detections of these protein markers performed along with the SCC test can significantly enhance the sensitivity of mastitis diagnosis. The other potential markers include lactose content, enzymes such as lactate dehydrogenase and *N*-acetyl-d-glucosaminidase (Adkins and Middleton 2018), and the activity of milk proteases such as plasmin (Bikker et al. 2014).

7.4.5 Mastitis Detection at Cow-Side

Various cow-side diagnostic tools have been developed for rapid, on-site detection of mastitis.

7.4.6 California Mastitis Test

The California Mastitis Test is the primary method used for the qualitative estimation of SCC in milk. It works on the principle of DNA and protein precipitation released from lysed cells. An anionic detergent combined with chromogenic dyes is used to lyse the cells, causing the precipitated DNA and proteins to alter the viscosity of the milk (Adkins and Middleton 2018).

7.4.7 Mammary Ultrasound

Ultrasonography is a widely used imaging tool that provides clinical information on various diseases in domestic animals. It is most commonly used to diagnose physiological and pathological conditions in the reproductive system (Quintela et al. 2012). However, several studies have shown that ultrasonography can be used to examine mammary glands and predict infections (Flöck and Winter 2006). For instance, Suzuki et al. (2020) used ultrasonography to study cow udders and observed structural changes in the milk ducts, hyperechoic spots in the parenchyma, and nonhomogeneous parenchyma. These abnormal sonographic findings can indicate disease conditions, aiding in the decision-making process for mastitis treatment (Suzuki et al. 2020).

7.4.8 Other Diagnostic Tools

In addition to the aforementioned cow-side diagnostic tools, several other methods are evaluated for detecting mastitis on site. These include measuring milk electric conductivity, infrared thermography, and on-farm culture, all of which can help to predict mammary gland infections (Tommasoni et al. 2023).

Cow-side diagnostic methods offer advantages over laboratory tools in terms of time, cost, and ease of use. However, owing to the high sensitivity and specificity of SCC, it remains the gold standard test for detecting mastitis in animals.

7.5 MAJOR TREATMENTS AND DRAWBACKS

Bacterial pathogens are the major microorganisms involved in mastitis infections, leading to significant use of antimicrobial therapies, such as antibiotics. Antibiotics include all narrow-spectrum and broad-spectrum antibiotics treatments administered via parental and intramammary routes. The intramammary application is a more popular route of treatment because the udder and mammary glands are the major sites of microbial infection. Systemic applications also avoid the undesired side effect of antibiotics on other healthy organs like kidney, liver, etc.

7.5.1 Antimicrobials

Antimicrobials are commonly used in animal husbandry to prevent and treat diseases as well as to promote growth. However, overuse of antimicrobials can lead to antimicrobial resistance, posing a risk to both animal and human health. It is vital to maintain the balance between advantages and threats of antibiotics applied for animals.

Mastitis involves microorganisms that remain on the skin and teat wounds known as contagious microorganisms and those that are not present on it known as environmental microorganisms (Ruegg 2017; Zeryehun and Abera 2017). There are about 140 studies that have reported major pathogens involved in mastitis (Amer et al. 2018). *Streptococcus agalactiae, Streptococcus uberis, Staphylococcus aureus,* another *Streptococcus* spp., and *E. coli,* are the microorganisms commonly found in mastitis (Keane et al. 2013; Poutrel et al. 2018; Krishnamoorthy et al. 2021). The pathogen types are now changed from major to minor pathogens like other bacilli and coagulase-negative *Staphylococcus* (Mpatswenumugabo et al. 2017; El Jakee et al. 2013) and play important roles in mastitis but differ from herd to herd (Vakkamäki et al. 2017; Zeryehun and Abera 2017).

The antibiotics like penicillin, streptomycin, tetracycline, ampicillin, and cloxacillin could be applied against mastitis directly through intramammary or parenteral routes of infusion (Bhosale et al. 2014). Pathogen sensitivity to antibiotics, kinds of mastitis, breed of cattle, and schedule of treatment are all factors that determine the efficacy of treating bovine mastitis (Barkema et al. 2006). Mastitis control is seriously impacted by the emergence of drug resistance, with resistance profiles often being specific to herds (Silveira-Filho et al. 2014).

There are no approved antimicrobial drugs for the systematic treatment of mastitis in the United States, and intramammary (IMM) infusion treatment of mastitis is only available for a limited number of antimicrobial drugs (Oliveira 2012). Since 2006, there have been no new antimicrobials approved for mastitis therapy despite the withdrawal of several products from the US market. There are only two classes of antimicrobial compounds that are approved by the US Food and Drug Administration among commercially available IMM infusion products. These classes feature six or seven IMM infusion products that are commercially available containing β-lactams like penicillin, cloxacillin, amoxicillin, ceftiofur, hetacillin, and cephapirin and another product containing lincosamide (pirlimycin) (Oliveira and Ruegg 2014).

Ceftiofur is a broad-spectrum third-generation cephalosporin used in most treatments of mastitis. The United States has a lot of systemic products that contain ceftiofur that do not require milk withdrawal, making them extensive (Erskine et al. 2003). Irrespective of the etiology of mastitis, milk production remained abnormal for about the same period after being treated for four to six days. IMM antimicrobials were used most commonly for the treatment of microbiologically negative cases (Oliveira and Ruegg 2014).

Prior to the approval of IMM ceftiofur in 2006, cephapirin was the most commonly used IMM drug for treating clinical mastitis (Raymond et al. 2006; Pol and Ruegg 2007). Cephapirin can only be found as an IMM preparation and requires a 96-hour withdrawal period, with short treatment duration of two doses for 12-hour interval compared with IMM ceftiofur. Cephapirin is intended to treat clinical mastitis caused by *Streptococcus agalactiae* and *Staphylococcus aureus,* but it is not considered a first-generation antibiotic. Yet, cephapirin, a cephalosporin of the first generation, is capable of assisting in the prevention of Gram-positive bacteria (Cortinhas et al. 2013).

Selection of the right antibiotic would be easier with a quick or on-farm bacteriological diagnosis (Raymond et al. 2006; Passantino 2007). For clinical mastitis treatment, narrow-spectrum antimicrobials are the preferred option in general (Pyörälä 2009). On-farm protocols designed specifically for mastitis treatment can ensure the precise and judicious use of antimicrobials. β-Lactams are the first choice for treating mastitis caused by Streptococci and penicillin-susceptible Staphylococci.

It is not recommended to use broad-spectrum antibiotics such as third- or fourth-generation cephalosporins as the first choice for mastitis because that will increase the risk of antibiotic resistance. In cases of *S. aureus* and severe coliform mastitis, systemic treatment is recommended, especially when accompanied by IMM therapy (Barkema et al. 2006). In many countries, it is recommended to treat clinical mastitis for at least three days.

There are 16 products available in the United States for treating bovine mastitis, six of which need a prescription. Cloxacillin, hetacillin, amoxicillin, pirlimycin, and novobiocin are the main prescribed drugs, while over-the-counter drugs are listed as dihydrostreptomycin, erythromycin, penicillin G, novobiocin, and sulfamethazine (for nonlactating dairy cows only) preparations.

7.5.1.1 Route of Administration of Antibiotics

The milk or udder tissue should have an accumulation of the antimicrobial agent (Erskine 2003), based on the etiological agent. The target site may differ depending on the cause of the infection, the causative agent: Streptococci are known to stay in the milk compartment, but *S. aureus* has the ability to penetrate the udder tissue resulting in severe infection (Table 7.2). In mastitis, the intramuscular route is the most commonly used method for administering antimicrobials. This approach is advantageous due to the ability to achieve high antimicrobial concentrations in the milk and requires less antibiotic overall, as the drug is injected directly into the infected area. The uneven distribution throughout the udder is the drawback of IMM administration, and infusing the product through the teat canal is highly risky for

TABLE 7.2
Representative Antibiotics Applied Against Mastitis Pathogens

Sr. No.	Antibiotic	Properties	Minimum Inhibitory Concentration [MIC] (Sensitive Bacteria)	Reference
1	Penicillin	Moderately lipid soluble, weak organic acid with pKa 2.7	0.12 to 0.19 μg/ml (*Staphylococcus aureus*) 0.5/1–4/8 μg/ml (Streptococci)	(Pengov and Ceru 2003) (Cameron et al. 2016)
2	Dihydrostreptomycin	A base, with pKa 8.0	0.2 and 5 μg/ml (*Staphylococcus aureus*) 5 to 50 μg/ml (Streptococci) ≤32 and >64 μg/ml (*Klebsiella pneumoniae*)	(Ziv 1969) (Ziv et al. 1970) (Saini et al. 2012)
3	Neomycin	Poor lipid solubility	≤1 to 64 μg/ml (Staphylococci and Streptococci)	(Gianneechini et al. 2002)
4	Erythromycin	Highly lipid-soluble base with pKa 8.8 Low degree of ionization at physiological pH	0.25 to 16 μg/ml (Staphylococci and Streptococci)	(MacDiarmid 1978) (Rubin et al. 2011)
5	Oxytretracycline	Ionized at physiological pH, form less polar organic complexes with body fluid constituents	0.5 to 64 μg/ml (Staphylococci and Streptococci) Average 2.44 μg/ml (*E. coli*)	(Gianneechini et al. 2002; Cameron et al. 2016) (Pilloud 1973)
6	Tylosin	Low degree of ionization at physiological pH and by its high lipid solubility	MIC 0.5 to ≥256 μg/ml (Staphylococci and CoNS) (85% strains inhibited at ≤2 μg/ml) and 0.12 to ≥256 μg/ml (Streptococci) (>85% strains inhibited at ≤2 μg/ml)	(Entorf et al. 2014) (Simjee et al. 2012)
7	Chloramphenicol	An alcohol with a high degree of lipid solubility	4 to 32 μg/ml (*Staphylococcus aureus*) ≤2 to >32 μg/ml (*Klebsiella* spp.)	(Rubin et al. 2011) (Sani et al. 2012)
8	Sulfonamides	Acidic with different pKa values	19 to 76 μg/ml (Staphylococci) 2 and 16 μg/ml (Streptococci)	(Rubin et al. 2011)
9	Trimethoprim	Short half-life	<0.06 to 8 μg/ml (*Staphylococcus aureus*) ≤0.06 to 2 μg/ml (Streptococci)	(Gianneechini et al. 2002)

infecting the quarter (Ehinger and Kietzmann 2000), It has been suggested that the systemic approach to clinical mastitis treatment is more efficient than the intramammary approach as this route allows the antimicrobial to better penetrate the udder tissue (Erskine 2003). However, obtaining and maintaining therapeutic levels in milk or udder tissue after systemic infusion is challenging.

The success of systemic mastitis treatment relies on the drugs' ability to move from the bloodstream into the infected area and into milk. The drugs' passage depends on the following factors: its degree of ionization, lipid solubility, molecular size, and protein-binding capacity (MacDiarmid 1978).

There are a limited number of drugs for treatment of systemic mastitis with optimal pharmacokinetic and pharmacodynamic properties. In milk, maintaining therapeutic levels of broad-spectrum antibiotics such as ceftiofur, trimethoprim-sulfonamide, and oxytetracycline is difficult (Kaartinen et al. 1999). Clinical trials have not shown that macrolides are effective when given systemically to treat clinical mastitis (Owens et al. 1999), despite the fact that they would have excellent pharmacokinetics (Franklin et al. 1986; Sanders et al. 1992). Tylosin and spiramycin expressed reasonable efficacy against mastitis caused by Streptococci (McDougall et al. 2007). Penicillin G is the antibiotic used in systemic treatment; as a weak acid, it enters the mammary gland slowly, but because sensitive organisms have very low minimum inhibitory concentrations (MICs), therapeutic doses can be reached in milk (Ziv and Storper 1985). Because of the more lipophilic nature of the penicillin G formulation known as pentamate, it disperses better in milk compared to penicillin G alone (Ziv and Storper 1985).

However, due to the increase of different groups of Gram-positive and Gram-negative opportunistic microorganisms, the therapeutic approach to IMM infusion has little demand for mastitis treatment. The treatment of clinical mastitis by farmers is mostly on the basis of symptoms and not on diagnosis of microorganisms, and hence, there is treatment without cause determination (Hoe and Ruegg 2006). This strategy often leads to the prescription of antibiotics in cases that may not work.

7.5.1.2 Pharmacokinetics and Pharmacodynamics Factors Affecting Antibiotic Use

The clinical efficacy and toxicity of any drug are directly related to its pharmacokinetic (PK)/pharmacodynamic (PD) parameters. Key PK parameters influencing antimicrobial activity include the maximum concentration achieved after the first dose, total drug exposure over the dosing interval, half-life, clearance, volume of distribution, and bioavailability. Changes in these interrelated parameters can significantly impact drug effectiveness. Additionally, the level of protein binding affects drug activity, as only the unbound fraction exerts antimicrobial effects. In mastitis treatment, achieving the desired drug concentration at the target site – the infected mammary – is a main challenge. When the drug is infused intramammarily, the pharmacokinetics of the drug such as solubility in lipids, degree of ionization, protein binding, and the type of vehicles used for delivery significantly influence drug absorption in milk and its distribution within the udder (Pyörälä 2009). Furthermore, eliminating residual antibiotics from milk after treating infected cows is a critical aspect of effective mastitis management (Wagner and Erskine 2006).

The PD process emphasizes the biochemical and physiological influence of drugs and their mechanisms of action. The PD of drugs is characterized by the concentration of drug at the site of action and is related to the magnitude of observed effects (Marino et al. 2024). Therefore, when treating mastitis in lactating animals, it is crucial to consider the interaction between milk and the drugs used, ensuring that this interaction does not interfere with the drug's antimicrobial activity. Previous studies have shown that various classes of antibiotics, including tetracyclines, macrolides, and trimethoprim-sulfonamides, exhibited reduced antimicrobial activity in the presence of milk (Louhi et al. 1992; Fang and Pyörälä 1996). Moreover, in infected cows, the natural immune response is diminished, and the phagocytic activity at the site of infection is weakened. Consequently, the choice of antibiotic should prioritize bactericidal rather than bacteriostatic agents for systemic use (Kehrli and Harp 2001). It is also important to select drugs with low protein binding, high bioavailability, and an extended half-life within the body (MacDiarmid 1978).

7.5.2 Antibiotic Resistance

For several decades, antibiotics have been the preferred treatment for mastitis. However, there is still no consensus on the efficacy, safety, and cost of treatment. Additionally, the development of antibiotic resistance among microbial pathogens poses a significant challenge to effectively controlling mastitis. The degree of resistance to the different classes of antibiotics often varies and is herd specific (Silveira-Filho et al. 2014). The extensive use of antibiotics in treating mastitis significantly increases the risk of transmitting antibiotic-resistant microorganisms and the horizontal spread of antibiotic resistance to humans. Among the various causative microorganisms, *Staphylococcus aureus* reported resistance to penicillin G (Olsen et al. 2006; Hendriksen et al. 2008).

Methicillin-resistant *Staphylococcus aureus* (MRSA) is a substantial public health alarm, as it has been proven that these strains can transmit antibiotic resistance to humans (Pascu et al. 2022). The emergence of resistance in *Staphylococcus* toward different β-lactam antibiotics may be due to expression of the β-lactamase gene, modification of targets, pharmacological issues, or virulence factors. Among *Staphylococcus* species, coagulase-negative Staphylococci (CNS) are generally more resistant than *Staphylococcus aureus* and tend to develop multidrug resistance more easily (Pitkälä et al. 2004; Sawant et al. 2009). Similarly, *Streptococcus* species isolated from bovine milk have been reported to show emergent resistance to lincosamides and macrolides, which is associated with the presence of the *ermB* gene (Pitkälä et al. 2004; Loch et al. 2005). The genetic determinants responsible for antibiotic resistance by mastitis pathogens and environmental contaminants are discussed in Table 7.3, which summarizes various mechanisms of antibiotic resistance.

7.6 NEED FOR ALTERNATIVE TREATMENT TO ANTIBIOTIC THERAPY

Public health is at high risk due to harmful antibiotic residues in milk resulted from the use of antimicrobial therapy for mastitis treatment in cattle's and the development of resistance caused by overexposure to antibiotics (Wagner and Erskine 2006;

TABLE 7.3
Genetic Determinants of Antibiotic Resistance in Mastitis Pathogens

Pathogen	Resistance to Antibiotic	Resistance Mechanism	Reference
Gram positive			
Staphylococcus aureus	Ciprofloxacin, daptomycin, gentamicin, linezolid, oxacillin, or vancomycin	Activation of intrinsic factors such as mprF, ndh, fmtA, graR, or dltA	(Blázquez et al. 2014)
	Quinolones or lincomycin	Activation of some chromosomally encoded multidrug efflux pumps such as NorA or LmrS	(Costa et al. 2013)
	Novobiocin	Expression of a novobiocin-resistant GyrB protein	
	β-Lactams, tetracyclines, aminoglycosides, amphenicols, macrolides, trimethoprim, lipopeptides, and lincosamides	Acquired resistance	(Barnett and Bushby 1970; Wendlandt et al. 2013; Nobrega et al. 2018)
S. aureus, S. epidermidis, S. haemolyticus, S. chromogenes, and CNS species	β-Lactam antibiotics: penicillin and methicillin	blaZ gene encodes the PC1 β-lactamase	(Llarrull and Mobashery 2012)
Staphylococcus spp. and CNS	Methicillin and oxacillin	mecA gene, which encodes PBP2a (with low β-lactam affinity in Staphylococcal cassette chromosome mec (SCCmec) elements	(Schwarz et al. 2018)
	Tetracycline	tet(K) and tet(L)genes, which code for membrane-associated efflux proteins of the major facilitator superfamily transferred by plasmid	(Schwarz et al. 2018)
	Streptomycin	Multidrug-resistant gene cluster: gene aadE, ant(6)-Ia, str, lsa(E) and lnu(B), aadE	(Antók et al. 2019)
Staphylococcus spp.	Aminoglycosides	Genes coding for inactivating enzymes	(Schwarz et al. 2018)
	Anamycin, neomycin, amikacin	Gene aphA3 codes for a phosphotransferase	
	Gentamicin, kanamycin, tobramycin, amikacin	aacA and aphD code for an acetyltransferase and phosphotransferase	

MRSA	Macrolides or lincosamides or streptogramins	Expression of one or more erm class A, B, C, T, 43, 44, or 48 genes can induce a modification of the target site at the rRNA and inhibit the binding to ribosome	(Schwarz et al. 2018)
	Macrolide and streptogramin B	msr, mph, ere, Inu, vga, lsa, or sal genes. msr(A) and msr(B), which code for an efflux pump from the ABCF subfamily protein, confer resistance to macrolide and streptogramin B	
	Lincosamides, pleuromutilins, streptogramin A	Expression of plasmid genes vga(A), vga(C), lsa(E), and sal(A), which code for ABC transporters and mediate resistance	(Wendlandt et al. 2013, 2015)
	Kanamycin, neomycin, and tobramycin	Gene aadD	(Qu et al. 2019)
Streptococci	Quinolones, aminoglycosides	Overexpression of ABC efflux pumps	(Garvey and Piddock, 2008)
	Macrolides or lincosamides or streptogramins	Expression of genes like erm(A), erm(B), resulting target protection, following methylation of the ribosome	(Haenni et al. 2018)
		Expression of genes like mef, msr, or mre families code for efflux pumps	(Haenni et al. 2018)
		Acquisition of an mph gene through the production of inactivating enzymes	(Chesneau et al. 2007)
	Tetracyclines	Membrane efflux systems [tet(K) tet(L)] and ribosomal protection enzymes [tet(M), tet(O), tet(S)]	(Haenni et al. 2018)
	β-Lactams	Mutation (substitutions) in the penicillin-binding protein	(McDougall et al. 2020)
	-Penicillin and ampicillin	Presence of the gene bl2b	(Vélez et al. 2017)
	Fluoroquinolones	Point mutations in the quinolone resistance determinant regions of the gyrA and parC genes	(Drlica and Zhao 1997)
	Chloramphenicol and sulfonamides	(cat1, cat2), or (sul1, sul2, sul3), respectively	(Tian et al. 2019)
	Aminoglycoside	aphA-3 and aad-6	(Ahmed et al. 2021)
Gram negative			

(*Continued*)

TABLE 7.3 (*Continued*)
Genetic Determinants of Antibiotic Resistance in Mastitis Pathogens

Pathogen	Resistance to Antibiotic	Resistance Mechanism	Reference
Enterobacteriaceae: *E. coli* and *K. pneumoniae*	Macrolides, aminocoumarins, or glycopeptides	Poor membrane permeability of these antimicrobial classes, unable to penetrate the outer membrane of Gram-negative bacteria	(Klobucar and Brown 2022)
	β-Lactams: Penicillins and narrow-spectrum cephalosporins	TEM-1 and SHV-1 and SHV-2	(EFSA 2011)
	Penicillins, cephalosporins, monobactams	Extended-spectrum β-lactamases (ESBL) and AmpC cephalosporinases β-Lactam inhibitors such as clavulanic acid, sulbactam, or tazobactam	(Padmini et al. 2017) (Poirel et al. 2018)
	Tetracyclines	tet(A), tet(B), tet(E), and tet(G) resistance genes	(Metzger and Hogan 2013)
	Aminoglycosides such as neomycin, streptomycin, gentamicin, or kanamycin	Target modification and enzymatic inactivation	(Poirel et al. 2018)
	Aminoglycosides and fluoroquinolones	Enzymatic modification Aminoglycoside *N*-acetyltransferases AAC(3)-II/IV and AAC(6)-Ib, part of gene cassettes in integrons Nucleotidyltransferases ANT(2″) and ANT(3″) coded by the aadB and aadA genes are contained within gene cassettes	(Poirel et al. 2018) (Majumder et al. 2021)
	Streptomycin	Phosphotransferases APH(6)-Iα and APH(6)-Id coded by strA and strB genes	(Poirel et al. 2018)
	Kanamycin	aph(3″)-I/II genes	(Massé et al. 2020)

Quinolones	Mutations of the gyrA/gyrB or parC/parE chromosomal genes that code for gyrases and topoisomerases, respectively	(Yang et al. 2018)
	Qnr such as proteins: AAC(6′)-Ib-cr acetyltransferase, or active efflux pumps such as QepA and OqxAB	(Poirel et al. 2018)
Sulfonamides	sul1, sul2, and sul3 resistance genes	(Nobrega et al. 2021)
	Mutations on the folP gene	(Zheng et al. 2022)
Trimethoprim	dfr genes located in gene cassettes inserted into integrons class 1 or 2	(Nobrega et al. 2021)

Abbreviations: CNS, coagulase-negative Staphylococci; MRSA, methicillin-resistant *Staphylococcus aureus*.

WHO 2010; Tomanić et al. 2023). Besides, public health risks caused by antibiotic residues in milk can significantly impact the dairy industry, potentially compromising milk quality. The rapid global spread of antibiotic resistance has become a pressing issue, contributing to an increasing number of deaths each year (Gajdács et al. 2021). To effectively control the rapid rise of antibiotic resistance and achieve optimal health for humans, animals, and the environment, we must adopt a multisectoral One Health approach to raise awareness and promote coordinated action (Lancet 2022). Taking these factors into account, treating mastitis with antibiotics has become one of the greatest challenges in recent times. Consequently, it is necessary to search for harmless and effective alternatives to antibiotics.

7.7 ALTERNATIVE TREATMENT FOR PREVENTION OF BOVINE MASTITIS

To avoid the consequences linked with chemotherapy, such as antibiotic resistance, contamination of antibiotics in milk, etc., various alternative strategies are successfully used, which are known as novel strategies for mastitis control. These include applications of bacteriophages, phage endolysins, bacteriocins, and probiotics microorganisms. Additionally, vaccination of the cattle is an effective and popular treatment to protect the cattle from mastitis. The majority of the available vaccines are encapsulated, nonencapsulated, and attenuated toxins surface extracts, exopolysaccharides etc., that target major mastitis pathogens such as *S. aureus* and *Streptococcus agalactiae*. Core antigen J5 of *E. coli* is widely used for Gram-negative pathogens and was found to be significant in arresting new infections of Gram-negative bacteria (Landin et al. 2015). The Startvac I is a popular and widely used vaccine that is effective for arresting new infections of *Staphylococcus* spp. and *Streptococcus* spp. as well as increasing milk production and protection as compared to nonvaccinated cows (Schukken et al. 2014; Bradley et al. 2015; Ismail 2017; Freick et al. 2016). Although vaccine therapy seems to be significant due to the wide range of effectivity, it does not offer full protection against all types of mastitis. Besides that, success also depends on other disease management strategies.

7.7.1 Phage Therapy

Bacteriophages are the bacteria infecting viruses, and now the application of these viruses against human and animal pathogens is becoming a choice of treatment because of its host specificity. *Staphylococcus* is one of the major mastitis pathogens, and *Staphylococcus* phages are proven to be an effective solution due to phages' infective range of the *Staphylococcus* strains involved in pathogenesis (Kutter et al. 2010). Researchers advocated that the similarity in the surface chemical composition of *Staphylococcus* makes phages widely active against the range of pathogenic *Staphylococcus* species found in mastitis, although O'Flaherty et al. (2005a, 2005b) reported phage K like *Staphylococcus* phages can multiply in raw milk, and they are unable to control *S. aureus* because of the clumping of *S. aureus*.

Antistaphylococcal phages against *S. aureus* are restricted due to host complement factors that prevent phages from attach to the host cell and multiplying. But a cocktail of phage K and CS1 treatment administered to healthy bovine teat was reported as successful protection from mastitis. In many reports, the mix treatment of phages proved to be beneficial for protection of teat infections (Son et al. 2010; Kwiatek et al. 2012; Dias et al. 2013). Besides these, the phage's SPW and SA proved significant lytic capability against *Staphylococcus* and other mastitis pathogens (Li and Zhang 2014; Hamza et al. 2016; Tahir et al. 2017). Phage enzymes known as endolysin released from host cells lyse *Staphylococcus* and other Gram-positive pathogens of mastitis by breaking cell walls and binding or inhibiting peptidoglycan (Schmelcher et al. 2015). The five groups of endolysin are identified based on their mode of action. Donovan et al. (2006) reported phi11 and CONs, phi11 endolysin have the capacity to lyse Staphylococci strains related to mastitis. Besides that, some efforts are also reported for applications of engineered endolysin against mastitis pathogen (Zhou et al. 2017).

7.7.2 Bacteriocins and Probiotics

Bacteriocins are small antimicrobial peptides synthesized by various species of both Gram-positive and Gram-negative bacteria. The potential for bacteriocin production is among the prominent characteristics of probiotic microorganisms. The probiotic microorganisms have been well known for their health benefits to the host. Probiotics have the potential to modify the gut microflora by eliminating harmful microbes, activating the host immune system, and enhancing its defense against infectious pathogens. It has been proven that bacteriocins have the potential to inhibit biofilm formation, which is essential for eradicating mastitis pathogens within biofilms (Cotter et al. 2013; Davies 2003; Høiby et al. 2010; Gomes et al. 2016). Mathur et al. (2018) reported the antibiofilm potential of lantibiotics. De Souza Duarte et al. (2018) reported the first sacribiotic, Hyicin 4244, which has shown potential in inhibiting biofilm formation by *Staphylococcus* strains originating from human and bovine mastitis (De Souza Duarte et al. 2018).

Aureocin 4181, bovicin HC5 like own product of *S. aureus* and *Streptococcus,* was found to be effective against some Gram-negative pathogens involved in mastitis (Oftedal et al. 2021; Godoy-Santos et al. 2019). These metabolites disrupt the cell membrane of other pathogens and therefore can be used as alternatives for mastitis treatment. Ubericin K like metabolites of *Enterococcus faecium* were found to be effective against the *S. pyogenes* and *L. monocytogenes* (Oftedal et al. 2021). Nisin is one of the well-known antibiotics produced by *Leuconostoc mesenteroides*, which is an approved antimicrobial against the food microbes and was also found to be effective against mastitis pathogens by its pore-forming action in the cell membrane. However, development of resistance was reported (Kitazaki et al. 2017). To avoid resistance, various nisin-engineered derivatives must be found effective against the mastitis pathogens (Pérez-Ibarreche et al. 2021; Vidal Amaral et al. 2022).

Probiotics are the live microorganisms, the presence of which has positive effects on health, through immunomodulation and an antagonistic effect. Such microorganisms are found to be effective against the inflammatory actions of various pathogens (Table 7.4). When such microorganisms are infused in the teat, they reach the mammary gland, show antagonistic effects against the mastitis pathogens, and subside the infection. It was observed that *Lactobacillus casei* and *Lactobacillus acidophilus* cure subclinical infections as proven by a reduction in SCCs in infected cows. Bouchard et al. (2013) and Armas et al. (2017) also reported *Lactobacillus brevis, Lactobacillus plantarum,* and *Lactobacillus lactis* are good competitors of *S. aureus, S. epidermis, S. agalactiae,* and a plethora of other mastitis pathogens. Souza et al. (2018) also reported some *Lactobacillus* spp. were isolated from the mammary gland to reduce the expression of cytokines induced by the *S. aureus* pathogen. In addition, the biofilm formation mechanism is the major mechanism of colonization of mastitis pathogen. It was observed that some probiotics microorganisms are also useful in inhibiting biofilm formation, replacing the pathogens from the mammary glands, and recovering from the infection (Wallis et al. 2019). In feeding cattle *Bacillus subtilis* like probiotics throughout lactation was found to reduce the SCC to normal and also reduced the incidence of subclinical mastitis (Urakawa et al. 2022). The majority of bifidobacterium, *Lactobacillus* spp., have been proven successful as alternative therapy for the control of mastitis by producing antibacterial peptides, antagonism, and competitive displacement like mechanisms (Table 7.4).

TABLE 7.4
Mastitis Treatment Using Probiotics Bacteria

Sr. No.	Probiotic Bacterial Strain	Mechanism of Action	References
1	*Lactobacillus lactis*	Produces an antibiotic peptide, nisin, active against antibiotic-resistant bacteria	(Pieterse 2008; Sang and Blecha 2008)
2	*Lactobacillus fermentum*	Prevents the growth of bacterial pathogens causing mastitis; strengthens the immune response in the mammary gland	(Jiménez et al. 2008)
3	*Bifidobacterium bifidum*	Strengthens the mucosal barriers, reduces pathogenic load, decreases somatic cell count	(Kober et al. 2022)
4	*Streptococcus thermophilus*	Enhances epithelial cell integrity, supports probiotic colonization in mammary gland	(Pieterse 2008; Luo et al. 2022)
5	*Lactobacillus salivarius*	Competitive displacement of pathogens (*Staphylococcus aureus*), anti-inflammatory effect	(Akhtar et al. 2022; Kang et al. 2017)

7.7.3 Herbal Treatment

Traditionally in various rural areas, the application of plant parts such as leaves and root decoctions are reported for mastitis control. Herbal formulations have comparatively less toxicity. The applications of herbal formulations are primarily topical applications for teat or udder infections. Although there are various reports of plant metabolites like limolin, thymol, and baicalein going against the mastitis pathogen such as *S. aureus* (Liang et al. 2014; Fu et al. 2014; Wei et al. 2014; He et al. 2015), there are no known effective herbal formulations available commercially for clinical and subclinical mastitis. Various essential oils are also reported against the mastitis pathogens, e.g., essential oil from *Origanum vulgare, Cinnamomum cassia, Minthostachys verticillata,* etc., reported against both Gram-negative and Gram-positive pathogens causing mastitis. The mechanism reported is inhibition of biofilm formation by reducing adhesions and invasions, and activation of macrophage-mediated phagocytosis in the mammary gland (Cho et al. 2015; Federman et al. 2016; Zhu et al. 2016).

7.8 CONCLUSION

The World Organization for Animal Health, the Food and Agriculture Organization, and the World Health Organization collaborate to implement strategies through the One Health Initiative. This association will take action for human health, animal health, and the environment. Subclinical mastitis requires more attention as it does not express visual symptoms. Subclinical mastitis needs to be detected on site for pathogens at species level with advanced molecular methods and treated with alternative methods instead of direct use of antibiotics depending on the severity of the mastitis. The profound use of antibiotics should be restricted in the animal sector as it turns to antibiotic resistance for human use. Clinical mastitis presently has antibiotics as the best treatment choice, but we search for effective alternative strategies that will reduce the use of antibiotics. Even if an antimicrobial sensitivity test is performed before the drug is prescribed at higher concentration and a narrow-spectrum antibiotic is applied, the efforts will be beneficial. The future perspective should be focused on emerging potential approaches like herbal medicines, probiotics, bacteriocins, bacteriophages, and vaccination, which have potential in the fight against antibiotic resistance. Novel approaches will overcome the limitations of traditional antimicrobial therapy. This strategy will not entirely replace antibiotics, but it can be used as a complementary strategy for the treatment of mastitis.

7.9 ACKNOWLEDGMENT

Dr. SVP is grateful to RGSTC, Mumbai, for financial support for working on a farmer's problem due to environmental changes (File No. KBCNMU/RGSTC/sanction order/155 dated 25/07/2023). All authors are thankful for the constant support and inspiration of Prof. Anthony I Okoh, Director, SAMRC Microbial Water Quality Monitoring Centre, University of Fort Hare, Alice, South Africa, during our research journey.

REFERENCES

Abebe, R., H. Hatiya, M. Abera, B. Megersa, and K. Asmare. 2016. Bovine mastitis: Prevalence, risk factors and isolation of *Staphylococcus aureus* in dairy herds at Hawassa milk shed, South Ethiopia. *BMC Vet Res* 12:1–11. https://doi.org/10.1186/s12917-016-0905-3

Addis, M. F., V. Tedde, G. M. G. Puggioni, S. Pisanu, A. Casula, C. Locatelli, N. Rota, V. Bronzo, P. Moroni, and S. Uzzau. 2016. Evaluation of milk cathelicidin for detection of bovine mastitis. *J Dairy Sci* 99:8250–8258.

Adkins, P. R., and J. R. Middleton. 2018. Methods for diagnosing mastitis. *Vet Clin Food Anim Pract* 34:479–491.

Ahmed, W., H. Neubauer, H. Tomaso, F. I. El Hofy, S. Monecke, A. A. Abd El-Tawab, and H. Hotzel. 2021. Characterization of *Enterococci*-and ESBL-producing *Escherichia coli* isolated from milk of bovides with mastitis in Egypt. *Pathogens* 10:97. https://doi.org/10.3390/pathogens10020097

Akhtar, M., S. U. A. S. Naqvi, Q. Liu, H. Pan, Z. Ma, N. Kong, Y. Chen, D. Shi, M. F. E. A. Kulyar, J. A. Khan, and H. Liu. 2022. Short chain fatty acids (SCFAs) are the potential immunomodulatory metabolites in controlling *Staphylococcus aureus*-mediated mastitis. *Nutrients* 14:3687. https://doi.org/10.3390/nu14183687

Algharib, S. A., A. S. Dawood, L. Huang, A. Guo, G. Zhao, K. Zhou, C. Li, J. Liu, X. Gao, W. Luo, and S. Xie. 2024. Basic concepts, recent advances, and future perspectives in the diagnosis of bovine mastitis. *J Vet Sci* 25(1):e18. https://doi.org/10.4142/jvs.23147

Amer, S., F. L. A. Gálvez, Y. Fukuda, C. Tada, I. L. Jimenez, W. F. M. Valle and Y. Nakai. 2018. Prevalence and etiology of mastitis in dairy cattle in El Oro Province, Ecuador. *J Vet Med Sci* 80:861–868.

Antók, F. I., R. Mayrhofer, H. Marbach, J. C. Masengesho, H. Keinprecht, V. Nyirimbuga, O. Fischer, S. Lepuschitz, W. Ruppitsch, M. Ehling-Schulz, and A. T. Feßler. 2019. Characterization of antibiotic and biocide resistance genes and virulence factors of *Staphylococcus* species associated with bovine mastitis in Rwanda. *Antibiotics* 9:1–17.

Armas, F., C. Camperio, and C. Marianelli. 2017. In vitro assessment of the probiotic potential of *Lactococcus lactis* LMG 7930 against ruminant mastitis-causing pathogens. *PLoS One* 12:e0169543. https://doi.org/10.1371/journal.pone.0169543

Ayers, S. H., and C. S. Mudge. 1922. The *Streptococci* of the Bovine Udder: IV. Studies of the *Streptococci*. *J Infect Dis* 31:40–50.

Barkema, H., Y. H. Schukken, and R. N. Zadoks. 2006. Invited review: The role of cow, pathogen, and treatment regimen in the therapeutic success of bovine *Staphylococcus aureus* mastitis. *J Dairy Sci* 89:1877–1895.

Barnett, M., and S. R. M. Bushby. 1970. Trimethoprim and the sulphonamides. *Vel Rec* 87:43–50.

Batavani, R. A., S. Asri, and H. Naebzadeh. 2007. The effect of subclinical mastitis on milk composition in dairy cows. *Iran J Vet Res* 8:205–211.

Bennur, T., A. Ravi Kumar, S. Zinjarde, V. Javdekar. 2015. Nocardiopsis species: Incidence, ecological roles and adaptations. *Microbiol Res* 174:33–47.

Bhat, A. M., J. S. Soodan, R. Singh, I. A. Dhobi, T. Hussain, M. Y. Dar, and M. Mir. 2017. Incidence of bovine clinical mastitis in Jammu region and antibiogram of isolated pathogens. *Vet World* 10:984–989.

Bhosale, R., R. A. Osmani, P. P. Ghodake, S. M. Shaikh, S. R. Chavan. 2014. Mastitis: An intensive crisis in veterinary science. *Int J Pharma Res Health Sci* 2:96–103.

Bikker, F. J., G. Koop, N. B. Leusink, K. Nazmi, W. E. Kaman, H. S. Brand, and E. C. Veerman. 2014. Tailor made plasmin substrates as potential diagnostic tool to test for mastitis. *Vet Res Commun* 38:271–277.

Bisutti, V., A. Vanzin, A. Toscano, S. Pegolo, D. Giannuzzi, F. Tagliapietra, and A. Cecchinato. 2022. Impact of somatic cell count combined with differential somatic cell count on milk protein fractions in holstein cattle. *J Dairy Sci* 105:6447–6459.

Bitew, M., A. Tafere, T. Tolosa. 2010. Study on bovine mastitis in dairy farms of Bahir dar and its environs. *J Anim Vet Adv* 9:2912–2917.

Blázquez, B., L. I. Llarrull, J. R. Luque-Ortega, C. Alfonso, B. Boggess, and S. Mobashery. 2014. Regulation of the expression of the β-lactam antibiotic-resistance determinants in methicillin-resistant *Staphylococcus aureus* (MRSA). *Biochemistry* 53:1548–1550.

Bouchard, D. S., L. Rault, N. Berkova, Y. Le Loir, and S. Even. 2013. Inhibition of *Staphylococcus aureus* invasion into bovine mammary epithelial cells by contact with live *Lactobacillus casei*. *Appl Environ Microbiol* 79:877–885.

Bradley, A. J., J. Breen, B. Payne, V. White, M. J. Green. 2015. An investigation of the efficacy of a polyvalent mastitis vaccine using different vaccination regimens under field conditions in the United Kingdom. *J Dairy Sci* 98:1706–1720. https://doi. org/10.3168/jds.2014–8332

Bytyqi, H., U. Zaugg, K. Sherifi, A. Hamidi, M. Gjonbalaj, S. Muji, and H. Mehmeti. 2010. Influence of management and physiological factors on somatic cell count in raw cow milk in Kosova. *Vet Arhiv* 80:173–183.

Cameron, M., M. Saab, L. Heider, J. T. McClure, J. C. Rodriguez-Lecompte, J. Sanchez. 2016. Antimicrobial susceptibility patterns of environmental Streptococci recovered from bovine milk Samples in the Maritime Provinces of Canada. *Front Vet Sci* 3:79.

Carpenter, C. M. 1922. Experimental production of bovine mastitis with *streptococci* and other bacteria. *J Infect Dis* 31:1–9.

Cheng, W. N., and G. H. Sung. 2020. Bovine mastitis: Risk factors, therapeutic strategies, and alternative treatments-A review. *Asian-Australas J Anim Sci* 33:1699–1713.

Chesneau, O., K. Tsvetkova, and P. Courvalin. 2007. Resistance phenotypes conferred by macrolide phosphotransferases. *FEMS Microbiol Lett* 269:317–322.

Cho, B. W., C. N. Cha, S. M. Lee, M. J. Kim, J. Y. Park, C. Y. Yoo, S. E. Son, S. Kim, and H. J. Lee. 2015. Therapeutic effect of oregano essential oil on subclinical bovine mastitis caused by *Staphylococcus aureus* and *Escherichia coli*. *Korean J Vet Res* 55:253–257.

Cobirka, M., V. Tancin, and P. Slama. 2020. Epidemiology and classification of mastitis. *Animals (Basel)* 10:2212. https://doi.org/10.3390/ani10122212.

Cortinhas, C. A., L. Oliveira, C. Hulland, M. V. Santos, and P. L. Ruegg. 2013. Minimum inhibitory concentrations of cephalosporin compounds and their active metabolites for selected mastitis pathogens. *Am J Vet Res* 74:683–690.

Costa, J. C. M., I. D. F. Espeschit, F. A. Pieri, L. A. Benjarim, and M. A. S. Moreira. 2014. Increase in biofilm formation by *Escherichia coli* under conditions that mimic the mastitic mammary gland. *Cienc Rural* 44:666–671.

Costa, S. S., M. Viveiros, L. Amaral, and I. Couto. 2013. Multidrug efflux pumps in *Staphylococcus aureus*: An update. *Open Microbiol J* 7:59–71.

Cotter, P. D., Ross, R. P., C. Hill. 2013. Bacteriocins – a viable alternative to antibiotics? *Nat Rev Microbiol* 11:95–105

Crowley, R. C., J. A. Leigh, P. N. Ward, H. M. Lappin-Scott and L. D. Bowler. 2011. Differential protein expression in *Streptococcus uberis* under planktonic and biofilm growth conditions. *Appl Environ Microbiol* 77:382–384.

Darwish, S. F., and H. A. E. Asfour. 2013. Investigation of biofilm forming ability in *Staphylococci* causing bovine mastitis using phenotypic and genotypic assays. *Sci World J* 378492. https://doi.org/10.1155/2013/378492

Davies, D. 2003. Understanding biofilm resistance to antibacterial agents. *Nat Rev Drug Discov* 2:114–122.

De Souza Duarte, A. F., H. Ceotto-Vigoder, E. S. Barrias, T. C. B. S. Souto-Padrón, I. F. Nes, and M. D. C. de Freire Bastos. 2018. Hyicin 4244, the first sactibiotic described in *staphylococci*, exhibits an anti-staphylococcal biofilm activity. *Int J Antimicrob Age* 51:349–356.

Defaala, E. N., and H. M. Gharib. 1958. Study of mastitis in a goat caused by *Nocardia asteroides*. *Brit Vet J* 114:143–145.

Dias, R. S., M. R. Eller, V. S. Duarte, A. L. Pereira, C. C. Silva, H. C. Mantovani, L. L. Oliveira, E. de A. M. Silva, and S. O. De Paula. 2013. Use of phages against antibiotic-resistant *Staphylococcus aureus* isolated from bovine mastitis. *J Anim Sci* 91:3930–3939.

Ditchfield, J., C. A. Butas, and R. J. Julian. 1959. Mastitis due to *Nocardia braziliensis*. *Canad J Comp Med Vet Sci* 23:93–96

Dogan, B., S. Klaessig, M. Rishniw, R. A. Almeida, S. P. Oliver, K. Simpson, and Y. H. Schukken. 2006. Adherent and invasive *Escherichia coli* are associated with persistent bovine mastitis. *Vet Microbiol* 116:270–282.

Donovan, D. M., M. Lardeo, and J. Foster-Frey. 2006. Lysis of *staphylococcal* mastitis pathogens by bacteriophage phi11 endolysin. *FEMS Microbiol Lett* 265:133–139.

Dosogne, H., E. Meyer, A. Struk, J. van Loon, A. M. Massart-Leen, and C. Burvenich. 2002. Effect of enrofloxacin treatment of plasma endotoxin during bovine *Escherichia coli* mastitis. *Inflamm Res* 51:201–205.

Drlica, K., and X. Zhao. 1997. DNA gyrase, topoisomerase IV, and the 4-quinolones. *Microbiol Mol Biol Rev* 61:377–392.

Eckel, E. F., and B. N. Ametaj. 2016. Invited review: Roles of bacterial endotoxins in the etiopathogenesis of periparturient diseases of transition dairy cows. *J Dairy Sci* 99:5967–5990.

EFSA. 2011. Scientific opinion on the public health risks of bacterial strains producing extended-spectrum β-lactamases and/or AmpCβ-lactamases in food and food-producing animals. *EFSA J* 9:2322.

Ehinger, A. M., and M. Kietzmann. 2000. Tissue distribution of oxacillin and ampicillin in the isolated perfused bovine udder. *J Vet Med A* 47:157–168.

El Jakee, J., N. E. Aref, A. Gomaa, M. ElHariri, H. M. Galal, S. A. Omar, A. Samir. 2013. Emerging of coagulase negative *staphylococci* as a cause of mastitis in dairy animals: An environmental hazard. *Int J Vet Sci Med* 1:74–78.

Elhadidy, M., and E. Zahran. 2014. Biofilm mediates Enterococcus faecalis adhesion, invasion, and survival into bovine mammary epithelial cells. *Lett Appl Microbiol* 58:248–254.

Entorf, M., A. T. Feßler, K. Kadlec, H. Kaspar, J. Mankertz, T. Peters, and S. Schwarz. 2014. Tylosin susceptibility of S*taphylococci* from bovine mastitis. *Vet Microbiol* 171:368–373.

Erskine, R. J. 2003. Antibacterial therapy of clinical mastitis-part I. Drug selection. Part II Administration. In *North American Veterinary Conference, Proceedings*, 13–16.

Etifu, M., and M. Tilahun. 2019. Prevalence of bovine mastitis, risk factors, isolation and antibio gram of major pathogens in Mid Rift Valley, Ethiopia. *Int J Livest Prod* 10:14–23.

Evans, A. C. 1918. A Streptothrix (*Nocardia*) infection of cows' udders. *J Infect Dis* 23:373–376.

Fabres-Klein, M. H., M. J. Caizer Santos, R. Contelli Klein, G. Nunes de Souza, and A. de Oliveira Barros. 2015. An association between milk and slime increases biofilm production by bovine *Staphylococcus aureus*. *BMC Vet Res* 11:1–8.

Fang, W., and S. Pyörälä. 1996. Mastitis causing Escherichia coli: Serum sensitivity and susceptibility to selected antibacterials in milk. *J Dairy Sci* 79:76–82.

FAO. 2014. *Impact of Mastitis in Small Scale Dairy Production Systems*. Animal Production and Health Working Paper. No. 13. Rome.

Federman, C., J. Joo, J. A. Almario, S. Salaheen, and D. Biswas. 2016. Citrus-derived oil inhibits *Staphylococcus aureus* growth and alters its interactions with bovine mammary cells. *J Dairy Sci* 99:3667–3674.

Fisseha, D., T. S. Tessema, and B. T. Birhanu. 2020. Aerobic bacterial isolates, incidence rate and associated risk factors of heifer and cow mastitis in and around Debre-Libanos district, Oromia, Ethiopia. *J Vet Med Anim Health* 12:7–13.

Flöck, M., and P. Winter. 2006. Diagnostic ultrasonography in cattle with diseases of the mammary gland. *The Vet J* 171:314–321.

Franklin, A., M. Horn af Rantzien, N. Obel, K. Ostensson, and G. Astrom. 1986. Concentrations of penicillin, streptomycin, and spiramycin in bovine udder tissue liquids. *Am J Vet Res* 47:804–807.

Freick, M., Y. Frank, K. Steinert, A Hamedy, O Passarge, and A. Sorbiraj. 2016. Mastitis vaccination using a commercial polyvalent vaccine or a herd-specific *Staphylococcus aureus* vaccine. *Tierarztl Prax Ausg G Grosstiere Nutztiere* 44:219–229. http://doi.org/10.15653/TPG150912

Fu, Y., R. Gao, Y. Cao, M. Guo, Z. Wei, E. Zhou, Y. Li, M. Yao, Z. Yang, and N. Zhang. 2014. Curcumin attenuates inflammatory responses by suppressing TLR4-mediated NF-κB signaling pathway in lipopolysaccharide-induced mastitis in mice. *Int Immunopharmacol* 20:54–58.

Gajdács, M., E. Urbán, A. Stájer, and Z. Baráth. 2021. Antimicrobial resistance in the context of the sustainable development goals: A brief review. *Eur J Investig Health Psychol Educ* 11:71–82.

Garvey, M. I., and L. J. V. Piddock. 2008. The efflux pump inhibitor reserpine selects multidrug-resistant *Streptococcus pneumoniae* strains that overexpress the ABC transporters PatA and PatB. *Antimicrob Agents Chemother* 52:1677–1685.

Ghose, B., R. Sharda, D. Chhabra, U. K. Garg, and S. Tiwari. 2001. Clinical mastitis in cows of Malwa region of Madhya Pradesh: Incidence, etiology and antibiogram of bacterial isolates. *Indian J* 349–352.

Gianneechini, R. E., C. Concha, and A. Franklin. 2002. Antimicrobial susceptibility of udder pathogens isolated from dairy herds in the west littoral region of Uruguay. *Acta Vet Scand* 43:1–10.

Giovannini, A. E. J., B. H. P. van den Borne, S. K. Wall, O. Wellnitz, R. M. Bruckmaier, and C. Spadavecchia. 2017. Experimentally induced subclinical mastitis: Are lipopolysaccharide and lipoteichoic acid eliciting similar pain responses? *Acta Vet Scand* 59:1–13. https://doi.org/10.1186/s13028-017-0306-z

Godoy-Santos, F., M. S. Pinto, A. A. Barbosa, M. A. Brito, H. C. Mantovani. 2019. Efficacy of a ruminal bacteriocin against pure and mixed cultures of bovine mastitis pathogens. *Indian J Microbiol* 59:304–312.

Gomes, F., M. J. Saavedra, M. Henriques. 2016. Bovine mastitis disease/pathogenicity: Evidence of the potential role of microbial biofilms. *Pathog Dis* 74:1–7.

Gonzalo, C., J. R. Martínez, J. A. Carriedo, and F. San Primitivo. 2003. Fossomatic cell-counting on ewe milk: Comparison with direct microscopy and study of variation factors. *J Dairy Sci* 86:138–145.

Gray, J. D. A. 1935. An aerobic acid-fast Actinomyces in cow's milk due to an udder infection. *J Dairy Res* 6:49.

Gruet, P., P. Maincent, X. Berthelot, and V. Kaltsatos. 2001. Bovine mastitis and intramammary drug delivery: Review and perspectives. *Ad Drug Delivery Rev* 50:245–259.

Gwatkin, R., S. Hadwen, and H. M. Legard. 1937. Bovine mastitis: Notes on incidence, aetiology and diagnosis. *Can J Comp Med* 1:7–16.

Haenni, M., A. Lupo, and J. Y. Madec. 2018. Antimicrobial resistance in *Streptococcus* spp. *Microbiol Spectr* 6:10–1128. https://doi.org/10.1128/microbiolspec.arba-0008–2017

Hakogi, E., H. Tamura, S. Tanaka, A. Kohata, Y. Shimada, K. Tabuchi. 1989. Endotoxin levels in milk and plasma of mastitis-affected cow measured with a chromogenic limulus text. *Vet Microbiol* 20:267–274.

Hale, H. H., C. F. Helmboldt, W. N. Plastridge, and E. F. Stula. 1962. Bovine mastitis caused by a *Mycoplasma* species. *Cornell Vet* 52:582–591.

Hamadani, H., A. A. Khan, M. T. Banday, I. Ashraf, N. Handoo, A. Bashir, and A. Hamadani. 2013. Bovine mastitis- A disease of serious concern for dairy farmers. *Int J Livest Res* 3:42–55.

Hamza, A., S. Perveen, Z. Abbas, S. Ur Rehman. 2016. The lytic SA phage demonstrate bactericidal activity against mastitis causing *Staphylococcus aureus*. *Open Life Sci* 11:39–45.

He, X., Z. Wei, E. Zhou, L. Chen, J. Kou, J. Wang, and Z. Yang. 2015. Baicalein attenuates inflammatory responses by suppressing TLR4 mediated NF-κB and MAPK signaling pathways in LPS-induced mastitis in mice. *Int Immunopharmacol* 28:470–476.

Heeschen, W. 1975. Determination of somatic cells in milk (technical aspects of counting). *Bull. Int Dairy Fed* 85:79–92.

Hendriksen, R. S., D. J. Mevius, A. Schroeter, C. Teale, D. Meunier, P. Butaye, A. Franco, A. Utinare, A. Amado, M. Moreno, and C. Greko. 2008. Prevalence of antimicrobial resistance among bacterial pathogens isolated from cattle in different European countries: 2002–2004. *Acta Vet Scand* 50:28. https://doi.org/10.1186/1751-0147-50-28

Hillerton, J. E., and A. J. Bramley. 1989. Infection following challenge of the lactating and dry udder of dairy cows with *Actinomyces pyogenes* and *Peptostreptococcus indolicus*. *Br Vet J* 145:148–158.

Hoe, F. G., and P. L. Ruegg. 2006. Opinions and practices of Wisconsin dairy producers about biosecurity and animal well-being. *J Dairy Sci* 89:2297–2308

Hogan, J., and K. L. Smith. 2012. Managing environmental mastitis. *Vet Clin Food Anim Pract* 28:217–224.

Høiby, N., T. Bjarnsholt, M. Givskov, S. Molin, and O. Ciofu. 2010. Antibiotic resistance of bacterial biofilms. *Int J Antimicrob Agents* 35:322–332.

Hu, H., Z. Fang, T. Mu, Z. Wang, Y. Ma, and Y. Ma. 2021. Application of metabolomics in diagnosis of cow mastitis: A review. *Front Vet Sci* 8:747519. https://doi.org/10.3389/fvets.2021.747519

International Dairy Federation. 2013. Guidelines for Use and Interpretation of Bovine Milk Somatic Cell Counts (SCC) in the Dairy Industry. *Bull Int Dairy Fed* 466/2013

Ismail, Z. B. 2017. Mastitis vaccines in dairy cows: Recent developments and recommendations of application. *Vet World* 10:1057–1062. http://doi.org/10.14202/vetworld.2017.1057–1062

Jaeger, S., F. Virchow, P. R. Torgerson, M. Bischoff, B. Biner, S. Hartnack, and S. R. Rüegg. 2017. Test characteristics of milk amyloid A ELISA, somatic cell count, and bacteriological culture for detection of intramammary pathogens that cause subclinical mastitis. *J Dairy Sci* 100:7419–7426.

Jagielski, T., K. Roeske, Z. Bakuła, T. Piech, Ł. Wlazło, M. Bochniarz, P. Woch, and H. Krukowski. 2019. A survey on the incidence of Prototheca mastitis in dairy herds in Lublin province, Poland. *J Dairy Sci* 102:619–628.

Jiménez, E., L. Fernández, A. Maldonado, R. Martín, M. Olivares, J. Xaus, and J. M. Rodríguez. 2008. Oral administration of Lactobacillus strains isolated from breast milk as an alternative for the treatment of infectious mastitis during lactation. *Appl Environ Microbiol* 74:4650–4655.

Jones, G. F., and Ward G. E. 1989. Cause, occurrence, and clinical signs of mastitis and anorexia in cows in a Wisconsin study. *J Am Vet Med Assoc* 195:1108–1113.

Kaartinen, L., K. Löhönen, B. Wiese, A. Franklin, and S. Pyörälä. 1999. Pharmacokinetics of sulphadiazine-trimethoprim in lactating dairy cows. *Acta Vet Scand* 40:271–278.

Kang, M. S., H. S., Lim, J. S., Oh, Y. J. Lim, K. Wuertz-Kozak, J. M. Harro, and Y. Achermann. 2017. Antimicrobial activity of *Lactobacillus salivarius* and *Lactobacillus fermentum* against *Staphylococcus aureus*. *Pathog Dis* 75:ftx009. https://doi.org/10.1093/femspd/ftx009

Keane, O. M., K. E. Budd, J. Flynn, and F. McCoy. 2013. Pathogen profile of clinical mastitis in Irish milk-recording herds reveals a complex aetiology. *Vet Record* 173:17–17.

Kehrli, M., and J. Harp. 2001. Immunity in the mammary gland. *Vet Clin North Am Food Anim Pract* 17:495–516.

Kerro Dego, O. J., E. van Dijk, and H. Nederbragt. 2002. Factors involved in the early pathogenesis of bovine Staphylococcus aureus mastitis with emphasis on bacterial adhesion and invasion. A Review. *Vet Q.* 24:181–198.

Kirkeby, C., T. Halasa, M. Farre, G. N. Chehabi, and K. Græsbøll. 2021. Transmission dynamics of *Corynebacterium* spp. Within two danish dairy cattle herds. *Front Vet Sci* 8:735345.

Kitazaki, K., S. Koga, K. Nagatoshi, K. Kuwano, T. Zendo, J. Nakayama, K. Sonomoto, H. Ano, H. Katamoto. 2017. In vitro synergistic activities of cefazolin and nisin A against mastitis pathogens. *J Vet Medical Sci* 79:1472–1479.

Klobucar, K., and E. D. Brown. 2022. New potentiators of ineffective antibiotics: Targeting the Gram-negative outer membrane to overcome intrinsic resistance. *Curr Opin Chem Biol* 66:102099.

Kober, A. H., S. Saha, M. A. Islam, M. S. R. Rajoka, K. Fukuyama, H. Aso, J. Villena, and H. Kitazawa. 2022. Immunomodulatory effects of probiotics: A novel preventive approach for the control of bovine mastitis. *Microorganisms* 10:2255. https://doi.org/10.3390/microorganisms10112255

Krishnamoorthy, P., A. L. Goudar, K. P. Suresh, and P. Roy. 2021. Global and countrywide prevalence of subclinical and clinical mastitis in dairy cattle and buffaloes by systematic review and meta-analysis. *Res Vet Sci* 136:561–586.

Krömker, V., F. Reinecke, J. H. Paduch, et al. 2014. Bovine *Streptococcus uberis* intramammary infections and mastitis. *Clin Microbiol* 3:157. https://doi.org/10.4172/2327–5073.1000157.

Kutter, E., D. De Vos, G. Gvasalia, Z. Alavidze, L. Gogokhia, S. Kuhl, and S. T. Abedon. 2010. Phage therapy in clinical practice: Treatment of human infections. *Curr Pharm Biotechnol* 11:69–86.

Kwiatek, M., S. Parasion, L. Mizak, R. Gryko, M. Bartoszcze, and J. Kocik. 2012. Characterization of a bacteriophage, isolated from a cow with mastitis, that is lytic against *Staphylococcus aureus* strains. *Arch Virol* 157:225–234.

Lancet, T. 2022. Antimicrobial resistance: Time to repurpose the Global Fund. *Lancet* 399:335. https://doi.org/10.1016/S0140–6736(22)00091–5

Landin, H., M. J. Mörk, M. Larsson, and K. P. Waller. 2015. Vaccination against *Staphylococcus aureus* mastitis in two Swedish dairy herds. *Acta Vet Scand* 57:81. https://doi.org/10.1186/s13028- 015-0171-6 74.

Li, L., and Z. Zhang. 2014. Isolation and characterization of a virulent bacteriophage SPW specific for *Staphylococcus aureus* isolated from bovine mastitis of lactating dairy cattle. *Mol Biol Rep* 41:5829–5838.

Liang, D., F. Li, Y. Fu, Y. Cao, X. Song, T. Wang, W. Wang, M. Guo, E. Zhou, D. Li, and Z. Yang. 2014. Thymol inhibits LPS-stimulated inflammatory response via down-regulation of NF-κB and MAPK signaling pathways in mouse mammary epithelial cells. *Inflammation* 37:214–222.

Llarrull, L. I., and S. Mobashery. 2012. Dissection of events in the resistance to β-lactam antibiotics mediated by the protein BlaR1 from *Staphylococcus aureus*. *Biochemistry* 51:4642–4649.

Loch, I. M., K. Glenn, and R. N. Zadoks. 2005. Macrolide and lincosamide resistance genes of environmental *streptococci* from bovine milk. *Vet Microb* 111:133–138.

Loong, S. K., H. Y. Lee, J. J. Khoo, F. S. Lim, S. N. Ahmad-Nasrah, A. S. Azman, C. Suntharalingam, C. Panchadcharam, and S. AbuBakar. 2019. Microbiological analysis of raw milk unveiled the presence of a dairy contaminant, *Corynebacterium lipophiloflavum*. *J Appl Biol Biotechnol* 7:41–44.

Louhi, M., K. Inkinen, V. Myllys, and M. Sandholm. 1992. Relevance of sensitivity testings (MIC) of *S. aureus* to predict the antibacterial action in milk. *J Vet Med B* 39:253–262.

Loy, J. D., M. L. Clawson, P. R. Adkins, and J. R. Middleton. 2023. Current and emerging diagnostic approaches to bacterial diseases of ruminants. *Vet Clin North Am Food Anim Pract* 39:93–114.

Lücken, A., N. Wente, Y. Zhang, S. Woudstra, and V. Krömker. 2021. *Corynebacteria* in bovine quarter milk samples-species and somatic cell counts. *Pathogens* 10:831. https://doi.org/10.3390/pathogens10070831

Luo, S., Y. Wang, X. Kang, P. Liu, and G. Wang. 2022. Research progress on the association between mastitis and gastrointestinal microbes in dairy cows and the effect of probiotics. *Microb Pathogenesis* 173:105809. https://doi:10.1016/j.micpath.2022.105809

MacDiarmid, S. C. 1978. Antibacterial drugs used against mastitis in cattle by the systemic route. *New Zeal Vet J* 26:290–295.

Majumder, S., D. Jung, J. Ronholm, and S. George. 2021. Prevalence and mechanisms of antibiotic resistance in *Escherichia coli* isolated from mastitic dairy cattle in Canada. *BMC Microbiol* 21:222.

Marino, M., Z. Jamal, and P. M. Zito. 2024. Pharmacodynamics. (Updated 2023 Jan 29). In. *StatPearls* [Internet]. Treasure Island (FL): StatPearls Publishing. January. Available from: https://www.ncbi.nlm.nih.gov/books/NBK507791/

Mashburn-Warren, L. R., R. J. McLean, and M. Whiteley. 2008. Gram-negative outer membrane vesicles: Beyond the cell surface. *Geobiology* 6:214–219

Massé, J., S. Dufour, and M. Archambault. 2020. Characterization of *Klebsiella* isolates obtained from clinical mastitis cases in dairy cattle. *J Dairy Sci* 103:3392–3400.

Mathers, G. 1916. Different types of *Streptococci* and their relation to bovine mastitis. *J Infect Dis* 19.222 235.

Mathur, H., D. Field, M. C. Rea, P. D. Cotter, C. Hill, and R. P. Ross. 2018. Fighting biofilms with lantibiotics and other groups of bacteriocins. *NPJ Biofilms Microbiomes* 4:9. https://doi.org/10.1038/s41522-018-0053-6

Mbindyo, C. M., G. C. Gitao, and C. M. Mulei. 2020. Prevalence, etiology, and risk factors of mastitis in dairy cattle in Embu and Kajiado Counties. Kenya. *Vet Med Int* 8831172. https://doi.org/10.1155/2020/8831172

McDougall, S., K. E. Agnew, R. Cursons, X. X. Hou, and C. R. W. Compton. 2007. Parenteral treatment of clinical mastitis with tylosin base or penethamate hydriodide in dairy cattle. *J Dairy Sci* 90:779–789.

McDougall, S., L. Clausen, H. J. Ha, I. Gibson, M. Bryan, N. Hadjirin, E. Lay, C. Raisen, X. Ba, O. Restif, J. Parkhill, and M. A. Holmes. 2020. Mechanisms of β-lactam resistance of *Streptococcus uberis* isolated from bovine mastitis cases. *Vet Microbiol* 242:108592.

Metzger, S. A., and J. S. Hogan. 2013. Short communication: Antimicrobial susceptibility and frequency of resistance genes in Escherichia coli isolated from bovine mastitis. *J Dairy Sci* 96:3044–3049.

Milanov, D., B. Prunic, M. Velhner, D. Todorovic, and V. Polacek. 2015. Investigation of biofilm formation and phylogenetic typing of *Escherichia coli* strains isolated from milk of cows with mastitis. *Acta Vet* 65:202–216.

More, S. J., T. A. Clegg, P. J. Lynch, and L. O'Grady. 2013. The effect of somatic cell count data adjustment and interpretation, as outlined in European Union legislation, on herd eligibility to supply raw milk for processing of dairy products. *J Dairy Sci* 96:3671–3681.

Mpatswenumugabo, J. P., L. C. Bebora, G. C. Gitao, V. A. Mobegi, B. Iraguha, O. Kamana, B. Shumbusho. 2017. Prevalence of subclinical mastitis and distribution of pathogens in dairy farms of Rubavu and Nyabihu Districts, Rwanda. *J Vet Med* 8456713. https://doi.org/10.1155/2017/8456713

Mtmch-Petersen, E. 1954. Actinomyces (*Nocardia*) Sp. From a bovine udder infection. *Aust Vet J* 30:97.

Myllys, V., K. Asplund., E. Brofeldt, V. Hirvelä-Koski, T. Honkanen-Buzalski, J. Junttila, L. Kulkas, O. Myllykangas, M. Niskanen, H. Saloniemi. 1998. Bovine mastitis in Finland in 1988 and 1995—Changes in prevalence and antimicrobial resistance. *Acta Vet Scand* 39:119.

Nickerson, S. C., W. E. Owens, and R. L. Boddie. 1995. Mastitis in dairy heifers: Initial studies on prevalence and control. *J Dairy Sci* 78:1607–1618.

Nobrega, D. B., A. P. Calarga, L. C. Nascimento, C. G. Chande Vasconcelos, E. M. de Lima, H. Langoni, and M. Brocchi. 2021. Molecular characterization of antimicrobial resistance in *Klebsiella pneumoniae* isolated from Brazilian dairy herds. *J Dairy Sci* 104:7210–7224.

Nobrega, D. B., S. Naushad, S. A. Naqvi, L. A. Z. Condas, V. Saini, J. P. Kastelic, C. Luby, J. De Buck, and H. W. Barkema. 2018. Prevalence and genetic basis of antimicrobial resistance in nonaureus *Staphylococci* isolated from Canadian dairy herds. *Front Microbiol* 9:256. https://doi.org/10.3389/fmicb.2018.00256

O'Flaherty, S., R. P. Ross, J. Flynn, W. J. Meaney, G. F. Fitzgerald, and A. Coffey. 2005a. Isolation and characterization of two antistaphylococcal bacteriophages specific for pathogenic *Staphylococcus aureus* associated with bovine infections. *Lett Appl Microbiol* 41:482–486.

O'Flaherty, S., R. P. Ross, W. Meaney, G. F. Fitzgerald, M. F. Elbreki, A. Coffey. 2005b. Potential of the polyvalent anti-staphylococcus bacteriophage K for control of antibiotic-resistant staphylococci from hospitals. *Appl Environ Microbiol* 71:1836–1842.

Oftedal, T. F., K. V. Ovchinnikov, K. A. Hestad, O. Goldbeck, D. Porcellato, J. Narvhus, C. U. Riedel, M. Kjos, and D. B. Diep. 2021. Ubericin K, a new pore-forming bacteriocin targeting mannose-PTS. *Microbiol Spectrum* 9:e0029921.

Ogola, H., A. Shitandi, and J. Nanua. 2007. Effect of mastitis on raw milk compositional quality. *J Vet Sci* 8:237–242.

Oliveira, L. 2012. *Characteristics of Clinical Mastitis Occurring in Cows on Large Dairy Herds in Wisconsin* (Dept. of Dairy Science. PhD dissertation, Madison: University of Wisconsin)

Oliveira, L., and P. L. Ruegg. 2014. Treatments of clinical mastitis occurring in cows on 51 large dairy herds in Wisconsin. *J Dairy Sci* 97:5426–5436 http://doi.org/10.3168/jds.2013–7756

Olsen, J. E., H. Christensen, and F. M. Aarestrup. 2006. Diversity and evolution of blaZ from *Staphylococcus aureus* and coagulase-negative *staphylococci*. *J Antimicrob Chemother* 57:450–460.

Olson, M. E., H. Ceri., D. W. Morck, A. G. Buret, and R. R. Read. 2002. Biofilm bacteria: Formation and comparative susceptibility to antibiotics. *Can J Vet Res* 66:86–92.

Owens, W. E., S. C. Nickerson, and C. H. Ray. 1999. Efficacy of parenterally or intramammarily administered tilmicosin or ceftiofur against *Staphylococcus aureus* mastitis during lactation. *J Dairy Sci* 82:645–647.

Padmini, N., A. A. K. Ajilda, N. Sivakumar, and G. Selvakumar. 2017. Extended spectrum β-lactamase producing *Escherichia coli* and *Klebsiella pneumoniae*: Critical tools for antibiotic resistance pattern. *J Basic Microbiol* 57:460–470.

Parker, A. M., P. A. Sheehy, M. S. Hazelton, K. L. Bosward, and J. K. House. 2018. A review of *Mycoplasma* diagnostics in cattle. *J Vet Intern Med* 32:1241–1252.

Pascu, C., V. Herman, I. Iancu, and L. Costinar. 2022. Etiology of mastitis and antimicrobial resistance in dairy cattle farms in the western part of Romania. *Antibiotics* 11:57. https://doi.org/10.3390/antibiotics11010057

Passantino, A. 2007. Ethical aspects for veterinarians regarding antimicrobial drug use in Italy. *Int J Antimicrob Agents* 29:240–244.

Pengov, A., and S. Ceru. 2003. Antimicrobial drug susceptibility of *Staphylococcus aureus* strains isolated from bovine and ovine mammary glands. *J Dairy Sci* 86:3157–3163.

Pérez-Ibarreche, M., D. Field, R. P. Ross, and C. Hill. 2021. A bioengineered nisin derivative to control *Streptococcus uberis* biofilms. *Appl and Environ Microbiol* 87:e0039121.

Pieterse, R. 2008. *Control of Bacterial Pathogens Associated With Mastitis in Dairy Cows With Natural Antimicrobial Peptides Produced by Lactic Acid Bacteria* (Doctoral dissertation, Stellenbosch: Stellenbosch University).

Pilloud, M. 1973. Pharmacokinetics, plasma protein binding and dosage of oxytetracycline in callie and horses. *Res Vet Sci* 15:224–230.

Pitkälä, A., M. Haveri, S. Pyörälä, V. Myllys, and T. Honkanen-Buzaski. 2004. Bovine mastitis in Finland 2001-prevalence, distribution of bacteria and antimicrobial resistance. *J Dairy Sci* 87:2433–2441.

Plastridge, W. N., and E. O. Anderson. 1936. Detection of shedders of streptococci responsible for infectious bovine mastitis. *Am J Public Health Nations Health* 26:711–715.

Poirel, L., J.-Y. Madec, A. Lupo, A.-K. Schink, N. Kieffer, P. Nordmann, and S. Schwarz. 2018. Antimicrobial resistance in *Escherichia coli*. *Microbiol Spectr* 6:10.1128. https://doi.org/10.1128/microbiolspec.ARBA-0026–2017

Pol, M., and P. L. Ruegg. 2007. Treatment practices and quantification of antimicrobial drug usage in conventional and organic dairy farms in Wisconsin. *J Dairy Sci* 90:249–261.

Poutrel, B., S. Bareille, G. Lequeux, and F. Leboeuf. 2018. Prevalence of mastitis pathogens in France: Antimicrobial susceptibility of *Staphylococcus aureus*, *Streptococcus uberis* and *Escherichia coli*. *J Vet Sci Technol* 9:1–3.

Pyörälä, S. 2009. Treatment of mastitis during lactation. *Irish Vet J* 62:1–5.

Qadri, K., S. Ganguly, P. A. Para, P. K. Praveen, and R. Wakchaure. 2015.Summer mastitis in cattle: A review. *J Biol Chem Res* 32:1006–1009.

Qu, Y., H. Zhao, D. B. Nobrega, E. R. Cobo, B. Han, Z. Zhao, S. Li, M. Li, H. W. Barkema, and J. Gao. 2019. Molecular epidemiology and distribution of antimicrobial resistance genes of *Staphylococcus* species isolated from Chinese dairy cows with clinical mastitis. *J Dairy Sci* 102:1571–1583.

Quintela, L. A., M. Barrio, A. I. Peña, J. J. Becerra, J. Cainzos, P. G. Herradón, and C. C. Díaz. 2012. Use of ultrasound in the reproductive management of dairy cattle. *Reprod Domest Anim* 47:34–44.

Radostits, O. M., C. C. Gay, K. W. Hinchcliff, and P. D. Constable. 2007. *Veterinary Medicine. A Textbook of the Diseases of Cattle, Sheep, Pigs, Goats and Horses*. 10th edn. Saunders Elsevier. pp. 1045–1046.

Rai, A. K., A. Nayak, J. Jogi, V. Gupta, R. V. Singh, K. K. Jadav, P. Shakya, and B. M. S. Dhakar. 2022. Prevalence of clinical and subclinical mastitis in dairy cows and buffaloes of Jabalpur district of Madhya Pradesh. *Pharma Innov J* 11:4771–4773.

Raymond, M. J., R. D. Wohrle, and D. R. Call. 2006. Assessment and promotion of judicious antibiotic use on dairy farms in Washington state. *J Dairy Sci* 89:3228–3240.

Rosell, J. M., and W. T. Miller. 1933. A biological and bacteriological study of bovine mastitis. *J Am Vet Med Assoc* 82:587–607.

Rosini, R., and I. Margarit. 2015. Biofilm formation by *Streptococcus* agalactiae: Influence of environmental conditions and implicated virulence factors. *Front Cell Infect Microbiol* 5:6.

Rowe, S., J. K. House, and R. N. Zadoks. 2024. Milk as diagnostic fluid for udder health management. *Australian Vet J* 102:5–10.

Rubin, J. E., K. R. Ball, M. Chirino-Trejo. 2011. Antimicrobial susceptibility of *Staphylococcus aureus* and *Staphylococcus pseudintermedius* isolated from various animals. *Can Vet J* 52:153–157.

Ruegg, P. L. 2017. A 100-year review: Mastitis detection, management, and prevention. *J Dairy Sci* 100:10381–10397.

Ryan, K. J., and C. G. Ray. 2004. *Sherris Medical Microbiology: An Introduction to Infectious Diseases*. 4th edn. McGraw Hill Publishers, New York, USA pp. 937–994.

Saini, V., J. T. McClure, D. Léger, G. P. Keefe, D. T. Scholl, D. W. Morck, and H. W. Barkema. 2012. Antimicrobial resistance profiles of common mastitis pathogens on Canadian dairy farms. *J Dairy Sci* 95:4319–4332.

Sanders, P., G. Moulin, P. Guillot, M. Dacom, P. Perjant, B. Delepine, C. Gaudiche, and D. Mourot. 1992. Pharmacokinetics of spiramycin after intravenous, intramuscular and subcutaneous administration in lactating cows. *J Vet Pharmacol Therap* 15:53–61.

Sang, Y., and F. Blecha. 2008. Antimicrobial peptides and bacteriocins: Alternatives to traditional antibiotics. *Anim Health Res Rev* 9:227–235.

Saravanan, S., and K. M. Palanivel. 2020. Acute gangrenous and haemorrhagic mastitis due to *Bacillus* Species in a preparturient cow. *Rumin Sci* 9:199–201.

Savage, W. G. 1920. Domestic animals as a factor in the spread of infection. *J R Sanit Inst* 41:285–291.

Sawant, A. A., B. E. Gillespie, and S. P. Oliver. 2009. Antimicrobial susceptibility of coagulase-negative *Staphylococcus* species isolated from bovine milk. *Vet Microbiol* 134:73–81.

Schmelcher, M., A. M. Powell, M. J. Camp, C. S. Pohl, and D. M. Donovan. 2015. Synergistic streptococcal phage lambda SA2 and B30 endolysins kill *Streptococci* in cow milk and in a mouse model of mastitis. *Appl Microbiol Biotechnol* 99:8475–8486.

Schukken, Y., V. Bronzo, C. Locatelli, C. Pollera, N. Rota, A. Casula, F. Testa, L. Scaccabarozzi, R. March, D. Zalduendo, and R. Guix. 2014. Efficacy of vaccination on *Staphylococcus aureus* and coagulase-negative *Staphylococci* intramammary infection dynamics in 2 dairy herds. *J Dairy Sci* 97:5250–5264. https://doi.org/10.3168/jds.2014–800872

Schwarz, S., A. T. Feßler, I. Loncaric, C. Wu, K. Kadlec, Y. Wang, and J. Shen. 2018. Antimicrobial resistance among *Staphylococci* of animal origin. *Microbiol Spectrum* 6:ARBA–0010–2017.

Shaheen, M., H. A. Tantary, and S. U. Nabi. 2016. A treatise on bovine mastitis: Disease and disease economics, etiological basis, risk factors, impacts on human health, therapeutic management, prevention and control strategy. *J Adv Dairy Res* 4:1000150.

Sharma, A., and N. Sindhu. 2007. Occurrence of clinical and subclinical mastitis in buffaloes in the State of Haryana (India). *Italian J Animl Sci* 6:965–967.

Sharma, N., N. K. Singh, and M. S. Bhadwal. 2011. Relationship of somatic cell count and mastitis: An overview. *Asian-Australas J Anim Sci* 24:429–438.

Sheldon, I. M., D. E. Noakes, A. N. Rycroft, D. U. Pfeiffer, and H. Dobson. 2002. Influence of uterine bacterial contamination after parturition on ovarian dominant follicle selection and follicle growth and function in cattle. *Reproduction* 123:837–845.

Silveira-Filho, V. M., I. S. Luz, A. P. F. Campos, W. M. Silva, M. P. S. Barros, E. Medeiros, M. F. L. Freitas, R. A. Mota, M. J. Sena, and T. C. Leal-Balbino. 2014. Antibiotic resistance and molecular analysis of *Staphylococcus aureus* isolated from cow's milk and dairy products in Northeast Brazil. *J Food Prot* 77:583–591.

Simjee, S., J. Amedeo, and A. M. Barletta. 2012. Use of Tylan 200® for the treatment of mastitis caused by *Staphylococcus aureus* and/or *Streptococcus uberis*. In *Udder Health and Communication*. Wageningen Academic. pp. 295–295.

Son, J. S., E. B. Kim, S. J. Lee, S. Y. Jun, S. J. Yoon, S. H. Kang, and Y. J. Choi. 2010. Characterization of *Staphylococcus aureus* derived from bovine mastitis and isolation of two lytic bacteriophages. *J Gen Appl Microbiol* 56:347–353.

Sordillo, L. M. 2005. Factors affecting mammary gland immunity and mastitis susceptibility. *Livest Prod Sci* 98:89–99.

Soucie, J. M. 1994. *Epidemiology of Kidney Stones in the United States*. Emory University, Atlanta, GA, USA. ISBN 9798208369999.

Souza, R. F. S., L. Rault, N. Seyffert, V. Azevedo, Y. Le Loir, and S. Even. 2018. Lactobacillus *casei* BL23 modulates the innate immune response in *Staphylococcus aureus*-stimulated bovine mammary epithelial cells. *Benef Microbes* 9:985–995.

Sundekilde, U. K., N. A. Poulsen, L. B. Larsen, and H. C. Bertram. 2013. Nuclear magnetic resonance metabonomics reveals strong association between milk metabolites and somatic cell count in bovine milk. *J Dairy Sci* 96(1):290–299.

Suzuki, N., T. Kurose, S. Kaneko, A. Haraguchi, and N. Isobe. 2020. Outcome prediction from the first examination in clinical mastitis using ultrasonography in dairy cows. *Anim Sci J* 91:e13452.

Tahir, A., M. Asif, Z. Abbas, and S. Rehman. 2017. Three bacteriophages SA, SA2 and SNAF can control growth of milk isolated *Staphylococcal* species. *Pakistan J Zool* 49:493–496.

Tian, X. Y., N. Zheng, R. W. Han, H. Ho, J. Wang, Y. T. Wang, S. Q. Wang, H. G. Li, H. W. Liu, and Z. N. Yu. 2019. Antimicrobial resistance and virulence genes of *Streptococcus* isolated from dairy cows with mastitis in China. *Microb Pathog* 131:33–39.

Tomanić, D., M. Samardžija, and Z. Kovačević. 2023. Alternatives to antimicrobial treatment in bovine mastitis therapy: A review. *Antibiotics* 12(4):683.

Tommasoni, C., E. Fiore, A. Lisuzzo, and M. Gianesella. 2023. Mastitis in dairy cattle: On-farm diagnostics and future perspectives. *Animals* 13:2538.

Urakawa, M., T. Zhuang, H. Sato, S. Takanashi, K. Yoshimura, Y. Endo, T. Katsura, T. Umino, K. Tanaka, H. Watanabe, and H. Kobayashi. 2022. Prevention of mastitis in multiparous dairy cows with a previous history of mastitis by oral feeding with probiotic *Bacillus subtilis*. *Anim Sci J* 93:e13764.

Vakkamäki, J., S. Taponen, A. M. Heikkilä, and S. Pyörälä. 2017. Bacteriological etiology and treatment of mastitis in Finnish dairy herds. *Acta Vet Scand* 59:33.

Vélez, J. R., M. Cameron, J. C. Rodríguez-Lecompte, F. Xia, L. C. Heider, M. Saab, J. T. McClure, and J. Sánchez. 2017. Whole-genome sequence analysis of antimicrobial resistance genes in *Streptococcus uberis* and *Streptococcus dysgalactiae* isolates from Canadian dairy herds. *Front Vet Sci* 4:63.

Vidal Amaral, J. R., R. T. Jucá Ramos, F. Almeida Araújo, R. Bentes Kato, F. Figueira Aburjaile, S. de Castro Soares, A. Góes-Neto, M. Matiuzzi da Costa, V. Azevedo, B. Brenig, and S. Soares de Oliveira. 2022. Bacteriocin producing-g Streptococcus agalactiae strains isolated from bovine mastitis in Brazil. *Microorganisms* 10:588. https://doi.org/10.3390/microorganisms10030588

Wagner, S., and R. Erskine. 2006. Antimicrobial drug use in bovine mastitis. In *Antimicrobial Therapy in Veterinary Medicine*. 4th edn. Giguère, S., Prescott, J. D., and Baggot, R. D. et al. (eds). Blackwell, Oxford.

Wallis, J. K., V. Kromker, and J.-H. Paduch. 2019. Biofilm challenge: Lactic acid bacteria isolated from bovine udders versus *Staphylococci*. *Foods* 8:79.

Wawron, W., M. Bochniarz, T. Piech, W. Łopuszyński, and J. Wysocki. 2013. Outbreak of protothecal mastitis in a herd of dairy cows in Poland. *Bull Vet Inst Pulawy* 57:335–339. https://doi.org/10.2478/bvip-2013–0058.

Wei, Z., E. Zhou, C. Guo, Y. Fu, Y. Yu, Y. Li, M. Yao, N. Zhang, and Z. Yang. 2014. Thymol inhibits *Staphylococcus aureus* internalization into bovine mammary epithelial cells by inhibiting NF-κB activation. *Microb Pathogenesis* 71:15–19.

Wellenberg, G. J., W. H. M. van der Poel, and J. T. Van Oirschot. 2002. Viral infections and bovine mastitis: A review. *Vet Microbiol* 88:27–45.

Wendlandt, S., A. T. Feßler, S. Monecke, R. Ehricht, S. Schwarz, and K. Kadlec. 2013. The diversity of antimicrobial resistance genes among *Staphylococci* of animal origin. *Int J Med Microbiol* 303:338–349.

Wendlandt, S., J. Shen, K. Kadlec, Y. Wang, B. Li, W. J. Zhang, A. T. Feßler, C. Wu, and S. Schwarz. 2015. Multidrug resistance genes in *Staphylococci* from animals that confer resistance to critically and highly important antimicrobial agents in human medicine. *Trends Microbiol* 23:44–54.

WHO. 2010. *Annex I: Report of the Tenth Meeting of the OIE Animal Production Food Safety Working Group*. World Organisation for Animal Health, Paris.

Workineh, S., M. Bayleyegn, H. Mekonnen, and L. N. D. Potgieter. 2002. Prevalence and aetiology of mastitis in cows from two major Ethiopian dairies. *Trop Anim Health Prod* 34:19–25.

Woudstra, S., A. Lücken, N. Wente, Y. Zhang, S. Leimbach, M. K. Gussmann, C. Kirkeby, and V. Krömker. 2023. Reservoirs of *Corynebacterium* spp. in the environment of dairy cows. *Pathogens* 12:139.

Yang, F., S. Zhang, X. Shang, L. Wang, H. Li, and X. Wang. 2018. Characteristics of quinolone-resistant *Escherichia coli* isolated from bovine mastitis in China. *J Dairy Sci* 101:6244–6252.

Yassin, A. F., H. Hupfer, C. Siering, and P. Schumann. 2011. Comparative chemotaxonomic and phylogenetic studies on the genus Arcanobacterium Collins et al. 1982 emend. Lehnen et al. 2006: Proposal for *Trueperella gen.* nov. and emended description of the genus *Arcanobacterium*. *Int J Syst Evol Microbiol* 61:1265–1274.

Yimana, M., and J. Bekele. 2022. Isolation, identification and antimicrobial profile of corynebacterium bovis from selected dairy farm in Bishoftu, central Ethiopia. *Arch Vet Med* 15:69–84.

Zadoks, R. N., E. Scholz, S. M. Rowe, J. M. Norris, H. B. Pooley, and J. House. 2023. A framework for evaluation of on-farm mastitis diagnostics in Australia. *Australian Vet J* 101:142–152.

Zeryehun, T., and G. Abera. 2017. Prevalence and bacterial isolates of mastitis in dairy farms in selected districts of Eastern Harrarghe Zone, Eastern Ethiopia. *J Vet Med* 6498618.

Zheng, Z., P. J. Gorden, X. Xia, Y. Zheng, and G. Li. 2022. Whole-genome analysis of *Klebsiella pneumoniae* from bovine mastitis milk in the U.S. *Environ Microbiol* 24:1183–1199.

Zhou, Y., H. Zhang, H. Bao, X. Wang, and R. Wang. 2017. The lytic activity of recombinant phage lysin LysKDeltaamidase against staphylococcal strains associated with bovine and human infections in the Jiangsu province of China. *Res Vet Sci* 111:113–119.

Zhu, H., M. Du, L. Fox, and M-J. Zhu. 2016. Bactericidal effects of Cinnamon cassia oil against bovine mastitis bacterial pathogens. *Food Control* 66:291–299. https://doi.org/10.1016/j.food cont.2016.02.013

Zigo, F., M. Vasil', S. Ondrašovičová, J. Výrostková, J. Bujok, and E. Pecka-Kielb. 2021. Maintaining optimal mammary gland health and prevention of mastitis. *Front Vet Sci* 8:607311.

Ziv, G. 1969. Antibiotic sensitivity of *Staphylococcus aureus* strains isolated from bovine udders in Israel. *Refuah Vet* 26:104–113.

Ziv, G., A. Rosenzuaig, and A. Schacked. 1970. Antibiotic sensitivity of streptococci isolated from bovine udders in Israel. *Refuah Vet* 27:55–59.

Ziv, G., and M. Storper. 1985. Intramuscular treatment of subclinical staphylococcal mastitis in lactating cows with penicillin G, methicillin and their esters. *J Vet Pharmacol Therap* 8:276–283.

Zygmunt, L., K. Jolanta, and B. Aneta. 2015. Factors determining the susceptibility of cows to mastitis and losses incurred by producers due to the disease-A review. *Annal Anim Sci* 15:1–24. https://doi.org/10.1515/aoas-2015–003

8 Natural Defenses
Exploring Alternatives to Antimicrobials in Animal Husbandry

Charu Gupta

8.1 INTRODUCTION

Antibiotic pollution significantly impacts the biosphere, affecting the structure and function of microbial communities, with adverse effects on plants and the carbon and nitrogen cycles. Antibiotic-resistant genes and bacteria have proliferated because of the extensive use of antibiotics, endangering both human health and ecosystems. Even minimal exposure can induce resistance in pathogenic bacteria, causing harm to disease prevention and control systems. Antibiotics can also directly harm the human body, leading to allergies, poisoning, and gastrointestinal issues.

Antimicrobial contamination within the environment has a critical effect on the life exercises of creatures and plants, especially the biosphere. It arises with detrimental effects on the ecosystem's trophic natural communities, particularly as a microbial inhibitor. This fundamentally affects the composition and activities of microbial communities within the biological system, causing changes in the carbon and nitrogen cycles that microorganisms participate in as well as the coordination of toxic effects on plants. The extensive use of antimicrobials has led to the emergence of antibiotic-resistant genes (ARGs) and bacteria (ARBs) in the environment, which could be harmful to both human health and the environment.

Indeed, negligible antimicrobial presentation within the environment has the potential to actuate resistance in pathogenic microscopic organisms. Upon entering the human body, these pathogenic microorganisms seriously damage the systems responsible for preventing and controlling infections. Apart from the possible harm posed by ARBs, antimicrobials also have coordinated negative effects and even toxic qualities for human health. The one-time or persistent admissions of antimicrobials past the greatest dose for the human body leads to hypersensitivities, harm, coma, shock, and indeed passing in extreme cases. The ingestion of an expansive number of antimicrobials can crush the adjustment of the microbial community within the gastrointestinal tract and actuate a series of intestinal infections caused by the lopsidedness of the microbial community (Yang et al., 2024).

DOI: 10.1201/9781003269847-8

The increasing prevalence of antimicrobial resistance (AMR) in animal husbandry has become a critical issue, with significant implications for both animal and human health. The use of antimicrobials in livestock production, primarily for growth promotion and disease prevention, has been widespread for decades. However, as a result of this technique, resistant bacterial strains have emerged and spread, and they now pose a threat to human health through direct contact, animal product consumption, and environmental pathways.

As the effectiveness of traditional antibiotics wanes, there is an urgent need to explore and implement alternative strategies that can maintain animal health and productivity without contributing to the AMR crisis (Stoica and Cox, 2021).

The European Medicine Agency (EMA) and the European Food Safety Authority (EFSA) have jointly examined actions taken within the European Union (EU) to reduce the need for and use of antibiotics in food-producing animals, as well as the ensuing effects on AMR. Reduction techniques have been successfully used in some Member States. These protocols include national reduction goals, benchmarking of antibiotic use, approval and restriction restrictions on the use of antibiotics that are very needed, together with improvements in animal husbandry and disease prevention and control techniques. Although there is evidence linking reduced antibiotic usage to decreased AMR, the impact of any one degree is difficult to quantify due to the multitude of factors that contribute to AMR.

Limiting the use of antibiotics requires the implementation of a diverse approach that is tailored to local conditions.

Suggested choices (nonprioritized) incorporate improving national techniques; orchestrating frameworks for checking antimicrobial utilization and AMR improvement; setting up national targets for antimicrobial utilization lessening; utilizing on-farm well-being plans; expanding the obligation of veterinarians for antimicrobial endorsing; preparing, instructing, and raising open-mindfulness; expanding the accessibility of fast and dependable diagnostics; making strides in cultivation and administration strategies for infection avoidance and control; and reconsidering animals' generation frameworks to decrease inalienable infection hazard. A constrained number of ideas give vigorous proof of alternatives to antimicrobials that positively impact well-being parameters. Potential solutions include nipple sealants, bacteriophages, immunomodulators, nipple probiotics and prebiotics, and competitive prohibition.

Improvement of an administrative system that licenses the utilization of items as options ought to be considered. Assisting in the investigation to assess the potential of elective cultivating systems on lessening AMR is additionally prescribed. Creatures enduring bacterial contaminations ought to be treated with antimicrobials based on veterinary determination and medicine (Murphy et al., 2017).

It may have been common practice for decades to utilize the subtherapeutic dosage of antimicrobials in food-animal bolsters to avoid illnesses and improve production in present-day animal production. In the meantime, efforts to develop so-called options for antimicrobials have been sparked by worries about the growing number of bacteria that are resistant to antibiotics as a result of the absurd use of antimicrobials and the disappearance of traditional antimicrobials.

Whether or not the alternatives seem able to truly supplant antimicrobials remains a questionable issue. This chapter summarizes later advancements and viewpoints of alternatives for antimicrobials. All of the following are discussed: plant extricates; inhibitors targeting pathogenicity (bacterial majority detection, biofilm formation, and destructiveness); nourishing chemicals; resistance balancing specialists; bacteriophages and their lysins; antimicrobial peptides; master-, pre-, and synbiotics; and so on. Last, a thorough analysis is done on the alternatives' achievability about antimicrobials. It is difficult to conclude that in the foreseeable future, alternatives could replace antimicrobials in veterinary medications.

At this time, judicious use of antimicrobials and the foundation of logical observing frameworks are the leading and quickest ways to restrain the antagonistic impacts of the manhandling of antimicrobials and to guarantee the security of animal-derived nourishment and environment (Cheng et al., 2014).

8.2 THE RISE OF ANTIMICROBIAL RESISTANCE

The use of antimicrobials in animal husbandry accounts for a substantial proportion of total antibiotic consumption globally. In some countries, up to 80% of all antibiotics are used in the agricultural sector. The World Health Organization has identified the misuse and overuse of antibiotics in agriculture as a key driver of AMR, which is estimated to cause 700,000 deaths annually worldwide, with projections suggesting this could rise to 10 million deaths per year by 2050 if no action is taken (WHO, 2018).

The mechanisms of resistance involve genetic mutations and the acquisition of resistance genes through horizontal gene transfer. The selection pressure exerted by antimicrobial use in animals facilitates the emergence of resistant strains, including multidrug-resistant (MDR) pathogens. Notable examples include methicillin-resistant *Staphylococcus aureus* (MRSA), extended-spectrum β-lactamase (ESBL) producing Enterobacteriaceae, and colistin-resistant Enterobacteriaceae, which pose significant treatment challenges in human medicine.

8.3 ALTERNATIVES TO ANTIMICROBIALS

To combat AMR, various alternatives to traditional antibiotics are being explored. These alternatives aim to maintain animal health and performance while reducing the reliance on antibiotics. Key strategies include the use of probiotics and prebiotics, phage therapy, plant-derived compounds, immunomodulatory approaches, and improved management practices (Laishevtsev et al., 2023). Other key alternatives to antimicrobial growth promoters for broilers include enzymes, (in)organic acids, etheric oils, and immunostimulants, which can improve feed efficiency and performance without causing antibiotic resistance. Alternatives to ah promoters in poultry and swine production can help slow the spread of antibiotic-resistant bacteria, but their widespread use requires careful monitoring and wise use of antimicrobials (Rahman et al., 2022).

8.4 IMMUNOMODULATORY STRATEGIES: ANTIBACTERIAL VACCINES

Vaccination and the use of immune-enhancing agents are crucial for disease prevention in livestock. Vaccines can provide specific protection against pathogens, reducing the need for therapeutic antibiotic use. The development of vaccines for bacterial pathogens such as *Mycoplasma hyopneumoniae* in pigs and *Clostridium perfringens* in poultry has been particularly successful (Richards and Pomorska-Mól, 2020). Moreover, immunomodulators, including cytokines and immunoglobulins, can boost the innate and adaptive immune responses of animals, enhancing their ability to resist infections.

Live attenuated bacteria, replicating transiently in the host, can express a full repertoire of antigens. Live attenuated bacteria, such as *Salmonella*, can express a full range of antigens and have been tested for efficacy. These strains, however, have a higher chance of returning to full virulence and remain in the animal body for a longer period of time. Certain live attenuated *Salmonella* vaccines have not been field tested, and they offer insufficient cross-protection against other serotypes that are not host-adapted.

Therefore, more research is needed to develop effective and safe live attenuated vaccines (Desin et al., 2013).

Because they are produced by killing in vitro–grown bacterial cultures and encasing them in oil-based adjuvants to boost immune responses, killed vaccines are considered safer than live ones. They are produced at a low cost and store well. *Pasteurella multocida, Salmonella, Mycoplasma, Ornithobacterium, Haemophilus, Staphylococcus, Escherichia coli*, and *Bordetella bacterins* are among the current inactivated autogenous poultry vaccines. Nevertheless, there are drawbacks to killed vaccinations, including broad-spectrum protection, antigenic competition, a lack of protective antigens, and safety issues.

Adjuvants are required, limiting delivery options, and most are injectable products not used in intensive broiler operations (Potter et al., 2008).

A subunit vaccine, consisting of single or multiple defined antigens, is less immunogenic and requires appropriate adjuvants. Despite the ongoing development of *Salmonella* subunit vaccines, it is difficult to determine the efficacy of each class. The low absorption in the intestines and antigen degradation make oral subunit vaccinations difficult in large animals.

Despite these challenges, the development of *Salmonella* subunit vaccines is ongoing (Desin et al., 2013).

DNA vaccines are a promising improvement over conventional vaccines, consisting of a genetically engineered plasmid containing the antigen sequences from a pathogen. When injected into the body, the host cells convert the DNA into pathogenic proteins, triggering immune responses. However, DNA vaccines have limited protective capacity and may risk integrating the vector's genetic elements into the host genome. Most literature on DNA vaccines has focused on viral antigens delivered in mouse models, but results are often disappointing in large animals. Therefore, DNA vaccines are unlikely to reach the market until plasmid dose control and effective delivery problems are solved (Potter et al., 2008).

The antibacterial vaccine market is still relatively rare, with no commercially available vaccines for *C. jejuni* or *E. coli O157*, despite immunization against these strains in chicken and cattle. Bacteriocins have not demonstrated any protective effect against *E. Coli O157*. Despite reports of vaccinations against *Brachyspira hyodysenteriae* and *E. coli*, the absence of safety and efficacy data has prevented these vaccines from being commercialized.

S. aureus vaccines are developed for bovine mastitis, but methodological differences and lack of rigid scientific criteria can hinder their efficacy assessment. Additionally, the protection rates of animals by antibacterial vaccines are low, highlighting the need for further research and development (Pereira et al., 2011).

The development of practical and affordable vaccines for poor countries remains a significant challenge. The cost of poultry vaccines and their ability to be used in most cases are the main challenges for mass immunization. Although vaccines may reduce antibiotic reliance, they are complementary rather than replacement (Zhang and Sack, 2012).

8.5 OTHER IMMUNOMODULATORS

Immunomodulators, primarily immunostimulants, can enhance innate immune function and enhance disease resistance. In infectious diseases, they can modulate the immune response to microbes, modify antigen-based responses, and minimize end-organ damage using nonspecific anti-inflammatory agents (Masihi, 2000). β-Glucans, bacterial products, and plant constituents can activate innate defense mechanisms, triggering intracellular gene(s) that produce antimicrobial molecules.

Immunostimulants, including nucleotides, thymosin, and oregano oil, have been used since the 1990s. Probiotics, herbs, and their extracts have also been studied (Thacker, 2010). β-1,3/1,6-Glucan derived from yeast cell walls is fed to animals as a means of defense against pathogens. According to studies, immunostimulants can strengthen an animal's innate defense, giving it resistance to pathogens during times of high stress, such as during grading, reproduction, transfer, and vaccination. Table 8.1 highlights some of the examples of immunostimulants.

Animal species respond differently to immunostimulants, and the connection between dose and effect is not linear. They can hinder the defensive functions of the human immune system and are more effective during or before infection. Adaptive immunologic events may result in immune-mediated processes that compromise systemic or organ-specific homeostasis when immunogenic stimulation endures or when autoregulatory immune mechanisms fall silent. Immunostimulants can strengthen innate defense in larval fish aquaculture without compromising fish development. Immunomodulating a neotenous animal before its immune system is fully developed, however, may harm the establishment of a normal immunologic response. Most immunomodulators enhance the immune system of animals, not directly killing bacteria (Thacker, 2010).

There is currently no uniform standards for evaluating the efficacy and safety of immunostimulants. Goats' immune systems were unaffected by a purported immunostimulant that included Proclin, *Ochrobactrum intermedium* lipopolysaccharides, and *Propionibacterium acnes* extract. It is uncertain how widely accepted the idea

TABLE 8.1
Classification of Immunostimulants

Category	Variety
Amino acids	Arginine, leucine, ubenimex, etc.
Bacterial extracts	β-Glucan, peptidoglycan, lipopolysaccharide, etc.
Biological (cytokines)	Interferon, transfer factor, interleukin, immune globulin, etc.
Chinese herbal remedies	Astragalus, Echinacea, etc.
Synthetic chemicals	Levomisole, cimetidine, sodium houttuyfonate, imiquimod, pidotimod, ubenimex, tilorone, polyinosinic acid, etc.
Hormones and hormone-like substances	Growth hormone, thymosin, metallothionein, thymopentin, etc.
Immunologic adjuvants	Aluminum adjuvant, propolis, liposome, Freund's adjuvant, etc.
Plant polysaccharides	Astragalus polysaccharide, lentinan, algal polysaccharides, ganoderan, *Polyporus* polysaccharide, chitosan, etc.
Vitamins	Vitamin A, vitamin E, vitamin C, etc.

of immunostimulatory plant natural products could be used therapeutically; it just implies that they are "tonic" without elaborating on the mechanisms involved.

Oral plant immunostimulants lack clinical evidence from the primary in vitro studies, and no conclusive data on orally administered immunostimulants are available in the scientific literature (Kak et al., 2012).

8.6 PHAGE THERAPY: BACTERIOPHAGES AND THEIR LYSINS

Bacteriophages and other antimicrobial agents show promise in reducing antibiotic resistance in animal husbandry, while food enzymes show promise for improving nutrient utilization (Laishevtsev et al., 2023).

Bacteriophages, viruses that specifically infect bacteria, offer a promising alternative to antibiotics. They can target specific bacterial pathogens without affecting the beneficial microbiota, reducing the likelihood of resistance development. Phage therapy has been successfully used to control bacterial infections in livestock, such as *Campylobacter* in poultry and *E. coli* in swine (Goodridge and Bisha, 2021). Recent animal studies show that phage therapy is effective in preventing pathogens like *E. coli O157:H7, Salmonella,* and *Campylobacter.* In 2006, the US Food and Drug Administration (FDA) approved a phage cocktail called LMP-102TM for food additives to prevent *Listeria* contamination in meat. In 2007, the US Department of Agriculture approved another phage product for disinfecting *E. coli* in cattle (Wittebole et al., 2014).

In treating bacterial infections, the use of bacteriophages, which may replicate in host cells and generate new lytic phages to keep up with pathogen mutations, has generated debate. Because of its stringent host strain specificity, it might be challenging to identify the precise microbe causing infection before utilizing phage therapy. Furthermore, the limited host range makes it difficult to employ a single type of phage to replace antibiotics.

Phages can be seen by the host's immune system as potential invaders, potentially leading to treatment failure. Phages may also transfer their DNA from one bacterial cell to another, potentially leading to the development of new microbes or more resistant bacteria. Therefore, phages that are unable to package extra host DNA or use host DNA to synthesize their DNA would be preferred (Allen et al., 2013).

Since the pharmacokinetic properties of phages are still not well understood, before conducting extensive clinical studies, the ideal dosage, mode of administration, frequency, and length of treatment must be determined. Since phage therapy has a short half-life, it may work better when used early. Phages can cause the release of toxins, such as endotoxin (LPS), from bacteria, which may cause side effects in the host, such as an inflammatory cascade leading to multiple organ failure (Drulis Kawa et al., 2012).

Through mutation, bacteria can develop resistance against phages; the rates of mutation for phages and antibiotics are 10–6 and 10–7, respectively. Bacterial resistance to a particular phage can be attributed to many mechanisms such as receptor loss or absence, structural alteration, and/or receptor masking.

Some phages can only survive in the intestines when the bacteria count reaches certain numbers, while preparation of phages should be done at a low temperature, as these biological products are not stable (Burrowes, 2011).

The promotion of phage preparations faces challenges due to the lack of large-scale clinical trial data and regulatory loopholes. The approval process used by the FDA and the European Medicines Agency is insufficient to handle the constantly shifting phage combinations, which is necessary for generating new treatments to remain ahead of the advancing MDR bacteria.

Most bacteriophage products are still in the research stage. The regulatory and logistic challenges associated with phage therapy, including specificity and phage-host interactions, need to be addressed for widespread adoption (Miedzybrodzki et al., 2012).

8.7 ENDOLYSINS

Enzymes produced during the late phage lytic cycle are known as endolysins, and they include glucosidase, amidase, endopeptidase, and transglycosylase. They break down the bacterial peptidoglycan to liberate fresh phages from contaminated bacteria. Initially identified in the 1950s, they have shown antibacterial action against *Staphylococcus, Bacillus anthracis, L. monocytogenes,* and *Clostridium butyricum.* They can treat sepsis and certain Gram-positive bacterial infections, such as *Enterococcus faecalis, C. perfringens,* and group B streptococcus. Pneumococcal illness incidence can be decreased by endolysins PAL and Cpl-1 from phage Cpl–1. Vancomycin-resistant *E. faecalis, E. faecium,* and *Staphylococcus* can all be killed by the phage K endolysins LysK and PlyV12. *Clostridium* infections can be treated using endolysins that were identified from phage phi3626 (Courchesne et al., 2009).

Endolysins are enzymes that have a wider antibacterial spectrum than phages, which makes their activities easier to detect. They work by targeting specific elements on bacterial cell walls, causing bacteria to lyse quickly and preventing the development of resistance. However, conventional methods for producing endolysins are complex and expensive, and recombinant DNA technology can lower costs but cannot guarantee the correct structure and full activity of the enzymes. Endolysins should be used to the full dosage because they are easily degraded and lose activity during use and storage. The efficacy of endolysins against Gram-negative bacteria limits the application of single-type endolysins. Future research will focus on chimeric endolysins, which may broaden the antibacterial spectrum (O'Flaherty et al., 2009).

There are not any published clinical trials on endolysins at the moment.

8.8 VIRION-ASSOCIATED PEPTIDOGLYCAN HYDROLASES

The phage lyases known as virion-associated peptidoglycan hydrolases (VAPGHs) hydrolyze bacterial peptidoglycan to enable phage entrance into bacterial cells. Antibacterial properties of many VAPGHs have been discovered; some of them have been reported to be more active against *S. aureus* during the early logarithmic growth phase. Other VAPGHs show lytic action against *S. aureus,* MRSA, *Pseudomonas, E. coli, S. typhimurium, Proteus vulgaris,* and other Gram-negative bacteria. Examples of these are protein 17 from phage P68 and protein gp61 from phage phiMR11. Phage KZ's protein Gp181 has lytic action against *Yersinia, Ralstonia solanacearum,* and these microorganisms. Phage KMV's protein gp36 has a high degree of heat stability and is efficient against both *P. aeruginosa* and *E. coli,* maintaining 26% of its activity after two hours of incubation at 100°C.

Currently, research on purified VAPGHs is limited, and VAPGHs from Gram-negative bacterial phages have a broad antimicrobial spectrum, while those from Gram-positive bacterial phages are only confined to specific host bacteria. VAPGHs also have antibacterial properties against some antibiotic-resistant pathogens, making them an alternative source for bacterial lyases, except for endolysins (Rodriguez-Rubio et al., 2013).

8.9 ANTIMICROBIAL PEPTIDES

Antimicrobial peptides show potential as suitable alternatives to conventional antibiotics in the swine and poultry industries due to their broad-spectrum activity against bacteria, fungi, and viruses, and their low resistance potential (Wang et al., 2016).

The use of antimicrobial peptides (AMPs) has been demonstrated to circumvent common resistance mechanisms, thereby decreasing the efficacy and safety of conventional antibiotics. The antibacterial activity of the novel AMP dendrimer G3KL has been demonstrated against drug-resistant and extensively drug-resistant *Acinetobacter baumannii* and *Pseudomonas aeruginosa*; drug-resistant *Staphylococcus aureus* is inhibited by AMP defensin derived from *Tribolium castaneum*; antimicrobial activity against drug-resistant *Acinetobacter baumannii* has also been demonstrated to be significant. Zhang et al. (2005) have shown that AMPs could inhibit MRSA and multidrug-resistant *Pseudomonas aeruginosa*. Additionally, in our laboratory, we discovered that the artificial antimicrobial peptide sublancin has potent inhibitory efficacy against MRSA.

However, pathogens have evolved mechanisms that resist damage by AMPs, such as proteolytic enzyme degradation, peptide repulsion, membrane efflux pumps, and bacterial-membrane fluidity reduction. Despite these defensive mechanisms, AMPs still provide protection, with a low propensity for resistance development compared to conventional antibiotics (Wang et al., 2016).

The two types of antimicrobial peptides are called ribosomally generated AMPs and nonribosomally synthesized AMPs. Nonribosomal AMPs, which are mostly made by bacteria, comprise sugar-peptide, bacitracin, polymyxin, and gramicidin. When used against Gram-negative bacteria, polymyxin effectively breaks down the cell membranes of the bacterium. Gram-positive bacteria are susceptible to bactericidal effects from bacitracin, a cyclic peptide that inhibits their cell wall peptidoglycans and glycoprotein core oligosaccharides. This action also targets Gram-negative cocci and spirochetes. Since 1960 and 1990, respectively, the United States and China have authorized the use of bacitracin zinc and bacitracin methylene salicylic acid as feed additives, with a broad range of drug compatibility.

Ribosomally synthesized AMPs, found in various organisms like mammals, amphibians, insects, plants, bacteria, and viruses, have various antibacterial, antimycotic, antiprotozoal, antiviral, and antineoplastic properties. On the membranes of bacteria, they can generate electrostatic adsorption with negatively charged phospholipid molecules, which can lead to structural damage. AMPs have been reported to have protective effects on humans and animals, particularly in the case of bacteriocins produced by bacteria. Bacteriocins are classified into four classes: lantibiotics, small heat-stable peptides, large heat-labile proteins, and undefined mixture proteins with lipids and carbohydrates (Bierbaum and Sahl, 2009).

Bacteriocins, including nisin, lactacin, lactocin, helveticin, fermenticin, sakacin, lacticin, plantacin, and subticin, have been identified for their strong killing and suppressive effects on various pathogens, including resistant ones. Although AMPs have good bactericidal effects and are easily digested without adverse effects on feed taste or environmental pollution, there are several constraints associated with their deepening research. Despite their potential, AMPs have been found to have limited authorized uses in foods (Field et al., 2011).

Bacteriocins (AMPs) are used as antibiotic alternatives to livestock due to their high production cost and complex purification process. Scientists are now using genetic engineering techniques to synthesize AMPs due to their peptide nature, which can be modified by protein engineering. Differential activity may result from the lack of consistency between the spatial structures produced by genetic engineering and those of natural AMPs.

Because of their hemolytic properties, some naturally occurring AMPs, including melittin, buthotoxin, and plant AMPs, may be harmful to eukaryotic cells.

A potent *E. coli* displaying multimeric AMPs on the cell surface can be converted into active AMP monomers by pepsin in the stomach of livestock (Molloy et al., 2013).

Most bacteriocins have a limited antibacterial spectrum and are only effective against closely related bacterial species. By lowering matching receptors, altering the structure of cell walls, or switching the main target of bacteriocin, bacteria can become resistant to AMPs. According to Cotter et al. (2013), some bacteria can also acquire resistance by using bacteriocin-hydrolase or immunologic genes.

Few research has been done in vivo on the pharmacodynamics, pharmacokinetics, and stability of AMPs, and it is unknown if AMPs or their metabolites are toxic to the body. Furthermore, during usage, AMPs are readily degraded by proteases in the alimentary canal and become unstable during transit (Nocek et al., 2012). The effects of antimicrobial peptides on the intestinal microbiota of poultry and pigs are shown in Table 8.2.

8.10 PRO-, PRE-, AND SYNBIOTICS – PROBIOTICS OR 'DIRECT-FED MICROBIALS'

The removal of antimicrobial growth promoters (AGP) has created a problem for animal nutrition, which has forced researchers to look at other strategies for managing and preventing the growth of harmful bacteria. Animal welfare, health, and productivity are all negatively impacted when antibacterials are removed from animal diets. The development of alternatives to antibiotics to preserve or enhance animal health and performance has been the subject of much research. Modulating the gut microbiota with zoo-technical feed additives such as prebiotics and probiotics to support animal husbandry, including livestock, poultry, and fish farming, is crucial for maximizing productivity and maintaining animal health and welfare (Anadón et al., 2019).

TABLE 8.2
Antimicrobial Peptide Effects on the Gut Microbiota of Poultry and Pigs

Antimicrobial Peptide	Visceral	Therapy Impacts
Antimicrobial peptide A3	Broilers Weanling pigs	Reduced coliform bacteria and *Clostridium* spp. counts in the feces of broilers and also in their ileum, caecum, and colon
Antimicrobial peptide P5	Broilers Weanling pigs	Reduced total anaerobic bacterial count including coliform bacteria in their excreta
Bovine lactoferrin-lactoferrampin	Pigs that are weanling	Decreased the number of *E. coli* in the colon, ileum, and caecum while increasing the number of bifidobacteria and lactobacilli in the feces, caecum, and ileum
Potato protein	Pigs that are weanling	Decreased viable counts of *Staphylococcus* spp., coliform, and total bacteria in the rectum and caecum
Lactoferrin	Pigs that are weanling	Decreased total viable count of *E. coli* and *Salmonella* in small intestine
Recombinant plectasin	Weanling pigs	Increased total viable count of bifidobacterium in ileum

Probiotics are bacteria that, when given in sufficient quantities, boost the host's health, according to the World Health Organization. By generating antimicrobial substances like bacteriocins and organic acids, these advantageous microbes suppress harmful germs (Fuller, 1992). By sticking to the mucosa, blocking pathogen adhesion, and competing for resources, they help enhance the microbial habitat in the gastrointestinal tract.

Probiotics can be categorized into colonizing species like *Lactobacillus* and *Enterococcus* spp. and non-colonizing species like *Bacillus* spp. and *Saccharomyces cerevisiae*. Competitive avoidance is used to treat day-old chicks with vague microbiota due to colonization resistance against pathogenic microorganisms. The ideal probiotic should have innate strains safe for preparation, corrosion, and the effects of bile salts and stomach-related chemicals. It should also be rapidly multiplying and be tested for antimicrobial resistance to prevent harmful antimicrobial resistance from entering the intestinal environment.

Commonly used probiotics include bacteria such as *Bacillus, Lactobacillus, Lactococcus, Streptococcus, Enterococcus, Pediococcus, Bifidobacterium, Bacteroides*, and *Pseudomonas*, as well as yeasts like *Saccharomyces* and fungi such as *Aspergillus* and *Trichoderma*. In the European Union, microbiological feed additives primarily consist of *Bacillus* spp. (e.g., *B. cereus* var. *toyoi, B. licheniformis, B. subtilis*), *Enterococcus (E. faecium), Lactobacillus (L. acidophilus, L. casei, L. farciminis, L. plantarum, L. rhamnosus), Pediococcus (P. acidilactici), Streptococcus (S. infantarius),* and fungi such as *Saccharomyces cerevisiae* and *Kluyveromyces*.

The intestinal microbiota plays a crucial role in a bird's immune system, enhancing resistance to enteropathogenic microorganisms like *C. perfringens*. The probiotic strain's activity may involve the generation of metabolites, interaction with receptors, and induction of the resistant framework. The most well-known group of probiotics is lactic corrosive microbes, which are used by anaerobic butyrate-producing microorganisms of clostridial clusters IV and XIVa to produce large concentrations of butyric acid. This mechanism, known as cross-feeding, is an advanced reason why lactic corrosive microbes organization can positively influence performance (Sherman et al., 2009). Studies have demonstrated that supplementation with probiotics and prebiotics can reduce the incidence of diseases such as necrotic enteritis and salmonellosis in poultry and improve feed efficiency and growth performance (Bajagai et al., 2020; Sun et al., 2019).

Similarly, the dietary supplementation with spent cider yeast positively influenced the gut microbiota diversity in weanling pigs, reducing *Salmonella* and *Escherichia* levels. The impact of dietary cider yeast on the gut microbial communities of pigs was investigated using 16S rRNA gene compositional sequencing. Dietary cider yeast may potentially modify the gut microbiota, but such changes depend on the pigs' existing microbiota, leading to variations in their relative response to that diet (Upadrasta et al., 2013).

Likewise, the lysed *Corynebacterium glutamicum* cell mass (CGCM) improves growth performance and reduces malondialdehyde in nursery pigs without affecting nutrient digestibility, intestinal morphology, or microbiota in the jejunal mucosa. Supplementing diets with 2.1% CGCM, replacing 1.5% blood plasma, enhanced growth performance and reduced MDA without impacting nutrient digestibility, intestinal morphology, and microbiota in the jejunal mucosa (Cheng et al., 2021).

The global use of probiotics has expanded significantly over the past few decades. Japan began using probiotics in the 1960s, followed by China in the 1980s. The FDA approved 42 probiotics by 1989 (Gaggia et al., 2010). By 2020, the global probiotic market had grown to approximately $48 billion, reflecting a compound annual growth rate of over 7% from 2014. Expanding consumer awareness of gut health, demand for functional foods, and the growing acceptance of natural health products have all contributed to this market development.

In China, the manufacture of probiotics is poorly supervised and managed, and the probiotic market is less structured. Although more than 50 probiotics are utilized in China, only 12 have received approval from the Ministry of Agriculture.

Animal poisoning, allergies, and diarrhea are reported due to a lack of standards. Probiotics can be harmful to congenitally immunodeficient animals and have inconsistent beneficial effects in animals. Studies have shown that probiotics can alter the gut microbiota, leading to urinary tract infections and other diseases.

Due to their ability to upset the ecological balance and cause urinary tract infections, lactobacillus and *Bacillus* have negative impacts on the normal gut flora when taken with probiotics.

Dietary cider yeast can alter the gut microbiota, but these changes depend on the endogenous microbiota of the organism.

Probiotics as feed additives face challenges such as limited safe bacterial species; easy inactivation in feed processing, transport, and storage; inability to withstand low pH in the gastrointestinal tract and bile acids; and difficulty reaching a high number of viable cells to colonize in the intestine. Additionally, there are no adequate regulations and standards, making it difficult to label probiotic products with proper doses and indicate suitable animal targets (Upadrasta et al., 2013).

8.11 PREBIOTICS

Prebiotics are indigestible dietary elements that selectively metabolize to benefit the digestive system. They consist of protein hydrolysates, natural plant extracts, oligosaccharides, and polyols. The most prevalent substances in food are called oligosaccharides, which are produced from several hexose monosaccharides, including mannose, glucose, fructose, and galactose. They have monosaccharides with a polymerization degree of 2–20. The primary sources are grain vegetables, particularly verbascose, stachyose, and raffinose.

Synthetic oligosaccharides are found through disaccharide coordinate polymerization or cell fractionation. Other oligosaccharides like arabinogalactose, arabinoxylan, and rhamno-galacturonose are found in soybean, wheat, and natural products (Van Craeyveld et al., 2009).

Prebiotics can specifically multiply gut flora, support immunologic responses, and exhibit antiviral properties. Prebiotics were first used as feed additives in China in the late 1990s, but their use dates to the late 1980s.

Multifunctional oligosaccharides and acidifiers are currently the most promising prebiotics.

Prebiotics as AGP replacers have a similar activity to probiotics due to their nondigestible nature. They stimulate the growth or metabolic movement of a few intestinal microbiota species, such as *Bifidobacteria* and *Lactobacillus* spp., and can have a similar effect as probiotics. Compounds derived from manufactured materials offer benefits such as using inulin and fructo-oligosaccharides (FOS) as substrates for desired microorganisms, but manno-oligosaccharides (MOS) have receptor properties for *Escherichia coli* and *Salmonella* spp., leading to their end with the digestive stream rather than a mucosal receptor. Oligosaccharide β-glucans of yeast cell dividers are thought to stimulate execution due to their immunomodulatory effects. They improve phagocytosis and expansion of monocytes and macrophages, which play a significant role in immunomodulation. Studies have reported significant health benefits from using safe tweaking β-1,3/1,6-glucan as a healthy fix to protect animals against microorganisms (Ramirez-Farias et al., 2009).

Prebiotics are stable compounds with no residue or induced resistance, but they are not authorized as feed additives in the EU under EC 1831/2003 due to their limitations. Prebiotics cannot inhibit or kill pathogens, making them unsuitable for treating bacterial infections like antibiotics. Because prebiotics ferment in the gastrointestinal system, large doses of them can have negative effects such as bloating, diarrhea, and other problems. Since both β-galactomannan and MOS from the yeast *Saccharomyces cerevisiae* inhibit *Salmonella*-induced production of IL6 and CXCL8, mannose's prebiotic effect is connected to its oligosaccharide structure. However, proinflammatory factors are similarly expressed in D-mannose cells. Prebiotic effectiveness varies according to animal species, age, and physical state, and antagonistic interactions between them might happen.

Additionally, the high cost of prebiotic production limits their application in the animal husbandry industry (Badia et al., 2013). Prebiotics and probiotics in feed can improve animal health and productivity by modulating gut microbiota and enhancing host protection in livestock, poultry, and fish farming (Anadón et al., 2019).

8.12 SYNBIOTICS

Synbiotics, which are a combination of probiotics and prebiotics, have dual roles in piglet health. They enhance immune function, improve daily gain and digestibility, reduce diarrhea morbidity and mortality, ease weaning stress response, and promote piglet performance. However, the beneficial effects of synbiotics on swine production are limited due to inadequate mixing proportions of probiotics and prebiotics. The synergy mechanism of probiotics and prebiotics has not been thoroughly understood, and the extensive application of synbiotics has a long way to go (Kolida and Gibson, 2011).

8.13 PLANT EXTRACTS

Plant materials are extensively used in traditional medicine and animal nutrition due to their antimicrobial, anti-inflammatory, anti-oxidative, and antiparasitic properties. Plant extracts, also known as phytobiotics, have beneficial multifunctional properties derived from their bioactive components. These bioactive constituents include secondary metabolites like terpenoids, phenolics, glycosides, and alkaloids. Among 109 new antibacterial drugs approved between 1981 and 2006, 69% originated from natural products, and 21% were natural derivatives or compounds mimicking natural products (Savoia, 2012).

Particularly in Asian, African, and South American nations, plant extracts are frequently utilized in feed as development accelerators and health protectors. They are thought to be secure and efficient against specific microorganisms. Plant extracts such as oregano, cinnamon, Mexican pepper, thyme, and *Camellia sinensis* have been shown in pig production to enhance body weight gain, reduce pathogenic microbial mass in the intestines, and improve pig performance.

Additionally, phytogenic feed additives have been reported to improve poultry production performance. These findings highlight the potential benefits of plant extracts in various industries (Abreu, 2012).

Recently, plant-derived nanoparticles have shown potential as biocontrol agents in food-producing animals, addressing antimicrobial resistance and ensuring food safety, human health, and environmental safety (Ajose et al., 2022). Recent studies have shown that nanoparticles (NPs) have enhanced antibacterial activity against resistant *S. aureus* from bovine mastitis. These NPs can contain medications without damaging the material's structure, enhancing the agent's pharmacological effectiveness. They offer advantages like consistent dosage, increased bioavailability, delivery to the infected site, reduced therapeutic time, and prevention of burst release and drug degradation, making them generally safe for public health and the environment (Algharib et al., 2020).

In in vitro bacterial susceptibility studies, plant extracts having minimum inhibitory concentrations (MICs) between 100 and 1,000 μg/ml are regarded as antibacterial. Alkaloids, phenolics/polyphenols, terpenoids/essential oils, and lectins/polypeptides are several categories into which these phytochemicals may be divided. There are several ways that phytochemicals work against bacteria. For example, tannins work by depriving the body of iron and interacting with essential proteins; cryptolepine functions as a DNA intercalator and topoisomerase inhibitor; and saponins combine with sterols to form complexes that damage membranes and cause cell collapse. The antibacterial capabilities of essential oils have long been known, but little is known about their precise mode of action.

Some in vivo observations suggest that phytogenic feed additives contribute to a reduction of intestinal pathogen pressure. As of 2008, only two types of human-used antibacterial plant extracts have completed clinical trials, with 13 more under clinical trials (Harvey, 2008).

Phytobiotics are a complex blend of bioactive components, with variations due to biological factors, manufacturing, and storage conditions. They can improve animal performance and control diseases under certain circumstances. However, bird growth responses to herbal plants are still controversial due to the lack of exact quality and quantity of active chemicals. Plant components, physical characteristics, genetic variance, age, dose, extraction technique, harvest period, and compatibility with other substances are some of the factors that influence the effectiveness of phytobiotics.

Dietary phytobiotics' beneficial effects can also be influenced by animal nutritional status, infection, diet composition, and environmental conditions. The exact quality and quantity of active chemicals in plant extracts are often lacking (Huyghebaert et al., 2011).

The complex composition of herbs and their extracts makes it challenging to conduct comprehensive toxicology studies and safety assessments. The goal is to recognize and measure several ways that herbs can enhance animal physiology, health, and feed use. Because they are frequently employed in substantial quantities, which may have an impact on feed nutrition, herbal feed additives currently do not adhere to the "trace and efficient" criterion.

Additionally, phytogenic feed additives may interact with other feed additives, such as enzyme preparations and protein, through partial denaturation. Therefore, research on their mechanisms of action, compatibility with diet, toxicity, and safety assessment is necessary before they can be widely applied in animal feeds (Sarica et al., 2005).

8.14 INHIBITORS TARGETING PATHOGENICITY – QUORUM SENSING INHIBITORS

The quorum sensing (QS) system, which includes downstream regulatory proteins, receptors, and self-induced signaling molecules (autoinducers [Ais]), regulates bacterial pathogenicity in part. Gram-negative bacteria produce AIs such as h-acyl homoserine lactones (AHLs) and autoinducing peptides (AIPs), whereas Gram-positive bacteria release AI-2, quinolones, esters, and fatty acids. QS-targeting inhibitors can inhibit QS system actions and stop bacterial pathogenicity. Three categories exist for QS inhibitors (QSIs): protein QSIs, peptides (mostly AIPs homologues), and nonpeptide small compounds. AHL analogs are nonpeptide QSIs that can obstruct the production of QS signal molecules or their ability to attach to receptors.

QS quenching enzymes and QS quenching antibodies are proteinaceous QSIs that degrade signaling molecules. Competing organisms can also clear the signal molecule to quench QS, such as *E. coli* ingesting AI2s to influence *Vibrio harveyi.* Rotifers, turbot or prawn larvae, and *Artemia* spp. are shielded from infection by bacteria that degrade AHL.

Apoprotein B (ApoB) binds with AIP1 molecules of *S. aureus*, effectively reducing its QS. The importance of QSIs is inferred from the growing number of patents related to this field and extensive research (Bhardwaj et al., 2013).

QSIs, except LED209, are not widely applicable due to their toxicity to eukaryotic cells, such as penicillic acid, patulin, and halogenated furanone. Although QSI does not interfere with bacterial growth, it can still cause new resistance through mutations in bacterial AHL synthases and QS signal receptors. Drugs based on AHL analogs may suffer from lactone ring hydrolysis, while proteins based on lactone cores may have stability issues. Therefore, caution is needed in developing synthetic anti-QS analogs (Amara et al., 2011).

Bacteria only use specific signal molecules, making it difficult to find broad-spectrum QSIs. Wide-ranging enzymes that quench QS have strong prospects; nonetheless, protein QSIs can trigger an immunologic response from the host. Degradation of QS signaling molecules affects the normal activities of the host intestinal flora. The expression of virulence factors in biofilm-forming bacteria is lower than in planktonic bacteria, so the actual value of using QSIs for treating pathogens with biofilms must be considered (Kalia and Purohit, 2011).

Combining antibiotics with QSIs can increase bacterial sensitivity, making it a better strategy to enhance antimicrobial effects and prevent bacterial resistance, as reported in the literature. Figure 8.1 depicts an overview of QSIs.

8.15 BACTERIAL VIRULENCE INHIBITORS

A new strategy to combat bacteria involves inhibiting bacterial virulence factors, such as the protein complex of anthrax toxin. This strategy involves compounds that inhibit the function and transmission of bacterial toxins. Anthrax toxin contains lethal factor, edema factor (EF), and PA, which can be combined with cisplatin to prevent anthrax infection. Cholestyramine can bind with clostridial toxin to weaken its toxicity. More than two dozen Gram-negative bacteria have the type three secretion system (T3SS), which delivers effects or proteins straight into the cytoplasm of the host cell.

One compound targeting the T3SS of *Yersinia pseudotuberculosis* prevents the translocation of effector molecules, attenuating pathogens. To stop *Salmonella* from invading intestinal epithelial cells, two T3SS effector molecules are inhibited by distinct acyl salicylaldehydes. Pilicides prevent the growth of bacterial pili; however, chaperone protein binding to pilus protein is competitively inhibited by pyridone and its *N*-terminal amino acid derivatives. Additionally, pyridinone prevents the production of biofilms and blood clotting, which prevents bacterial adherence. Controlling the expression of virulence factors in bacteria is crucial because virstatin prevents the development of the pilus and toxin of *Vibrio cholerae,* which prevents the bacterium from colonizing the intestines. One way to control virulence is to tamper with the bacterial QS. It is possible to reduce the negative effects of toxins by using inhibitors of Cl^- secretion, which can lessen diarrhea. Toxin inhibitors do not directly kill bacteria; therefore, resistance selection pressure is not produced; instead, they alter host reactions to toxins or interfere with the function of bacterial toxins.

Without these substances, bacteria can still grow and reproduce, potentially causing a recurrence of virulence (Clatworthy et al., 2007; Young and Collier, 2007)

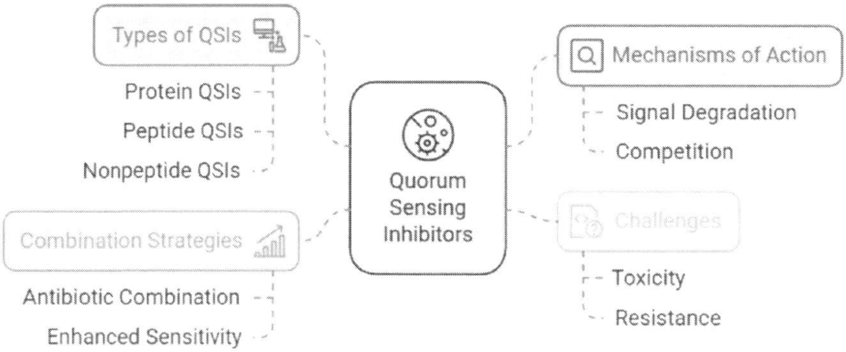

FIGURE 8.1 Quorum sensing inhibitors.

8.16 FEED ENZYMES

Dietary components are the main source of nutrients for the growth and proliferation of intestinal tract bacteria. These nutrients are either slowly absorbed or not broken down by digestive enzymes, which causes competition.

Exogenous enzymes influence nutrient absorption and produce nutrients for specific bacteria populations, directly impacting microflora populations (Bedford and Cowieson, 2012).

Feed enzymes, including phytase, are widely used in monogastric food-animal production. The creation of intestinal mucin and endogenous losses, as well as the digestibility of calcium, phosphorus, and other minerals, are all greatly impacted by these recombinantly generated enzymes. They also change the intestinal environment and nutrition supply, which modifies the selection pressures on different bacterial species. When added to a wheat-based diet, xylanase can help broiler hens suffering from the pathogenic effects of *C. perfringens.*

Encapsulated lysozyme, used as a feed additive in chickens, significantly reduced the concentration of *C. perfringens* and gastrointestinal lesions due to the organism in the ilium (Bedford and Cowieson, 2012).

Despite having a well-established market, feed enzymes have several drawbacks, including low activity, high costs, and a lack of production and quality control guidelines. The production method and how animals react to enzyme additions determine the stability and activity of feed enzymes.

Inconsistencies can be attributed to factors such as enzyme type, amount applied, enzyme side activities, diet composition, and animal variation. Feed enzymes improve the delivery of nutrients and provide fermentable oligosaccharides, which have an impact on the gut flora. Nevertheless, it is still unclear which main microbial species are involved and how much of this impact adds to the overall advantages of using enzymes. According to recent research, if prophylactic antibiotics are replaced with enzymes, the goal should be to increase the abundance of *E. coli* and *Clostridium* cluster XIVa species in the ileum and cecum while reducing the amount of *Lactobacillus* spp. present until a precise description of a desired flora can be recorded (Bedford and Cowieson, 2012).

Enzymes, such as glucose oxidase (GOD) from *Aspergillus niger* and *Penicillium* can reduce substrates for bacterial growth. GOD oxidizes D-glucose, producing gluconic acid and hydrogen peroxide. This can harm intestinal bacteria, inhibit their propagation, and consume oxygen in the intestine. An enzyme algino gel made of GOD, lacto-peroxidase, and guaiacol can prevent biofilm formation and inhibit established biofilms. However, the enzyme's mechanism of action differs from that of other feed enzymes, making further study of its feeding effects important (Cooper, 2013).

8.17 EXOGENOUS PROTEINS – ANTIMICROBIAL GROWTH PERFORMANCE PROMOTERS

By concentrating on intestinal bile salt hydrolase (BSH), an interesting microbiome target for the development of innovative AGP alternatives, AGPs increase animal productivity. By inhibiting the action of BSH, an enzyme generated by intestinal bacteria that has a detrimental effect on the digestion and utilization of fat, AGPs increase the development of animals. Research has indicated that the use of AGPs reduces the number of *Lactobacillus* species, which are the primary bacteria in the gut that produce BSH.

Some AGPs, such as tetracycline and roxarsone, also directly inhibit BSH activity. Therefore, BSH is a promising target for developing alternatives to AGPs, and dietary supplementation of BSH inhibitors may promote host lipid metabolism and energy harvest, leading to improved feed efficiency and body weight gain in food animals (Lin, 2014).

In animal feedstuffs, nonstarch polysaccharides (NSPs) are a complex category with a range of physiological effects, physical characteristics, and chemical compositions.

They include celluloses, pectin, oligosaccharides, arabinoxylans, and β-glucans. Different cereal types have different levels of NSPs, with maize containing mostly insoluble NSPs, wheat and grain having a solvent-to-insoluble ratio of 1/6, and rye having a high level of dissolvable NSPs, as noted by Choct (2002).

NSPs, which are substances found in the intestinal tract, have a significant antinutritive effect in poultry due to their thick nature. This is because the thick and dense substance slows down the dissemination of substrates and stomach-related proteins, preventing their interaction at the mucosal surface. Additionally, NSPs thicken the mucous layer on the intestinal mucosa, which is contrary to their metabolizable vitality (MEn) values in broiler chickens (Hedemann et al., 2009).

Viscous NSPs have a significant impact on intestinal physiology and morphology, but there are also negative effects on chickens' use of supplements. One adverse consequence is the aging of NSPs by the microbiota in the gut, which causes the small digestive tract to produce volatile fatty acids (VFAs). The whole caecal microbiota is made up of stringent anaerobes, whereas facultative anaerobes predominate in the chicken's tiny digestive tract under normal circumstances.

In an NSP-rich diet, VFA concentration increases primarily in the distal ileal lumen due to excess maturation and the growth of fermentative microflora. This competition for digestible supplements is evident in enzyme-free diets containing soluble-NSP-rich cereals like wheat, which initiate lymphocyte penetration and apoptosis of epithelial cells more than cereals with high levels of solvent NSPs (Teirlynck et al., 2009).

Diets with high NSP levels can be mitigated by incorporating AGPs. Without these, supplementing NSP-rich diets with chemicals leads to decreased ileal VFA concentration and increased caecal VFA concentration. Unpalatable substances are changed into rapidly absorbed VFAs throughout the aging process of the bowel. Dietary NSP enzymes increase the digesta section and supplement the absorption rate by decreasing the thickness of the digesta in the small intestine.

This reduces substrate and time for aging living beings to multiply, reestablishing normal and efficient endogenous enzymatic absorption of nutrients. Proteins counterbalance the negative effects of dissolvable NSP on performance (Teirlynck et al., 2009).

The impact of NSP-mediated decreased absorption on intestinal microbiota is more significant in the absence of antimicrobial development promoters, such as in the European Union. This is due to the debasement of stomach-related chemicals and bile salts, as well as colonization of the absorptive surface area. In less well-digested diets, there is a greater reaction to chemicals. NSP-debasing proteins also reduce the growth of pathogenic microorganisms like *Clostridium perfringens*. Broilers today contain proteins like xylanases and β-glucanases that break down NSPs (Jackson et al., 2003).

8.18 ORGANIC ACIDS

Natural acids, such as butyric corrosive, have been found to improve poultry performance and reduce the frequency of subclinical necrotic enteritis caused by *C. perfringens*. These acids are found in plants and animal tissues and are formed through microbial aging of carbohydrates in poultry caeca (Timbermont, 2009).

Natural acids with varying properties are used in drinking water supplements and as acidifiers. Salts, such as sodium, potassium, or calcium, are often available and are often odorless and easier to handle in food production due to their strong, less unstable structure. They are also less destructive and may be more solvent in water.

Natural acids have antibacterial properties, such as decreasing the pH of drinking water and decreasing buffering capacity, impacting the physiology of the gut and proventriculus. They can change from undissociated to separated frames, depending on the natural pH, increasing their antimicrobial impact. When in the undissociated frame, the corrosive can diffuse through the microorganisms' semipermeable film into the cell cytoplasm. The corrosive separates and suffocates bacterial cell chemistry once it is within the cell, when the pH is almost 7. It also strengthens the transport systems.

The effectiveness of a corrosive in repressing organisms is dependent on its pKa value, which is the pH at which 50% of the corrosive is separated (Van Immerseel et al., 2006).

Antibacterial action is influenced by several factors, including chemical equation, pKa value, chemical structure, atomic weight, microorganism-related MIC-value, microorganism nature, species, and buffering capacity of the food. Based on pH, film structure, and the microbiota species' in-cell physiology, each corrosive has a different spectrum of microbial mobility. Because of their wider spectrum of action, mixes of acids with a cluster of pKa values are employed. The impact of AGP substitution is also influenced by the physical form of the acids. Greasy acids' antibacterial mobility in the distal portion of the gastrointestinal system depends on their coating or microencapsulation with a dynamic "low release" lattice.

Additionally, added substance impacts of acids are possible. Medium-chain greasy acids, with a chain length between 8 and 12 carbon particles, may improve the viability of short-chain greasy acids. Blends of natural acids are used in the field to broaden their range and combine the benefits of different acids (Huyghebaert et al., 2011).

Extra natural acids, particularly greasy acids like butyric corrosive, may act as a quick-ingested vitality source and increase the development of the gastrointestinal mucosa near them. Butyric corrosive is a critical vitality source for intestine epithelial cells, invigorating cell multiplication and separation. It also has anti-inflammatory effects and fortifies the intestinal mucosal boundary by expanding the generation of antimicrobial peptides in mucous and fortifying the expression of tight intersection proteins. These acids are not only antibacterial but also contribute to the AGP-replacement effect. However, the impact of medium-chain greasy acids on cell exercises has not been explored (Dalmasso et al., 2008).

8.19 PLANT-DERIVED COMPOUNDS – HERBS AND ETHERIC OILS

The use of phytochemicals, including essential oils and plant extracts, is gaining interest due to their broad-spectrum antimicrobial properties. Compounds such as thymol, carvacrol, and eugenol have demonstrated effectiveness against a range of pathogens, including MRSA and *Salmonella* (Elisha et al., 2021). Additionally, these natural products can enhance animal growth and immune function, providing a dual benefit. The variability in the composition and potency of plant-derived products, however, poses challenges for standardization and efficacy. Plants possess multifunctional properties due to their bioactive components, including auxiliary metabolites like terpenoids, phenolics, glycosides, and alkaloids. The composition varies due to natural variables, production, and capacity conditions. The challenge is to identify and evaluate these activities and claims for food use, animal physiology, and overall well-being. The challenge lies in identifying and evaluating the vast array of benefits and claims in plant nutrition. Figure 8.2 depicts an overview of plant-derived compounds.

The specific components of etheric oil products that may stimulate stomach-related proteins and act as antioxidants, antimicrobial agents, or immunomodulators remain unclear due to the potential synergy between constituents. Test results show in vitro antimicrobial effects with specific MIC values and range of movement. Antimicrobial action is weak for ginger and pepper, medium for cumin, coriander, oregano, rosemary, sage, and thyme, and strong for clove, mustard, cinnamon, and garlic (Barbosa et al., 2009).

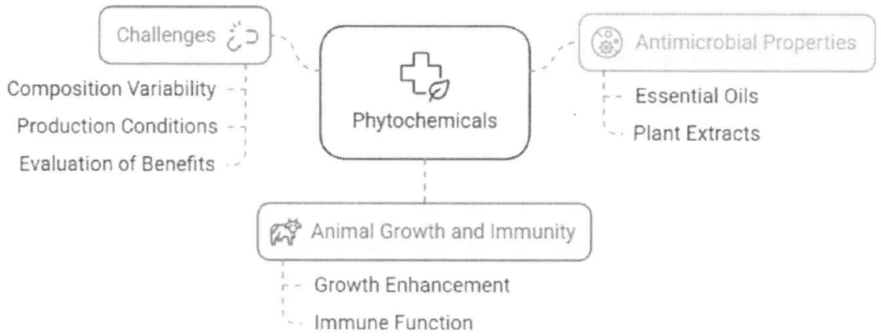

FIGURE 8.2 Plant-derived compounds (phytochemicals).

8.20 IMPROVED MANAGEMENT PRACTICES

Alongside biological alternatives, improved husbandry practices are essential in reducing the occurrence of diseases and the need for antibiotics. These include biosecurity measures, optimal nutrition, and appropriate housing conditions. For example, enhancing ventilation and sanitation can reduce the incidence of respiratory diseases in poultry, while tailored nutritional programs can improve gut health and immune function in livestock (Roura et al., 2019).

8.21 RECENT ADVANCES AND CHALLENGES

Recent research has made significant strides in developing and validating these alternatives. For instance, the use of next-generation sequencing and metagenomics has advanced our understanding of the gut microbiome's role in animal health and disease. This knowledge is pivotal in designing targeted probiotic and prebiotic interventions. Additionally, advancements in phage engineering have enabled the development of phage cocktails that can target multiple bacterial strains, enhancing the efficacy of phage therapy.

However, challenges remain in the implementation and adoption of these alternatives. The variability in the efficacy of natural products, regulatory hurdles for novel therapies, and the economic feasibility of large-scale implementation are significant barriers. Furthermore, there is a need for comprehensive monitoring and surveillance systems to assess the impact of these alternatives on AMR dynamics.

8.22 PERCEPTIONS

Alternatives to antibiotics should be safe, be simple to remove, not cause bacterial resistance, be stable in the gastrointestinal tract of animals and feed, be decomposable, be harmless to the environment, not interfere with palatability, not destroy healthy intestinal flora, not kill or inhibit the growth of pathogenic bacteria, increase immunity to disease, improve feed efficiency and encourage animal growth, and be well-compatible. These criteria are not currently met by any alternatives, nor are these needs fully met by any current alternatives.

There is still a big difference between antibiotics and alternatives in terms of how well they promote growth and prevent sickness. Bacterial infections are treated using antibacterial vaccinations, although few of them are curable. While feed enzymes and immunomodulators do not directly kill or suppress germs, they do maintain animal health. The safety of bacteriophages, which are only found in food, is uncertain. Probiotics and plant extracts have different impacts and safety hazards since they are unstable and complex substances. No products that target QS and bacterial virulence have been authorized; these inhibitors are still under study. Combining biofilm inhibitors with antibiotics yields positive outcomes. Although they have a limited antibacterial range and are expensive, AMPs can cure bacterial infections, which can lead to bacterial resistance.

Proteinaceous compounds, like feed enzymes and AMPs, are unstable and easily degraded in the digestive tract. Antibiotics, made from a single active ingredient, have high stability, consistency, and quality. Humans have not found a more effective way to treat indicated bacteria than selecting appropriate antibiotics.

The EU bans low doses of antibiotics as feed additives due to antimicrobial resistance. Penicillin, produced by *Penicillium,* has coexisted with bacteria for thousands of years. The development of penicillin resistance was discovered after extensive use. Drug pressure increases the likelihood of drug resistance development. Although antibiotic alternatives have not been linked to bacterial resistance, the widespread use of these alternatives may lead to mutated bacteria. The irregular and illegal use of these alternatives may also pose negative issues.

Antibiotics are still effective in preventing and controlling animal diseases in developing countries due to poor farming environments and high disease incidence. However, the ban on growth promoters in the EU requires improvements in farm hygiene. Without significant changes to the production settings, the frequency of bacterial illnesses in target animals may rise if the prohibition lowers the use of outdated antibiotics in feed. This may result in a rise in the therapeutic use of sophisticated antibiotics, posing significant risks to public health. There is insufficient scientific data to distinguish between the preventive and therapeutic usage of antibiotics with the development of resistance.

The decision to use in-feed antibiotics should be based on scientific approaches, as bacteria may not always listen to policymakers. The ban on antibiotics as growth promoters cannot be replicated globally.

Traditional antibiotics can still improve their efficacy by finding new bacterial targets and reinforcing anti-infectious therapy against MDR bacteria. For the treatment of *E. coli* infections that produce extended-spectrum β-lactamases, noncarbapenem options are frequently accessible. Novel formulations can enhance the antibacterial efficacy of antibiotics by using nanoparticles for drug delivery.

Broad-spectrum bactericidal agents should be used in empirical therapy to eradicate presumed MDR microorganisms. Once an infection is under control, it is crucial to switch to the most suitable narrow-spectrum agent, reducing the risk of adverse drug effects and antibiotic-induced resistance.

8.23 CONCLUSION

To ensure long-term sustainability in animal husbandry, it is crucial to regulate the use of antibiotics and develop alternatives. This includes defining target animals, treatment duration, and withdrawal periods, and strengthening laws to control antibiotic resistance and residues. Improving animal nutrition and production hygiene is also essential.

The shift toward alternatives to antimicrobials in animal husbandry is not merely a choice but a necessity in the fight against AMR. Bacteriophages, antimicrobial peptides, immunoglobulins, organic acids, metal nanoparticles, probiotics, prebiotics, synbiotics, and food enzymes are promising alternatives to antibiotics in animal husbandry. The development and implementation of effective alternatives require a multidisciplinary approach, encompassing microbiology, immunology, veterinary medicine, and public health. Collaborative efforts between researchers, industry stakeholders, and policymakers are essential to promote sustainable and safe livestock production practices. Recent European developments have shown a positive outcome from the ban on AGPs, resulting in improved animal welfare. Stronger research on the impacts of combined antibiotic usage and their alternatives is vital for sustaining a healthy agricultural economy and retaining effective antibiotics for human medicine. Research on antibiotic alternatives is likely to be a lengthy process.

The ongoing research and development in this field hold promise for a future where antibiotics are used judiciously, and the threat of AMR is significantly mitigated.

REFERENCES

Abreu, A.C.; McBain, A.J.; Simoes, M. Plants as sources of new antimicrobials and resistance-modifying agents. *Nat. Prod. Rep.* 2012, 29, 1007–1021. https://doi.org/10.1039/c2np20035j

Ajose, D.; Abolarinwa, T.; Oluwarinde, B.; Montso, P.; Fayemi, O.; Aremu, A.; Ateba, C. Application of plant-derived nanoparticles (PDNP) in food-producing animals as a bio-control agent against antimicrobial-resistant pathogens. *Biomedicines*. 2022, 10. https://doi.org/10.3390/biomedicines10102426.

Algharib, S.A.; Dawood, A.; Xie, S. Nanoparticles for treatment of bovine *Staphylococcus aureus* mastitis. *Drug Deliv.* 2020, 27, 292–308.

Allen, H.K.; Levine, U.Y.; Looft, T.; Bandrick, M.; Casey, T.A. Treatment, promotion, commotion: Antibiotic alternatives in food-producing animals. *Trends Microbiol.* 2013, 21, 114–119. https://doi.org/10.1016/j.tim.2012.11.001

Amara, N.; Krom, B.P.; Kaufmann, G.F.; Meijler, M.M. Macromolecular inhibition of quorum sensing: Enzymes, antibodies, and beyond. *Chem. Rev.* 2011, 111, 195–208. https://doi.org/10.1021/cr100101c

Anadón, A.; Ares, I.; Martínez-Larrañaga, M.; Martínez, M. Prebiotics and probiotics in feed and animal health. In: Gupta, R., Srivastava, A., Lall, R. (eds.), *Nutraceuticals in Veterinary Medicine*. Springer, Cham. 2019. https://doi.org/10.1007/978-3-030-04624-8_19.

Badia, R.; Lizardo, R.; Martinez, P.; Brufau, J. Oligosaccharide structure determines prebiotic role of beta-galactomannan against *Salmonella enterica* ser. Typhimurium in vitro. *Gut Microbes*. 2013, 4, 72–75. https://doi.org/10.4161/gmic.22728

Bajagai, Y.S.; Klieve, A.V.; Dart, P. J.; Bryden, W.L. *Probiotics in Animal Nutrition*. Springer. 2020.

Barbosa, L.N.; Rall, V.L.; Fernandes, A.A.; Ushimaru, P.I.; da Sliva Probst, I.; Fernandes Jr., A. Essential oils against foodborne pathogens and spoilage bacteria in minced meat. *Foodborne Pathog. Dis.* 2009, 6, 725–728.

Bedford, M.R.; Cowieson, A.J. Exogenous enzymes and their effects on intestinal microbiology. Anim. *Feed Sci. Technol.* 2012; 173, 76–85. https://doi.org/10.1016/j.anifeedsci.2011.12.018

Bhardwaj, A.K.; Vinothkumar, K.; Rajpara, N. Bacterial quorum sensing inhibitors: Attractive alternatives for control of infectious pathogens showing multiple drug resistance. *Recent Pat. Antiinfect. Drug Discov.* 2013, 8, 68–83. https://doi.org/10.2174/1574891X11308010012

Bierbaum, G.; Sahl, H.G. Lantibiotics: Mode of action, biosynthesis and bioengineering. *Curr. Pharm. Biotechnol.* 2009, 10, 2–18. https://doi.org/10.2174/138920109787048616

Burrowes, B.; Harper, D.R.; Anderson, J.; McConville, M.; Enright, M.C. Bacteriophage therapy: Potential uses in the control of antibiotic-resistant pathogens. *Expert Rev. Anti-Infect. Ther.* 2011, 9, 775–785. https://doi.org/10.1586/eri.11.90

Cheng, G.; Hao, H.; Xie, S.; Wang, X.; Dai, M.; Huang, L.; Yuan, Z. Antibiotic alternatives: The substitution of antibiotics in animal husbandry? *Front. Microbiol.* 2014, 5. https://doi.org/10.3389/fmicb.2014.00217.

Cheng, Y.; Duarte, M.; Kim, S. Nutritional and functional values of lysed *Corynebacterium glutamicum* cell mass for intestinal health and growth of nursery pigs. *J. Anim. Sci.* 2021, 99. https://doi.org/10.1093/jas/skab331.

Choct, M. Non-starch polysaccharides: Effect on nutritive value. In: McNab, J.M., Boorman, K.N. (eds.), *Poultry Feedstuffs*. CABI Publishing, New York, US. 2002.

Clatworthy, A. E.; Pierson, E.; Hung, D.T. Targeting virulence: A new paradigm for antimicrobial therapy. *Nat. Chem. Biol.* 2007, 3, 541–548. https://doi.org/10.1038/nchembio.2007.24

Cooper, R.A. Inhibition of biofilms by glucose oxidase, lactoperoxidase and guaiacol: The active antibacterial component in an enzyme alginogel. *Int. Wound J.* 2013, 10(6), 630–637. https://doi.org/10.1111/iwj.12083.

Cotter, P.D.; Ross, R.P.; Hill, C. Bacteriocins–a viable alternative to antibiotics? *Nat. Rev. Microbiol.* 2013, 11, 95–105. https://doi.org/10.1038/nrmicro2937

Courchesne, N.M.; Parisien, A.; Lan, C.Q. Production and application of bacteriophage and bacteriophage-encoded lysins. *Recent Pat. Biotechnol.* 2009, 3, 37–45. https://doi.org/10.2174/187220809787172678

Dalmasso, G.; Nguyen, H.T.; Yan, Y.; Charrier-Hisamuddin, L.; Sitaraman, S.V.; Merlin, D. Butyrate transcriptionally enhances peptide transporter PepT1 expression and activity. *PLoS One.* 2008, 3, e2476.

Desin, T.S.; Koster, W.; Potter, A.A. Salmonella vaccines in poultry: Past, present and future. *Expert Rev. Vaccines.* 2013, 12, 87–96. https://doi.org/10.1586/erv.12.138

Drulis-Kawa, Z.; Majkowska-Skrobek, G.; Maciejewska, B.; Delattre, A. S.; Lavigne, R. Learning from bacteriophages– advantages and limitations of phage and phage-encoded protein applications. *Curr. Protein Pept. Sci.* 2012, 13, 699–722. https://doi.org/10.2174/138920312804871193

Elisha, I. L.; Dzoyem, J. P.; McGaw, L. J.; Eloff, J.N. The antimicrobial activities of essential oils from selected South African medicinal plants. *BMC Compl. Med. Therap.* 2021, 21(1), 1–11.

Field, D.; Hill, C.; Cotter, P.D.; Ross, R.P. The dawning of a ‘Golden era’ in lantibiotic bioengineering. *Mol. Microbiol.* 2011, 78, 1077–1087. https://doi.org/10.1111/j.13652958.2010.07406.x

Fuller, R. History and development of probiotics. In: Fuller, R. (ed.), *Probiotics– The Scientific Basis*. Chapman and Hall. London. 1992.

Gaggia, F.; Mattarelli, P.; Biavati, B. Probiotics and prebiotics in animal feeding for safe food production. *Int. J. Food Microbiol.* 2010, 141(Suppl. 1), S15–S28. https://doi.org/10.1016/j.ijfoodmicro.2010.02.031

Goodridge, L.D.; Bisha, B. Phage-based biocontrol strategies to reduce foodborne pathogens. *Antibiotics.* 2021, 10(2), 120.

Harvey, A.L. Natural products in drug discovery. *Drug Discov. Today.* 2008, 13, 894–901. https://doi.org/10.1016/j.drudis.2008.07.004

Hedemann, M.S.; Theil, P.K.; Bach-Knudsen, K.E. The thickness of the intestinal mucous layer in the colon of rats fed various sources of nondigestible carbohydrates is positively correlated with the pool of SCFA but negatively correlated with the proportion of butyric acid in digesta. *British J. Nutr.* 2009, 102, 117–125.

Huyghebaert, G.; Ducatelle, R.; Immerseel, F.V. An update on alternatives to antimicrobial growth promoters for broilers. *Vet. J.* 2011, 187(2), 182–188. https://doi.org/10.1016/j.tvjl.2010.03.003

Jackson, M.E.; Anderson, D.M.; Hsiao, H.Y.; Mathis, G.F.; Fodge, D.W. Beneficial effect of b-mannanase feed enzyme on performance of chicks challenged with Eimeria sp. and Clostridium perfringens. *Avian Dis.* 2003, 47, 759–763.

Kak, V.; Sundareshan, V.; Modi, J.; Khardori, N.M. Immunotherapies in infectious diseases. *Med. Clin. North Am.* 2012, 96, 455–474. https://doi.org/10.1016/j.mcna.2012.04.002

Kalia, V.C.; Purohit, H.J. Quenching the quorum sensing system: Potential antibacterial drug targets. *Crit. Rev. Microbiol.* 2011, 37, 121–140. https://doi.org/10.3109/1040841X.2010.532479

Kolida, S.; Gibson, G.R. Synbiotics in health and disease. *Annu. Rev. Food Sci. Technol.* 2011, 2, 373–393. https://doi.org/10.1146/annurev-food-022510–133739

Laishevtsev, A.; Smirnov, D.; Ezhova, E.; Pimenov, N.; Oleynik, N. Antibiotic substitution programs in animal husbandry. *Veterinariya, Zootekhniya i Biotekhnologiya.* 2023. https://doi.org/10.36871/vet.zoo.bio.202302015.

Lin, J. Antibiotic growth promoters enhance animal production by targeting intestinal bile salt hydrolase and its producers. *Front. Microbiol.* 2014, 5. https://doi.org/10.3389/fmicb.2014.00033.

Masihi, K.N. Immunomodulators in infectious diseases; Panoply of possibilities. *Int. J. Immunopharmacol.* 2000, 22, 1083–1091. https://doi.org/10.1016/S01920561(00)00074–6

Miedzybrodzki, R.; Borysowski, J.; Weber-Dabrowska, B.; Fortuna, W.; Letkiewicz, S.; Szufnarowski, K., et al. Clinical aspects of phage therapy. *Adv. Virus Res.* 2012, 83, 73–121. https://doi.org/10.1016/B978-0-12-394438-2.00003-7

Molloy, E.M.; Field, D.; O'Connor, P.M.; Cotter, P.D.; Hill, C.; Ross, R.P. Saturation mutagenesis of lysine 12 leads to the identification of derivatives of nisin A with enhanced antimicrobial activity. *PLoS One.* 2013, 8, e58530. https://doi.org/10.1371/journal.pone.0058530

Murphy, D.; Ricci, A.; Auce, Z.; Beechinor, J.; Bergendahl, H.; Breathnach, R., et al. EMA and EFSA joint scientific opinion on measures to reduce the need to use antimicrobial agents in animal husbandry in the European Union, and the resulting impacts on food safety (RONAFA). *EFSA J.* 2017, 15. https://doi.org/10.2903/j.efsa.2017.4666.

Nocek, B.; Tikhonov, A.; Babnigg, G.; Gu, M.; Zhou, M.; Makarova, K.S., et al. Structural and functional characterization of microcin C resistance peptidase MccF from *Bacillus anthracis*. *J. Mol. Biol.* 2012, 420, 366–383. https://doi.org/10.1016/j.jmb.2012.04.011

O'Flaherty, S.; Ross, R.P.; Coffey, A. Bacteriophage and their lysins for elimination of infectious bacteria. *FEMS Microbiol. Rev.* 2009, 33, 801–819. https://doi.org/10.1111/j.1574–6976.2009.00176.x

Pereira, U.P.; Oliveira, D.G.; Mesquita, L.R.; Costa, G.M.; Pereira, L.J. Efficacy of *Staphylococcus aureus* vaccines for bovine mastitis: A systematic review. *Vet. Microbiol.* 2011, 148, 117–124. https://doi.org/10.1016/j.vetmic.2010.10.003

Potter, A.; Gerdts, V.; Littel-van den Hurk, S. Veterinary vaccines: Alternatives to antibiotics? *Anim. Health Res. Rev.* 2008, 9, 187–199. https://doi.org/10.1017/S1466252308001606

Rahman, M.; Fliss, I.; Biron, É. Insights in the development and uses of alternatives to antibiotic growth promoters in poultry and swine production. *Antibiotics.* 2022, 11. https://doi.org/10.3390/antibiotics11060766.

Ramirez-Farias, C.; Slezk, K.; Fuller, Z.; Duncan, A.; Holtrop, G.; Louis, P. Effect on inulin on the human gut microbiota: Stimulation of *Bifidobacterium adolescentis* and *Faecalibacterium prausnitzii. British J Nutr.* 2009, 101, 541–550.

Richards, P.; Pomorska-Mól, M. Immunomodulatory effects of probiotics on the host immune system. *Acta Vet. Scand.* 2020, 62(1), 1–9.

Rodriguez-Rubio, L.; Martinez, B.; Donovan, D.M.; Rodriguez, A.; Garcia, P. Bacteriophage virion-associated peptidoglycan hydrolases: Potential new enzybiotics. *Crit. Rev. Microbiol.* 2013, 39, 427–434. https://doi.org/10.3109/1040841X.2012.723675

Roura, E.; Homedes, J.; Klasing, K.C. Prevention of health disorders in broiler chickens by early feeding. *World's Poultry Sci. J.* 2019, 75(4), 567–582.

Sarica, S.; Ciftci, A.; Demir, E.; Kilinc, K.; Yildirim, Y. Use an antibiotic growth promoter and two herbal natural feed additives with and without exogenous enzymes in wheat-based broiler diets. S. *Afr. J. Anim. Sci.* 2005, 35, 61–72.

Savoia, D. Plant-derived antimicrobial compounds: Alternatives to antibiotics. *Future Microbiol.* 2012, 7, 979–990. https://doi.org/10.2217/fmb.12.68

Sherman, P.M.; Ossa, J.C.; Johnson-Henry, K. Unravelling mechanisms of action of probiotics. *Nutr. Clin. Pract.* 2009, 21, 10–14.

Stoica, C.; Cox, G. Old problems and new solutions: Antibiotic alternatives in food animal production. *Can. J. Microbiol.* 2021, 67(6), 427–444. https://doi.org/10.1139/cjm-2020–0601.

Sun, H.; Tang, J.W.; Fang, C.L.; Yao, X.H.; Wu, Y.F.; Wang, X., et al. Molecular analysis of intestinal bacterial microbiota of broiler chickens fed diets supplemented with different levels of chlorogenic acid. *Poultry Sci.* 2019, 98(2), 564–578.

Teirlynck, E.; Bjerrum, L.; Eeckhaut, V.; Huyghebaert, G.; Pasmans, F.; Haesebrouck, F., et al. The cereal type in feed influences gut wall morphology and intestinal immune cell infiltration in broiler chickens. *British J. Nutr.* 2009, 102, 1453–1461.

Thacker, E.L. Immunomodulators, immunostimulants, and immunotherapies in small animal veterinary medicine. *Vet. Clin. North Am. Small Anim. Pract.* 2010, 40, 473–483. https://doi.org/10.1016/j.cvsm.2010.01.004

Timbermont, L. *A Contribution to the Pathogenesis and Treatment of Clostridium Perfringens Associated Necrotic Enteritis in Broilers.* PhD thesis, Faculty of Veterinary Medicine, Ghent University, 2009.

Upadrasta, A.; O'Sullivan, L.; O'Sullivan, Ó.; Sexton, N.; Lawlor, P.; Hill, C., et al. The effect of dietary supplementation with spent cider yeast on the Swine distal gut microbiome. *PLoS One.* 2013, 8. https://doi.org/10.1371/journal.pone.0075714.

Van Craeyveld, V.; Holopainen, U.; Selinheimo, E.; Poutanen, K.; Delcour, J., Courtin, C.M. Extensive dry ball milling of wheat and rye bran leads to in situ production of arabinoxylan oligosaccharides through nanoscale fragmentation. *J. Agric. Food Chem.* 2009, 57, 8467–8473.

Van Immerseel, F.; Russell, J.B.; Flythe, M.D.; Gantois, I.; Timbermont, L.; Pasmans, F., et al. The use of organic acids to combat *Salmonella* in poultry: A mechanistic explanation of the efficacy. *Avian Patholo.* 2006, 35, 182–188.

Wang, S.; Zeng, X.; Yang, Q.; Qiao, S. Antimicrobial peptides as potential alternatives to antibiotics in food animal industry. *Int. J. Mol. Sci.* 2016, 17. https://doi.org/10.3390/ijms17050603.

WHO. 2018. https://www.who.int/news/item/07-11-2017-stop-using-antibiotics-in-healthy-animals-to-prevent-the-spread-of-antibiotic-resistance

Wittebole, X.; De Roock, S.; Opal, S.M. A historical overview of bacteriophage therapy as an alternative to antibiotics for the treatment of bacterial pathogens. *Virulence.* 2014, 5, 226–235. https://doi.org/10.4161/viru.25991

Yang, W.; Li, J.; Yao, Z.; Li, M. A review on the alternatives to antibiotics and the treatment of antibiotic pollution: Current development and future prospects. *Sci. Total Environ.* 2024, 926, 171757.

Young, J.A.; Collier, R.J. Anthrax toxin: Receptor binding, internalization, pore formation, and translocation. *Annu. Rev. Biochem.* 2007; 76, 243–265. https://doi.org/10.1146/annurev.biochem.75.103004.142728

Zhang, L.J.; Parente, J.; Harris, S.A.; Woods, D.E.; Hancock, R.; Fallal, T.J. Antimicrobial peptide therapeutics for cystic fibrosis. *Antimicrob. Agents Chemother.* 2005, 49, 2921–2927.

Zhang, W.; Sack, D. A. Progress and hurdles in the development of vaccines against enterotoxigenic *Escherichia coli* in humans. *Expert Rev. Vaccines* 2012, 11, 677–694. https://doi.org/10.1586/erv.12.37

9 Comprehensive Review of Recent Advancements and Future Perspectives of Antimicrobials in Animal Husbandry

Surendra Kumar Gautam

9.1 INTRODUCTION

Antimicrobials have played a very important role in the evolution of animal husbandry. They have been significant in improving the health, productivity, and welfare of animals. Their use ranges from treatment to preventive measures and growth promotion. However, the growing concern over antimicrobial resistance (AMR) has necessitated new strategies and innovations to ensure sustainable use of these agents in animal agriculture.

9.1.1 Importance of Antimicrobials in Animal Husbandry

The role of antimicrobial agents is significant to the control of infectious diseases in livestock. They will severely impact the agriculture sector from economic and health sides. Such agents are necessary for the assurance of safety in food and the good health status of animals alongside the economic viability of agriculture. Antimicrobial and antibacterial therapies prevent and manage bacterial infections, ensuring production and animal welfare, indicating higher yield and quality from animal sources (Aarestrup, 2015; Van Boeckel et al., 2019).

9.1.2 Historical Background and Traditional Application

The application of antimicrobials in animal husbandry dates back to the early 20th century, shortly after the discovery of antibiotics like penicillin. Traditionally, antimicrobials were used primarily to treat bacterial infections in animals. The applications later expanded to include prophylactic uses, which included growth promotion for low levels of antibiotics to promote acceleration of growth and improve feed efficiency in healthy animals, greatly improving livestock productivity and profitability.

 DOI: 10.1201/9781003269847-9

9.1.3 Current Challenges: Antimicrobial Resistance

9.1.3.1 Emergence and Spread of Antimicrobial Resistance

One of the most severe outcomes of the use of antimicrobials in animal agriculture is the development and spread of antimicrobial resistance (AMR). AMR is occurs when microorganisms such as bacteria, viruses, fungi, and parasites become tolerant to the action of previously effective antimicrobials that had killed them or otherwise retarded their growth. Overutilization and misutilization of antimicrobials in human and veterinary medicines have further worsened matters by proliferating resistant strains (Marshall & Levy, 2011; O'Neill, 2016).

9.1.4 One Health Approach

One Health promotes human, animal, and environmental health. AMR in livestock has implications in terms of animal health as well as human health and the environment. The transfer of resistant bacteria from animals to humans occurs through direct contact, consumption of animal products, and environmental pathways (Woolhouse et al., 2015). This has resulted in heightened efforts to devise comprehensive approaches to the control of antimicrobial use and spread of resistance.

9.2 OBJECTIVES OF THE REVIEW

This review provides a comprehensive overview of the recent developments and future outlook on antimicrobials in animal husbandry. This includes the following key areas:

- New antimicrobial agents and alternatives
- Antimicrobial delivery and use technologies
- Policy and regulatory amendments affecting antimicrobial practices
- Impact of these advancements on animal health and production
- State-of-play and way-forward strategies on managing AMR
- Future prospects and ways forward in sustainable antimicrobial use

The review does this with these topics as its core, discussing progress toward facing the challenge of antimicrobial use in animal husbandry and pointing to future directions for sustainability.

9.3 ANTIMICROBIALS AND THEIR MECHANISMS

Antimicrobials include drugs that result in either the killing of microorganisms or the halt of microorganism proliferation. In general, various mechanisms exist in antimicrobial agents where the target cell structures of the microbes encompass a considerable number of events that govern microbial survival and multiplication, and therefore form the bedrock of constructing more antimicrobial agents to contend with an increasingly emerging situation of AMR.

9.3.1 Classes of Antimicrobials and Recent Advances

9.3.1.1 β-Lactams

β-Lactam antibiotics encompass penicillins, cephalosporins, carbapenems, and monobactams. This class of antimicrobials comprises the largest group used by both human and veterinary medicine. β-Lactam antibiotics work by binding to and inhibiting penicillin-binding proteins, which are essential proteins for cross-linking peptidoglycan, thus resulting in lysis and death of cells in actively growing bacteria (Bush & Bradford, 2016).

9.3.1.1.1 Recent Breakthroughs

To overcome the resistance mechanism against β-lactam antibiotics, it has been discovered that certain compounds could inhibit the activity of β-lactamase and include clavulanic acid and sulbactam, which inhibit the enzymatic inactivation of β-lactam antibiotics by resistant bacteria (Drawz & Bonomo, 2010).

9.3.1.2 Aminoglycosides

Aminoglycosides include gentamicin, streptomycin, and neomycin. Such compounds work by binding onto the 30S ribosomal subunit, misreading the mRNA, and thereby preventing the synthesis of a protein. Eventually, these lead to the production of faulty proteins and to bacterial cell death (Kotra et al., 2000).

9.3.1.2.1 Recent Advancements

Liposomes of aminoglycosides are now tested for enhanced drug delivery and reduction in toxicity (Sercombe et al., 2015).

9.3.1.3 Tetracyclines

Tetracyclines include tetracycline, doxycycline, and minocycline, which inhibit protein synthesis by binding to the 30S ribosomal subunit and preventing the ribosome from accepting aminoacyl-tRNA. This causes the elongation of proteins to be blocked and bacteria to grow (Chopra & Roberts, 2001).

9.3.1.3.1 Recent Developments

There are modified tetracyclines, like tigecycline, to avoid resistance mechanisms due to efflux pumps and ribosomal protection proteins (Grossman, 2016).

9.3.1.4 Macrolides

Macrolides include erythromycin, azithromycin, and tylosin that act by binding to the 50S ribosomal subunit and inhibiting translocation of the growing peptide chain, thus inhibiting protein synthesis. The action is bacteriostatic but can be bactericidal in high concentrations (Leclercq, 2002).

9.3.1.4.1 Advances

Semisynthetic macrolides like telithromycin have enhanced pharmacokinetic profiles with superior activity against resistant organisms as well (Zhanel et al., 2001).

9.3.1.5 Fluoroquinolones

These drugs include, among others, ciprofloxacin, enrofloxacin, and danofloxacin, all of which exhibit inhibitory effects on bacterial DNA gyrase and topoisomerase IV. These two enzymes are concerned with replication and transcription of DNA. The inhibition caused by fluoroquinolones breaks DNA and finally leads to the death of the cell (Hooper, 1999).

9.3.1.5.1 Emerging Trends

New fluoroquinolones that have gained prominence are much more active but less susceptible to developing resistance (Dalhoff, 2012).

9.3.1.6 Sulfonamides

Sulfonamides, such as sulfamethoxazole, are antifolate drugs that act by inhibiting dihydropteroate synthase, an enzyme of folate synthesis. This blocks the synthesis of bacterial DNA and RNA. They are often combined with trimethoprim, which acts as an inhibitor of dihydrofolate reductase to enhance the bactericidal effect (Sköld, 2000).

9.3.1.6.1 Recent Advances

There is a trend toward the development of combination therapies and formulations to achieve greater efficacy and to minimize resistance (Mullard, 2017).

9.3.1.7 Polypeptides

Polypeptide antibiotics such as polymyxins and bacitracin interact with phospholipids causing cell membrane damage, which leads to leakage of the cellular components and eventual death of the cells (Storm et al., 1977).

9.3.1.7.1 Newer Trends

Some derivatives of polymyxin are nontoxic compared to the older versions and have been applied in the fight against multidrug-resistant infections (Falagas et al., 2010).

9.3.1.8 Ionophores

Ionophores, such as monensin and salinomycin, interact with the microbial cell membrane to interfere with ion gradients, thus causing energy metabolism disruption and death of the cells. They have mainly been used as growth promoters and coccidiostats in animals (Smith & Straughn, 1983).

9.3.1.8.1 Current Developments

Research into how ionophores may influence the gut microbiota for feed efficiency improvements is underway (Russell & Houlihan, 2003).

9.3.2 New Mechanisms and Strategies

9.3.2.1 Antimicrobial Peptides

Antimicrobial peptides are short, naturally occurring proteins that have broad-spectrum antimicrobial activity. They cause microbial membrane disruption, suppress the formation of biofilm, and regulate the immune response of the host. Antimicrobial peptides offer potential alternatives to traditional antibiotics, because they have alternative mechanisms and are less prone to the development of resistance (Hancock & Sahl, 2006).

9.3.2.1.1 Recent Advancement

More stable and more effective engineered antimicrobial peptides are being developed for use in veterinary applications (Mahlapuu et al., 2016).

9.3.2.2 Bacteriophage Therapy

Bacteriophages are viruses that are specific for bacteria and cause lysis. Phage therapy involves these natural pathogens as the antimicrobial tool against the infecting pathogen. This method has turned out to be an outstanding precision instrument against the concerning pathogen while not harming the desirable microbiota (Abedon et al., 2011).

9.3.2.2.1 Developing Currents

Genetically improved lytic activity and host range in phages are being researched (Cisek et al., 2017).

9.3.2.3 Probiotics and Prebiotics

Probiotics are described as beneficial bacteria, but prebiotics involve nondigestive components of food that promote growth for probiotics. Those increase immunities in their host and also inhibit growth in pathogenic bacteria. Animal diseases are prevented, and hence, no antibiotics are used in such methods (Gaggìa et al., 2010).

9.3.2.3.1 Recent Innovation

Synbiotics is the result of a synergistic combination of probiotics with prebiotics in order to achieve maximum beneficial health (Pandey et al., 2015).

9.3.2.4 Nanotechnology

Nanotechnology is the technique that applies nanoparticles for delivery purposes of antimicrobials. Nanoparticles enhance the solubility and stability of drugs. Nanoparticles help target delivery, and a reduced amount of drug required minimizes the adverse effects of drugs (Hajipour et al., 2012).

9.3.2.4.1 New Developments

Nanoparticles loaded with antimicrobials are being researched to increase efficacy and reduce resistance (Rana et al., 2016).

9.3.3 Challenges and Future Perspectives on Overcoming Antimicrobial Resistance

The persistent issue of AMR requires continuous innovation in the discovery of antimicrobials and stewardship practices. Solutions to counter AMR are discussed in the following sections.

9.3.3.1 Antimicrobial Stewardship Programs

They establish guidelines to improve the responsible use of antimicrobials in animals.

9.3.3.2 Surveillance and Monitoring

The use of antimicrobials and resistance patterns is tracked.

9.3.3.3 Research and Development

New antimicrobials, alternatives, and diagnostic tools are researched and developed (WHO, 2017).

9.3.4 Sustainable Use

Sustainable use of antimicrobials entails a balance between the need for effective infection control and the need to prevent resistance. This is achieved through the methods discussed in the following sections.

9.3.4.1 Integrated Disease Management

Integrated disease management is a combination of vaccination, biosecurity, and management practices with antimicrobials.

9.3.4.2 Precision Farming

Precision farming involves the application of technology in monitoring the health status of animals and enhancing the appropriate use of antimicrobials (Bailey et al., 2014b).

Animal husbandry antimicrobial mechanisms are multiple, encompassing various dimensions of microbial physiology and function. Recent scientific advancements in antimicrobial development, delivery, and alternative technologies present tremendous opportunities for furthering improvements in animal health and productivity while reducing the serious AMR concern. More innovation and research and development will be required in the future for sustainable utilization.

9.4 TECHNOLOGICAL ADVANCES IN ANIMAL HUSBANDRY ANTIMICROBIALS

Technological advancements in antimicrobials for animal husbandry are the key factors for improved animal health, productivity, and food safety. Simultaneously, these have addressed the serious challenge of AMR. The subsequent sections detail a wide range of technological advances, such as diagnostic tools, drug delivery systems, genetic modifications, alternative therapies, and precision farming techniques that shape the future of antimicrobials in livestock production.

9.4.1 Diagnostic Tools

9.4.1.1 Rapid Diagnostic Tests

Rapid diagnostic tests have changed veterinary practice with the possibility of rapid and accurate detection of pathogens for timely use of antimicrobials. Most of these tests utilize technologies such as polymerase chain reaction, lateral flow assays, and next-generation sequencing (Goldstein et al., 2013).

9.4.1.2 Polymerase Chain Reaction–Based Assays

Polymerase chain reaction–based assays can detect specific genetic markers of pathogens within hours. This can be viewed to provide a significant advantage when compared to the conventional culture methods (Rodríguez-Lázaro et al., 2014).

9.4.1.3 Lateral Flow Assays

These are point-of-care, portable, and user-friendly tests that will produce a result within minutes and are suitable for on-farm diagnostics (Posthuma-Trumpie et al., 2009).

9.4.2 Biosensors

Biosensors are analytical tools that contain a biological recognition element integrated into a physical transducer with the ability to identify pathogens or antimicrobial residues. Some of the key breakthroughs in biosensor technology result in the highly sensitive and specific devices that will enable real-time monitoring (Bahadir & Sezgintürk, 2015).

9.4.2.1 Electrochemical Biosensors

Measure electrical signals' change, occurring when the target molecules combine with the sensor for rapid, precise detection (Dhayal & Ratner, 2009).

9.4.2.2 Optical Biosensors

These are light techniques–based biosensors including surface plasmon resonance for detecting pathogens or residues. The sensitivity and multiplexing capabilities are very high (Homola, 2008).

9.4.3 Drug Delivery Systems

9.4.3.1 Nanotechnology

Nanotechnology has revolutionized drug delivery through targeted and controlled release of antimicrobials. This leads to increased efficacy while lowering the side effects. Drug solubility, stability, and bioavailability may be improved through nanoparticles (Kushwaha et al., 2012).

9.4.3.2 Liposomal Formulations

Liposomes are spherical vesicles that can encapsulate the antimicrobials, protecting it from degradation, and provide targeted delivery to infected tissues (Sercombe et al., 2015).

9.4.3.3 Polymeric Nanoparticles

These particles could be engineered to provide the controlled release of drugs in order to reduce the number of administrations, and compliance could be improved (Kumar et al., 2007).

9.4.4 Sustained-Release Implants

Controlled and extended-release implants gradually and constantly deliver antimicrobials over a considerable duration of time. This enables therapeutic levels of drugs to be maintained, which minimizes resistance development (Anderson & Shive, 2012).

9.4.4.1 Degradable Implants

These are developed using polymers such as polylactic acid (PLA) or polyglycolic acid (PGA). Here, it degrades to give up the drug steadily while releasing it as degradation takes place (Vert et al., 2012).

9.4.4.2 Osmotic Pumps

These operate on the basis of osmotic pressure for the release of drugs at a constant and controlled rate so that effective drug concentrations can be maintained for several weeks or even months (Santus & Baker, 1995).

9.4.5 Genetic Modifications

9.4.5.1 Genetically Engineered Livestock

Genetic engineering in livestock aims to enhance disease resistance, reduce the need for antimicrobials, and improve overall productivity. Techniques such as CRISPR-Cas9 and transgenic modifications are being employed to achieve these goals (Van Eenennaam, 2019).

9.4.5.2 CRISPR-Cas9

This genome editing tool allows precise modification of genes associated with disease resistance, offering the potential to create livestock with enhanced immunity (Doudna & Charpentier, 2014).

9.4.5.3 Transgenic Animals

Transgenic techniques involve introducing foreign genes into the genome of livestock to confer desirable traits, such as resistance to specific pathogens (Palmiter & Brinster, 1986).

9.4.6 Microbiome Engineering

Modifying the gut microbiome of livestock can improve health and reduce dependence on antimicrobials. Probiotics, prebiotics, and fecal microbiota transplantation are strategies used to manipulate the microbiome (Gaggìa et al., 2010).

9.4.6.1 Probiotics

Probiotics are beneficial bacteria that, when administered in adequate amounts, confer health benefits to the host by inhibiting pathogenic bacteria and modulating the immune system (Fuller, 1989).

9.4.6.2 Prebiotics

Prebiotics are nondigestible food ingredients that selectively stimulate the growth of beneficial bacteria in the gut (Gibson & Roberfroid, 1995).

9.4.6.3 Fecal Microbiota Transplantation

This is the transfer of fecal matter from healthy donors to recipient animals to restore a balanced microbiome and improve disease resistance (Borody & Khoruts, 2012).

9.4.7 Alternative Therapies

9.4.7.1 Phage Therapy

Bacteriophage therapy uses viruses that specifically infect and kill bacteria, offering a targeted approach to combating bacterial infections. Phages can be used alone or in combination with antibiotics to enhance efficacy (Salmond & Fineran, 2015).

9.4.7.1.1 Advantages

Phages are highly specific, reducing the impact on beneficial microbiota and lowering the risk of resistance development (Nobrega et al., 2015).

9.4.7.1.2 Challenges

Issues such as phage resistance and the regulatory landscape need to be addressed for widespread adoption (Hagens & Loessner, 2010).

9.4.8 Antimicrobial Peptides

Antimicrobial peptides (AMPs) are naturally occurring molecules with broad-spectrum activity against bacteria, fungi, and viruses. AMPs disrupt microbial membranes and modulate the host immune response (Hancock & Sahl, 2006).

9.4.8.1 Synthetic Antimicrobial Peptides

Advances in peptide synthesis have enabled the design of AMPs with enhanced stability, potency, and reduced toxicity (Mahlapuu et al., 2016).

9.4.8.2 Applications

AMPs are being explored as feed additives, topical treatments, and components of combination therapies (Wang et al., 2016).

9.4.9 Precision Farming Techniques

9.4.9.1 Automated Monitoring Systems

Precision farming involves the use of technology to monitor and manage livestock health and productivity. Automated systems collect and analyze data to optimize antimicrobial use and improve animal welfare (Bailey et al., 2014a).

9.4.9.2 Sensors and Internet of Things

Internet of Things devices and sensors monitor parameters such as temperature, humidity, and animal behavior in real time, enabling early detection of health issues (Wolfert et al., 2014).

9.4.9.3 Data Analytics

Advanced analytics and machine learning algorithms analyze data from sensors to predict disease outbreaks and inform treatment decisions (Kamphuis et al., 2010).

9.4.10 Smart Drug Delivery Systems

Smart drug delivery systems integrate sensors and automated dispensing mechanisms to deliver antimicrobials precisely when and where needed. These systems can reduce overuse and ensure optimal dosing (Hameed et al., 2020).

9.4.10.1 Wearable Devices

Wearable devices such as smart collars and implants monitor physiological parameters and administer drugs in response to specific triggers (Schukken et al., 2015).

9.4.10.2 Automated Injectors

These devices deliver precise doses of antimicrobials based on real-time data, reducing the risk of under- or overdosing (Kumar et al., 2018).

9.4.11 Challenges and Future Directions

9.4.11.1 Regulatory and Ethical Considerations

The adoption of new technologies in animal husbandry is often hindered by regulatory and ethical concerns. Ensuring the safety and efficacy of innovative antimicrobials and delivery systems is paramount (Mellor et al., 2018).

9.4.11.2 Regulatory Frameworks

Harmonizing global regulations and establishing clear guidelines for the approval and use of new technologies are critical for their adoption (Mossialos et al., 2010).

9.4.11.3 Ethical Considerations

Addressing ethical issues related to genetic modifications, animal welfare, and environmental impact is essential for public acceptance (Rollin, 2004).

9.4.12 Economic and Logistic Challenges

Implementing advanced technologies in animal husbandry can be costly and logistically challenging, particularly for small-scale farmers (Bowman et al., 2017).

9.4.12.1 Cost-Effectiveness

Developing affordable and scalable solutions is necessary to ensure widespread adoption and benefit to all stakeholders (Van Boeckel et al., 2015).

9.4.12.2 Infrastructure

Improving infrastructure and providing training and support to farmers are crucial for successful implementation of new technologies (Blake et al., 2016).

9.4.13 Research and Development

Ongoing research and development are essential to address the limitations of current technologies and explore new avenues for innovation (Laxminarayan et al., 2016).

9.4.13.1 Collaborative Efforts

Collaboration between academia, industry, and government agencies can drive progress and ensure that new technologies meet the needs of the livestock industry (O'Neill, 2016).

9.4.13.2 Interdisciplinary Research

Integrating knowledge from fields such as microbiology, engineering, and data science can lead to breakthrough innovations in antimicrobials and animal husbandry (Paphitou, 2013).

Technological innovations in antimicrobials for animal husbandry hold great promise for improving animal health, productivity, and food safety while combating antimicrobial resistance. Advances in diagnostic tools, drug delivery systems, genetic modifications, alternative therapies, and precision farming techniques are transforming the way antimicrobials are used in livestock production. However, addressing regulatory, ethical, economic, and logistic challenges is crucial for the successful adoption and sustainable use of these technologies. Continued research and collaboration across disciplines will be essential to harness the full potential of these innovations and secure a healthy future for animal husbandry.

9.4.14 Genetic Engineering

9.4.14.1 Genetically Engineered Livestock

Genetic engineering for livestock aims at increasing their resistance to diseases, the phasing out of antimicrobial use, and enhanced productivity. This involves the application of CRISPR-Cas9 and transgenic approaches (Van Eenennaam, 2019).

9.4.14.2 CRISPR-Cas9

This genome editing technology enables precise modification of genes associated with disease resistance; it can be used for the design of more resilient immune livestock (Doudna & Charpentier, 2014).

9.4.14.3 Transgenic Animals

Transgenic approaches involve the introduction of foreign genes into the genome of livestock, giving them desirable traits, like resistance to certain pathogens (Palmiter & Brinster, 1986).

9.4.15 Microbiome Engineering

The gut microbiome can be manipulated to enhance health and reduce reliance on antimicrobials. Probiotics, prebiotics, and fecal microbiota transplantation are strategies for managing the microbiome (Gaggìa et al., 2010).

9.4.15.1 Probiotics

Probiotics are beneficial bacteria that, when administered in adequate amounts, confer health benefits to the host by inhibiting pathogenic bacteria and modulating the immune system (Fuller, 1989).

9.4.15.2 Prebiotics

Prebiotics are nondigestible food ingredients that selectively stimulate the growth of beneficial bacteria in the gut (Gibson & Roberfroid, 1995).

9.4.15.3 Fecal Microbiota Transplantation

This involves mobilization of the fecal microbial material from donor animals toward the recipient animals for balance restoration and immune system boosting and enhancement in disease resistance among recipients (Borody & Khoruts, 2012).

9.4.16 Alternative Therapies

9.4.16.1 Phage Therapy

Bacteriophage therapy uses viruses that precisely target bacteria to infect it and cause its death due to specificity, thus giving one a focused approach on curbing the bacterial infection from within. These phages could be administered as a stand-alone dose or alongside antibiotics for more intense effectivity (Salmond & Fineran, 2015).

9.4.16.1.1 Advantage

Phages are species specific, which provides less chance of impact on the resident beneficial microbiota and reduces the risk of resistance (Nobrega et al., 2015).

9.4.16.1.2 Disadvantages

There are still challenges related to phage resistance, and the regulatory framework needs to be overcome before there is widespread use (Hagens & Loessner, 2010).

9.4.17 Antimicrobial Peptides

AMPs are naturally occurring molecules exhibiting a broad spectrum of activities against bacteria, fungi, and viruses. They disturb microbial membranes and adjust the host's immune response (Hancock & Sahl, 2006).

9.4.17.1 Synthetic Atimicrobial Peptides

The development of peptide synthesis enables researchers to design more stable AMPs having more potent activities with a less toxic impact (Mahlapuu et al., 2016).

9.4.17.2 Applications

AMPs are considered for use as feed additives, topical application, and in combination therapy (Wang et al., 2016).

9.4.18 Precision Farming Technologies

9.4.18.1 Automated Monitoring Systems

Precision farming is the application of technology to monitor and manage livestock health and productivity. Automated systems collect and analyze data to optimize antimicrobial use and improve animal welfare (Bailey et al., 2014a).

9.4.18.2 Sensors and Internet of Things

The real-time monitoring of temperatures, humidity, and changes in animal behavior can be carried out using Internet of Things devices and sensors. This will lead to the early detection of health issues.

9.4.18.3 Data Analytics

Advanced analytics and machine learning algorithms predict disease outbreaks and inform the treatment (Kamphuis et al., 2010).

9.4.19 Smart Drug Delivery Systems

Smart drug delivery systems consist of sensors and automated dispensing mechanisms that deliver antimicrobials precisely where and when they are required. This system may also avoid overuse and allow the optimal dosing of drugs (Hameed et al., 2020).

9.4.19.1 Wearable Devices

Wearable devices, which include smart collars and implants, monitor physiological parameters and administer drugs once specific triggers are detected (Schukken et al., 2015).

9.4.19.2 Automated Injectors

They provide real-time accurate doses of antimicrobials that can minimize under- or overdosing.

9.4.20 Challenges and Future Directions

9.4.20.1 Regulatory and Ethical Considerations

Regulatory and ethical considerations have often slowed the use of new technologies in animal husbandry. The focus here should be on safety and efficacy for innovative antimicrobials and their delivery systems (Mellor et al., 2018).

9.4.20.2 Regulatory Frameworks

Standardized global regulations and detailed guidelines for approval and implementation of new technologies must be implemented in order to be accepted (Mossialos et al., 2010).

9.4.20.3 Ethical Implications

An ethical concern would be brought about regarding genetic alterations, welfare of animals, and the environmental implications that will make it more acceptable by the public (Rollin, 2004).

9.4.21 Economic and Logistic Issues

Modern technologies in animal husbandry are costly and hard to adopt, especially among small-scale farmers (Bowman et al., 2017).

9.4.21.1 Cost-Effectiveness

The innovations should be cheap and scalable so that the technology can reach many people and help everyone (Van Boeckel et al., 2015).

9.4.21.2 Infrastructure

Improvement of infrastructure and training and assistance to the farmers are critical to the proper implementation of new technologies (Blake et al., 2016).

9.4.22 Research and Development

The requirement of the hour is continuous research and development to overcome the shortcomings of existing technologies and find new innovations (Laxminarayan et al., 2016).

9.4.22.1 Synergy

Synergy among academia, industry, and government agencies can foster change and ensure that new technologies fulfill the needs of the livestock industry (O'Neill, 2016).

9.4.22.2 Interdisciplinary Research

Innovations breakthroughs in antimicrobials as well as animal husbandry may arise from knowledge brought into this field from such disciplines as microbiology, engineering, and data science (Paphitou, 2013).

These innovations of antimicrobial technologies on animal husbandry have promises of enhancing animal health and productivity with improved food safety while lowering antimicrobial resistance. Diagnostic tools, drug delivery systems, genetic modifications, alternative therapies, and precision farming techniques now change the way antimicrobials are used on livestock production. For full-scale implementation and use that is sustainable, the existing regulatory, ethical, economic, and logistic hurdles are to be overcome. It is through continued research and the integration of experts from multiple disciplines that it will become possible to adequately realize these innovations and develop a healthy future for animal husbandry.

9.5 GLOBAL REGULATORY AND POLICY CHANGES

9.5.1 United States

The US Food and Drug Administation adopted the Veterinary Feed Directive as well as the Guidance for Industry 213 to provide impetus for responsible antibiotics use in food-producing animals.

9.5.2 European Union

The European Union has come up with more strict regulatory measures in developing the Veterinary Medicinal Products Regulation and the Regulation on Veterinary Medicinal Products and Medicated Feed, setting boundaries to restrict the application of antibiotics for growth promotion and prophylaxis.

9.5.3 China

The National Action Plan on Antimicrobial Resistance and Regulation on Veterinary Drug Management implemented by China has controlled antimicrobial use in livestock.

9.5.4 Regulatory and Policy Changes in India

9.5.4.1 Legislation and Guidelines

India set regulatory steps to restrict antimicrobial usage in animal rearing.

9.5.4.2 Drugs and Cosmetics Act, 1940

This act regulates the sale and distribution of veterinary drugs, including antimicrobials.

9.5.4.3 Prevention of Cruelty to Animals Act, 1960

This act governs the care and treatment of animals, which includes the administration of healthcare.

9.5.4.4 Food Safety and Standards Act, 2006

This act establishes the safety limits for foodstuffs and defines the residual limits of antimicrobials in products from animals.

9.5.5 Recent Additions and Initiatives

9.5.5.1 National Action Plan on Antimicrobial Resistance (NAP-AMR)

This was launched in 2017 to strengthen surveillance, regulation, and research against AMR.

9.5.5.2 Bureau of Indian Standards

The Bureau of Indian Standards sets standards for the use of antimicrobials in animal feeds and veterinary drugs.

9.5.5.3 Department of Animal Husbandry and Dairying

This department formulates policy and guidelines on antimicrobial stewardship in livestock.

9.5.6 Issues and Future Scope

Despite these regulatory efforts, numerous challenges remain in implementing and enforcing antimicrobial stewardship practices across the world. Future research should develop sustainable alternatives and improve the surveillance systems for monitoring patterns of resistance.

9.5.7 Role of Regulatory and Policy Changes toward Reducing Consequences of the Use of Antibiotics in Animal Husbandry on Global Health

Much needs to be done as more responsible use of antibiotics continues to be ensured to maintain its effectiveness to generations yet unborn.

Future research should be focused on sustainable antimicrobial alternatives, improvement of surveillance systems, and educational campaigns aimed at enhancing awareness about AMR and responsible antimicrobial use in Indian animal agriculture.

9.6 IMPACT OF ANTIMICROBIALS ON ANIMAL HEALTH AND PRODUCTION

Antimicrobials have revolutionized animal production as their application has dramatically increased disease prevention, increased infection control, and enhanced production efficiency. This systematic review explores the multidimensional impact of antimicrobials on animal health and production and the benefits and economic impacts on farmers. Case studies and field experiments serve as the source of information in this chapter.

9.6.1 Use of Antimicrobials for Disease Prevention and Control

Maintaining the animals' health through the role played by antimicrobials is key through control mechanisms that come with preventing mechanisms implicated by disease. A decline in morbidity and mortality through the proper application of antimicrobials among animals at appropriate times allows for a decrease in cases of wide outbreaks of disease. This is due to a reduction in the incidence of sickness and death in animals since cases have been dramatically reduced. This implies, for example, that research that has provided conclusive evidence regarding the applicability of antibiotics use in chickens and pigs has helped remarkably in averting the high possibility of them catching diseases that affect their respiratory and gastrointestinal systems.

9.6.1.1 Better Animal Welfare

Effective disease control supports good animal welfare since the minimal discomforts and sufferings caused by infections are avoided.

Effective antibiotic therapy eradicates the symptoms of diseases and, at the same time, accelerates recovery and improved health of animals (Johnson et al., 2018).

9.6.1.2 Improved Production

Antimicrobials improve production productivity through accelerated growth and enhanced feed conversion, thus ensuring proper performance of intensive farming systems.

Evidence shows that the addition of antimicrobials in feed will result in increased weight gain and feed efficiency, which means higher yields and returns for farmers (FDA, 2020).

9.6.2 Economic Implications for Farmers

The use of antimicrobials in animal agriculture has serious economic implications.

9.6.2.1 Cost of Antimicrobials

The cost of buying antimicrobials is a concern for farmers and should be weighed against the benefits arising from improved animal health and productivity.

However, cost-effectiveness in the use of antimicrobials is affected by factors including dosage regimen, disease prevalence, and market prices for agricultural products, such as those proposed by the European Commission (2019).

9.6.2.2 Market Pressure

With consumer demand for antibiotic-free and organic products, changing market trends have shifted production and have influenced decisions made about antimicrobial usage by farmers.

9.6.3 Challenges and Opportunities

Consumer demand for environmentally friendly and responsibly produced food products has pushed agricultural producers into the complex world of issues that come with these changes (Ministry of Agriculture and Rural Affairs of the People's Republic of China, 2022).

9.6.3.1 Case Studies and Field Trials

From the case studies, information is provided on how the application works effectively or fails to do the same when antimicrobial drugs are used in animal agriculture.

9.6.3.2 Best Practice: Success Stories

As illustrated in the following examples of successful antimicrobial programs, there are many instances in which strategic and responsible use contributed to positive outcomes for animal health and the farm's profitability.

9.6.3.2.1 Best Practice: Example

As reported in Case Study (Schukken & Deluyker, 1995), the adoption of a targeted protocol of antimicrobials for control of mastitis led the incidence rate to decrease followed by enhanced milk quality and better volumes of production.

9.6.4 Challenges and Lessons Learned

Praxis proves challenges and risks on antimicrobials use are complex.

9.6.4.1 Example: Case Study

Prolonged and non-discriminated use of antimicrobial drugs and its consequent treatment failure and economic burden-case study by Datta (2021).

9.6.4.2 Example: Case Study

A changing regulation impacts access and use of antimicrobials; by extension, livestock production must change to meet a new legal standard and, in doing so, be able to gain sustainability (Da Silva et al., 2023).

9.6.5 Regulatory and Policy Environment

To date, regulatory policies governing antimicrobial use in animal agriculture have been implemented by governments and international organizations.

9.6.5.1 Veterinary Feed Directives

Regulations bar the use of medically important antimicrobials for growth promotion purposes, and they must be administered through a veterinary prescription.

Veterinary feed directives aim at ensuring responsible use of antimicrobials, reducing the emergence of antimicrobial resistance, according to the FDA (2020).

9.6.5.2 International Activities

International organizations, among them, the World Health Organization and the World Organization for Animal Health, have developed worldwide plans regarding action against AMR related to monitoring, stewardship, and research.

Global activities include coordinating antimicrobial usage trend monitoring of resistance, researching alternate therapies, as well as upgrading veterinary capacity (WHO, 2020).

9.6.6 Future Horizons and Recommendations

Research and business directions in the future should emphasize alternative approaches to the use of antimicrobials.

9.6.6.1 Development and Validation of Sustainable Alternatives to Antimicrobials

Alternatives to treatment, such as probiotics, bacteriophage therapy, and vaccines, will help reduce reliance on conventional antibiotics as much as possible, and risks attributed to AMR can be regulated (Johnson et al., 2018).

9.6.6.2 Educational Programs

Programs should promote stewardship on antimicrobials in veterinarians, farmers, and stakeholders in general toward responsible use practices with risk mitigation on AMR.

Educational efforts to make people aware of appropriate antimicrobial use and the ethos of stewardship in agricultural livestock production are very important, according to the European Commission (2019).

9.6.6.3 Integrated Approach

An integrated approach should be taken for pest management or application of biosecurity for decreasing disease incidence and dependence on antimicrobials.

Integrated approaches are aimed toward holistic management of diseases, stressing prevention rather than therapeutic antimicrobial treatment, as much as possible.

Antimicrobials are critical tools in animal agriculture for health, welfare, and productivity. It is, however, an important challenge to manage these tools judiciously for the mitigation of risks that lead to antimicrobial resistance and to meet the changing consumers' demands for sustainable, safe food production. All stakeholders can protect animal health and preserve the efficacy of antimicrobial drugs by using alternative therapy advancements and responsible use.

9.7 FUTURE PERSPECTIVES IN ANTIMICROBIALS FOR ANIMAL HUSBANDRY

Addressing AMR while maintaining sustainable agricultural methods is critical to the future of antibiotics in animal husbandry. The research gaps, creative avenues, and cooperative strategies that are crucial for improving antimicrobial stewardship in animal agriculture are examined in this study.

9.7.1 Research Deficits and Prospects

9.7.1.1 Novel Research

Novel research investigating novel antimicrobial agents and substitutes.

Research must be advanced in order to find new antibacterial agents and substitutes.

9.7.1.2 Phage Therapy

Specific bacterial infections can be targeted with bacteriophages as antimicrobial agents Golkar and associates (2011).

9.7.1.3 Natural Chemicals

Examine the antibacterial properties of probiotics, essential oils, and plant-based chemicals Callaway and associates (2012).

9.7.1.4 Nano-Formations

Delivery methods based on nanotechnology may enhance antibacterial effectiveness and reduce environmental impact (Golkar et al., 2011).

9.7.1.5 Interdisciplinary Collaboration

Coordinating professionals from many sectors through interdisciplinary means is necessary to find solutions for AMR.

9.7.1.6 Microbiology and Immunology

Comprehend host-microbe interactions and resistance mechanisms at the microbial level.

9.7.1.7 Biotechnology and Engineering

In this case, developing resistant animals and boosting the efficacy of antimicrobial medications may be accomplished through genetic and microbial engineering.

9.7.1.8 Data Science and Analytics

Use AI and large data for AMR trend monitoring and predictive modeling.

9.7.2 Emerging Technologies and Interventions

9.7.2.1 Advanced Diagnostics

New diagnostics for early infection management and AMR containment, such as point-of-infection or resistance detection, should be implemented.

9.7.2.2 Point-of-Care Testing

Develop miniaturized, field-ready diagnostic equipment for the farm.

9.7.2.3 Genomic Sequencing

Whole-genome sequencing can help to determine resistance-conferring genes and transmission modes (EMA, 2018).

9.7.2.4 Biotechnological Innovations

Biotechnology including genetic and microbial engineering is an emerging area of the AMR solution.

9.7.2.5 Gene Editing

CRISPR/Cas9 can modify microbial genomes to increase sensitivity to antimicrobials.

9.7.2.6 Microbial Manipulation

Probiotics and beneficial bacteria can be engineered to competitively exclude pathogens and create healthy intestines (FDA, 2020).

9.7.3 Stakeholder Partnerships

9.7.3.1 Public-Private Partnerships

Public-private partnership building from strengths and expertise. It accelerates development.

9.7.3.2 Collaborative Research Initiatives

Collaborations may support funding for new, innovative research and development in novel antimicrobial therapies.

9.7.3.3 Technology Transfer

Assist in transposing scientific discovery to commercial use (FDA, 2020).

9.7.3.4 International Collaboration

Joint efforts should be made to tackle the difficulty of AMR. International collaboration is undertaken in a bid to transcend global difficulties concerning AMR.

9.7.3.5 Harmonization of Policies

Regulatory guidelines and regulations should be standardized to enhance prudent use of antimicrobials around the globe.

9.7.3.6 Global Surveillance Systems

Global surveillance systems should be set up to monitor the patterns of AMR and for the exchange of information (WHO, 2020).

The future of antimicrobials in animal husbandry will be determined by improving innovative research, adopting new technology, and developing collaboration that extends across disciplines and international boundaries. Improving biotechnological solutions, research gaps, and global cooperation will ultimately solve the problems of AMR while maintaining sustainable antimicrobial use in agriculture.

9.8 CONCLUSION

This review focuses on more recent developments regarding the current progress and challenges in the use of antimicrobials in animal husbandry, presenting benefits and risks from use. They are indeed critical parts of enhanced animal health and therefore provide less morbidity and mortality and enhanced productivity through greater weight gains and feed efficiencies. However, these advantages are accompanied by challenges that include the most notable: the development and spread of AMR among animal pathogens and the significant threats these present to public health and food security.

9.8.1 Balancing Antimicrobial Use and Resistance Management

This calls for a balancing act. The aim is to continue ensuring the efficacy of antimicrobials while avoiding the development and spread of resistance. This calls for integrated strategies that incorporate regulatory measures as well as on-farm practices.

9.8.2 Sustainable Practices: Long-Term Strategies for Responsible Use

Long-term strategies for sustainable use of antimicrobials in animal husbandry must be adopted.

9.8.2.1 Regulation

Strengthen the prescription of judicious use of antimicrobial agents by prohibiting nonessential uses, such as growth promotion.

9.8.2.2 Education and Training

Educate all stakeholders on the need for antimicrobial stewardship and guidelines for the appropriate use of antimicrobial agents.

9.8.2.3 Alternatives

Promote adoption of alternative therapies and preventive measures, such as vaccination and improved biosecurity measures.

9.8.2.4 Monitoring and Surveillance

Improve the surveillance systems to track use patterns of antimicrobials, developing resistance, and antimicrobial residues in the food products.

9.8.3 Future Vision: Innovation and Collaboration into the Future

The future of antimicrobials in animal husbandry looks forward to innovation and collaborative thinking.

9.8.3.1 Research and Development

Invest in research about finding new agents and different alternatives that would be viable and sustainable.

9.8.3.2 Technologies

Apply advanced diagnostics, genomic technology, and biotechnologies to combat AMR.

9.8.3.3 International Cooperation

Engage in international collaboration to harmonize regulations on the use of antimicrobial agents, exchange data on AMR trends, and develop global plans to manage AMR.

9.8.4 Sustainable Antimicrobial Use in the Future

Sustainable antimicrobial use will be attained by addressing health, environment, and economic considerations with holistic methods.

9.8.4.1 One Health Approach

The human-animal-environment interface has to be recognized as essential for solving the problem of AMR.

9.8.4.2 Economic Viability

Antimicrobial use can save for the economy and manage all the costs of the controls of AMR plus potential social consequences.

9.8.4.3 Environmental Aspects

Reducing resistant bacteria and its resultant antimicrobial agents in the environment. In this regard, more work needs to be focused on waste management and related treatment.

9.8.4.4 Continuous Improvement

The proper use of antimicrobials will, therefore, play a crucial role in reversing this growing threat. It would be possible to preserve the efficacy of antimicrobial therapies by all stakeholders for the future, ensure food security, and preserve animal health through intersectoral as well as transborder cooperation and through sustainability and innovation.

REFERENCES

Aarestrup, F. M. (2015). The livestock reservoir for antimicrobial resistance: A personal view on changing patterns of risks, effects of interventions and the way forward. *Philosophical Transactions of the Royal Society B: Biological Sciences*, 370(1670), 20140085.

Abedon, S. T., Kuhl, S. J., Blasdel, B. G., & Kutter, E. M. (2011). Phage treatment of human infections. *Bacteriophage*, 1(2), 66–85.

Anderson, J. M., & Shive, M. S. (2012). Biodegradation and biocompatibility of PLA and PLGA microspheres. *Advanced Drug Delivery Reviews*, 64, 72–82.

Bahadir, E. B., & Sezgintürk, M. K. (2015). Applications of commercial biosensors in clinical, food, environmental, and biothreat/biowarfare analyses. *Analytical and Bioanalytical Chemistry*, 407(3), 827–842.

Bailey, K. D., Case, L. A., & Miller, S. P. (2014a). The potential of precision dairy farming technologies to increase the efficiency of dairy farms. *Journal of Dairy Science*, 97(1), 340–345.

Bailey, K. L., Boyles, S. L., & Denman, S. E. (2014b). Precision livestock farming: Applications in ruminant production systems. *Animal Frontiers*, 4(3), 11–15.

Blake, R. W., Fernandez-Rivera, S., & Thorpe, W. (2016). Sustainable intensification in smallholder agriculture: An integrated systems research approach. *Outlook on Agriculture*, 45(1), 83–88.

Borody, T. J., & Khoruts, A. (2012). Fecal microbiota transplantation and emerging applications. *Nature Reviews Gastroenterology & Hepatology*, 9(2), 88–96.

Bowman, M. E., Robertson, R. M., & Redding, L. E. (2017). A framework for antimicrobial stewardship in veterinary medicine. *Journal of the American Veterinary Medical Association*, 251(5), 531–536.

Bush, K., & Bradford, P. A. (2016). β-lactams and β-lactamase inhibitors: An overview. *Cold Spring Harbor Perspectives in Medicine*, 6(8), a025247.

Callaway TR, et al. (2012). Phytogenic feed additives to young cattle diets modulate immune function and thrombocyte response. *Journal of Animal Science*, 90(7), 3039–3043.

Chopra, I., & Roberts, M. (2001). Tetracycline antibiotics: Mode of action, applications, molecular biology, and epidemiology of bacterial resistance. *Microbiology and Molecular Biology Reviews*, 65(2), 232–260.

Cisek, A. A., Dąbrowska, I., Gregorczyk, K. P., & Wyzewski, Z. (2017). Phage therapy in bacterial infections treatment: One hundred years after the discovery of bacteriophages. *Current Microbiology*, 74(2), 277–283.

Dalhoff, A. (2012). Global fluoroquinolone resistance epidemiology and implictions for clinical use. *Interdisciplinary Perspectives on Infectious Diseases*, 2012, 976273.

Da Silva, R. A., Arenas, N. E., Luiza, V. L., Bermudez, J. A. Z., & Clarke, S. E. (2023). Regulations on the use of antibiotics in livestock production in South America: A comparative literature analysis. *Antibiotics (Basel, Switzerland),* 12(8), 1303.

Datta, S. (2021). *Poultry as a livelihood opportunity – Two case studies – Kesla and Suguna: Kesla poultry: A journey of tribal women – From earning daily wages to owning India's largest poultry feed-mill.* Rajiv Gandhi Institute for Contemporary Studies, Rajiv Gandhi Foundation. https://www.rgics.org/wp-content/uploads/Kesla-Poultry-Case-Study-Ver-4-Jun-2021.pdf

Dhayal, B., & Ratner, D. M. (2009). Electrochemical biosensors for pathogen detection. *Annual Review of Analytical Chemistry*, 2(1), 297–319.

Doudna, J. A., & Charpentier, E. (2014). The new frontier of genome engineering with CRISPR-Cas9. *Science*, 346(6213), 1258096.

Drawz, S. M., & Bonomo, R. A. (2010). Three decades of β-lactamase inhibitors. *Clinical Microbiology Reviews*, 23(1), 160–201.

European Commission. Regulation (EU) 2019/6 on veterinary medicinal products. Available at: https://eur-lex.europa.eu/legal-content/EN/TXT/?uri=CELEX:32019R0006.

European Medicines Agency (EMA). EMA annual report 2018. Available at: http://www.ema.europa.eu/ema/index.jsp?curl=pages/special_topics/document_listing/document_listing_000109.jsp.

Falagas, M. E., Rafailidis, P. I., Ioannidou, E., Alexiou, V. G., & Matthaiou, D. K. (2010). Colistin therapy for microbiologically documented multidrug-resistant Gram-negative bacterial infections: A retrospective cohort study of 258 patients. *International Journal of Antimicrobial Agents*, 35(2), 194–199.

Food and Drug Administration (FDA). (2020). Veterinary Feed Directive (VFD). Available at: https://www.fda.gov/animal-veterinary/development-approval-process/veterinary-feed-directive-vfd.

Fuller, R. (1989). Probiotics in man and animals. *Journal of Applied Bacteriology*, 66(5), 365–378.

Gaggìa, F., Mattarelli, P., & Biavati, B. (2010). Probiotics and prebiotics in animal feeding for safe food production. *International Journal of Food Microbiology*, 141, S15–S28.

Gibson, G. R., & Roberfroid, M. B. (1995). Dietary modulation of the human colonic microbiota: Introducing the concept of prebiotics. *Journal of Nutrition*, 125(6), 1401–1412.

Goldstein, E. J., Citron, D. M., Merriam, C. V., Warren, Y. A., Tyrrell, K. L., & Fernandez, H. T. (2013). In vitro activities of the experimental aminoglycoside ACHN-490 against 1,031 Gram-negative and Gram-positive aerobic and anaerobic bacteria. *Antimicrobial Agents and Chemotherapy*, 57(10), 4848–4853.

Golkar Z, et al. (2011). Applications of nanotechnology in agriculture and food science. *International Journal of Food Science*, 46(7), 1489–1503.

Grossman, T. H. (2016). Tetracycline antibiotics and resistance. *Cold Spring Harbor Perspectives in Medicine*, 6(4), a025387.

Hagens, S., & Loessner, M. J. (2010). Bacteriophage for biocontrol of foodborne pathogens: Calculations and considerations. *Current Pharmaceutical Biotechnology*, 11(1), 58–68.

Hajipour, M. J., Fromm, K. M., Ashkarran, A. A., Jimenez de Aberasturi, D., Larramendi, I. R., Rojo, T., . . . & Parak, W. J. (2012). Antibacterial properties of nanoparticles. *Trends in Biotechnology*, 30(10), 499–511.

Hameed, N., Najam, F., Raza, A., Amin, F. U., Abbasi, H. T., & Aslam, Z. (2020). Smart drug delivery system: A model approach for the treatment of infectious diseases. *Drug Development and Industrial Pharmacy*, 46(3), 407–418.

Hancock, R. E., & Sahl, H. G. (2006). Antimicrobial and host-defense peptides as new anti-infective therapeutic strategies. *Nature Biotechnology*, 24(12), 1551–1557.

Homola, J. (2008). Surface plasmon resonance sensors for detection of chemical and biological species. *Chemical Reviews*, 108(2), 462–493.

Hooper, D. C. (1999). Mechanisms of fluoroquinolone resistance. *Drug Resistance Updates*, 2(1), 38–55.

Johnson T, et al. (2018) Antimicrobial use in animal agriculture: What have we learned? *Annual Review of Animal Biosciences*, 6, 249–269.

Kamphuis, C., Sherlock, R., Jago, J., & Mein, G. (2010). Automatic cow behaviour classification in grazing systems using accelerometers. *Precision Livestock Farming*, 97–105.

Kotra, L. P., Haddad, J., & Mobashery, S. (2000). Aminoglycosides: Perspectives on mechanisms of action and resistance and strategies to counter resistance. *Antimicrobial Agents and Chemotherapy*, 44(12), 3249–3256.

Kumar, M., Curtis, A., & Hoskins, C. (2018). Applications of nanotechnology in the delivery of drugs to the brain. *Therapeutic Delivery*, 9(8), 617–635.

Kumar, N., Long, S. R., & Smith, R. (2007). Polymers for drug delivery. *Journal of Controlled Release*, 120(1–2), 127–135.

Kushwaha, S. K., Keshari, T. K., & Dubey, S. K. (2012). Recent advances in nanotechnology-based drug delivery systems for the treatment of infectious diseases. *Drug Delivery and Translational Research*, 2(6), 481–497.

Laxminarayan, R., Matsoso, P., Pant, S., Brower, C., Røttingen, J. A., Klugman, K., & Davies, S. (2016). Access to effective antimicrobials: A worldwide challenge. *The Lancet*, 387(10014), 168–175.

Leclercq, R. (2002). Mechanisms of resistance to macrolides and lincosamides: Nature of the resistance elements and their clinical implications. *Clinical Infectious Diseases*, 34(4), 482–492.

Mahlapuu, M., Håkansson, J., Ringstad, L., & Björn, C. (2016). Antimicrobial peptides: An emerging category of therapeutic agents. *Frontiers in Cellular and Infection Microbiology*, 6, 194.

Marshall, B. M., & Levy, S. B. (2011). Food animals and antimicrobials: Impacts on human health. *Clinical Microbiology Reviews*, 24(4), 718–733.

Mellor, D. J., Patterson-Kane, E., & Stafford, K. J. (2018). *The sciences of animal welfare*. John Wiley & Sons.

Ministry of Agriculture and Rural Affairs of the People's Republic of China. (2022). Regulation on veterinary drug management. Available at: http://english.moa.gov.cn/policies/201910/t20191009_297822.html

Mossialos, E., Morel, C. M., Edwards, S., Berenson, J., Gemmill-Toyama, M., & Brogan, D. (2010). *Policies and incentives for promoting innovation in antibiotic research*. World Health Organization.

Mullard, A. (2017). (2016). FDA drug approvals. *Nature Reviews Drug Discovery*, 16(2), 73–76.
Nobrega, F. L., Costa, A. R., Kluskens, L. D., & Azeredo, J. (2015). Revisiting phage therapy: New applications for old resources. *Trends in Microbiology*, 23(4), 185–191.
O'Neill, J. (2016). *Tackling drug-resistant infections globally: Final report and recommendations*. Review on Antimicrobial Resistance. Retrieved from https://amr-review.org
Palmiter, R. D., & Brinster, R. L. (1986). Germ-line transformation of mice. *Annual Review of Genetics*, 20(1), 465–499.
Pandey, K. R., Naik, S. R., & Vakil, B. V. (2015). Probiotics, prebiotics and synbiotics: A review. *Journal of Food Science and Technology*, 52(12), 7577–7587.
Paphitou, N. I. (2013). Antimicrobial resistance: Action to combat the rising microbial challenges. *International Journal of Antimicrobial Agents*, 42(S1), S25–S28.
Posthuma-Trumpie, G. A., Korf, J., & van Amerongen, A. (2009). Lateral flow (immuno) assay: Its strengths, weaknesses, opportunities and threats. A review. *Analytical and Bioanalytical Chemistry*, 393(2), 569–582.
Rana, V. K., Sharma, R., & Kumari, A. (2016). Role of nanoparticles in antibacterial activity of silver nanoparticles against *E. coli* and *P. aeruginosa*. *Journal of Nanoscience and Nanotechnology*, 16(3), 2762–2770.
Rodríguez-Lázaro, D., Lombard, B., Smith, H., Rzezutka, A., D'Agostino, M., Helmuth, R., . . . & Cook, N. (2014). Trends in analytical methodology in food safety and quality: Monitoring microorganisms and genetically modified organisms. *Trends in Food Science & Technology*, 41(2), 104–113.
Rollin, B. E. (2004). Animal welfare and ethics. *The Veterinary Clinics of North America: Small Animal Practice*, 34(2), 329–340.
Russell, J. B., & Houlihan, A. J. (2003). Ionophore resistance of ruminal bacteria and its potential impact on human health. *FEMS Microbiology Reviews*, 27(1), 65–74.
Salmond, G. P., & Fineran, P. C. (2015). A century of the phage: Past, present and future. *Nature Reviews Microbiology*, 13(12), 777–786.
Santus, G., & Baker, R. W. (1995). Osmotic drug delivery: A review of the patent literature. *Journal of Controlled Release*, 35(1), 1–21.
Schukken, Y. H., & Deluyker, H. A. (1995). Design of field trials for the evaluation of antibacterial products for therapy of bovine clinical mastitis. *Journal of Veterinary Pharmacology and Therapeutics*, 18, 274–283.
Schukken, Y. H., Gunther, J., Fitzpatrick, J., Fontaine, M. C., Goetze, L., Holst, O., . . . & Rainard, P. (2015). Host-response patterns of intramammary infections in dairy cows. *Veterinary Immunology and Immunopathology*, 167(1–2), 1–17.
Sercombe, L., Veerati, T., Moheimani, F., Wu, S. Y., Sood, A. K., & Hua, S. (2015). Advances and challenges of liposome assisted drug delivery. *Frontiers in Pharmacology*, 6, 286.
Sköld, O. (2000). Sulfonamide resistance: Mechanisms and trends. *Drug Resistance Updates*, 3(3), 155–160.
Smith, L. W., & Straughn, W. D. (1983). Ionophores: Their effect on production in livestock. *Journal of Animal Science*, 57(6), 1489–1501.
Storm, D. R., Rosenthal, K. S., & Swanson, P. E. (1977). Polymyxin and related peptide antibiotics. *Annual Review of Biochemistry*, 46(1), 723–763.
Van Boeckel, T. P., Brower, C., Gilbert, M., Grenfell, B. T., Levin, S. A., Robinson, T. P., . . . & Laxminarayan, R. (2015). Global trends in antimicrobial use in food animals. *Proceedings of the National Academy of Sciences*, 112(18), 5649–5654.
Van Boeckel, T. P., Brower, C., Gilbert, M., Grenfell, B. T., Levin, S. A., Robinson, T. P., . . . & Laxminarayan, R. (2019). Global trends in antimicrobial use in food animals. *Proceedings of the National Academy of Sciences*, 116(9), 3463–3470.

Van Eenennaam, A. L. (2019). Application of genome editing in farm animals: Cattle. *Transgenic Research*, 28(2), 93–100.

Vert, M., Doi, Y., Hellwich, K. H., Hess, M., Hodge, P., Kubisa, P., . . . & Rinaudo, M. (2012). Terminology for biorelated polymers and applications (IUPAC recommendations 2012). *Pure and Applied Chemistry*, 84(2), 377–410.

Wang, G., Li, X., & Wang, Z. (2016). APD3: The antimicrobial peptide database as a tool for research and education. *Nucleic Acids Research*, 44(D1), D1087–D1093.

Wolfert, S., Goense, D., & Sørensen, C. A. G. (2014). A future Internet collaboration platform for safe and healthy food from farm to fork. *2014 Annual SRII Global Conference*. IEEE, 266–273.

Woolhouse, M., Ward, M., van Bunnik, B., & Farrar, J. (2015). Antimicrobial resistance in humans, livestock and the wider environment. *Philosophical Transactions of the Royal Society B: Biological Sciences*, 370(1670), 20140083.

World Health Organization. (2017). Stop using antibiotics in healthy animals to prevent the spread of antibiotic resistance. Retrieved from https://www.who.int

World Health Organization (WHO). (2020). Global action plan on antimicrobial resistance. Available at: https://www.who.int/antimicrobial-resistance/global-action-plan/en/.

Zhanel, G. G., Walters, M., Noreddin, A., & Vercaigne, L. (2001). Review of telithromycin: A novel ketolide. *Expert Review of Anti-Infective Therapy*, 10(1), 51–65.

10 Novel Approaches to Combat Microbial Infections in Animal Husbandry

P. Prakash, V. Ambikapathy, S. Babu, K. Vijayalakshmi, M. Ratha, N. Ganesh, and A. Panneerselvam

10.1 INTRODUCTION

In the past few decades, animal protein needs increased [OECD, 2021]. Hence, society utilized antimicrobial drugs in animal husbandry for animal health and rapid production. Antimicrobial resistance (AMR) in animal husbandry is now considered a major problem. The World Health Organization (WHO) stated that the bacteria, viruses, fungi, and parasites not controlled by antimicrobial drugs are a result of AMR [Tiseo et al., 2020]. According to WHO, the bacterial AMR was responsible for 1.27 million deaths globally in 2019 [WHO, 2024]. According to most of the investigations, these global death rates will increase to 10 million in 2050 [He et al., 2020].

Antimicrobial agents were used to prevent or cure microbial infections in animals as well as humans. Antimicrobial drugs were discovered after the Second World War, and they can be considered revolutionary for the treatment of many microbial infections in animals and for agriculture as well as veterinary medicine [Prescott, 2018].

Generally, antimicrobial agents were obtained from living organisms or were produced synthetically in laboratories. These antimicrobial agents were not only used as microbial infection treatments, they promoted the development of animal livestock [Hosain et al., 2021]. Regular utilization of antimicrobial agents in veterinary practices leads to some significant problems in human health [Islam et al., 2016; Bacanli and Basaran, 2019; Cully, 2014; Rushton, 2015].

Humans faced some health hazards due to the use of antimicrobial substances in animal husbandry. Livestock consumption led to the possibilities of allergies, carcinoma, and mutations in the human body [Kirbis, 2007]. However, misuse and overdosage of antimicrobials in humans, animals, and plants are the major factors in the development of antimicrobial-resistant pathogens.

DOI: 10.1201/9781003269847-10

Globally, the animal husbandry industry is projected to account for nearly half of all antibiotic use, reaching approximately 200,000 tons by 2030. In this regard, approximately 10% of research publications presented studies about antimicrobial resistance in animal husbandry [Van Boeckel et al., 2015].

Antimicrobial drug alternatives are needed. Hence, antimicrobial peptides, probiotics, prebiotics, and postbiotics were promising microbial infection control strategies [Ghosh et al., 2019]. Food supplements have also been used to promote antimicrobial efficiency and growth in both humans and animal husbandry. Especially, postbiotics have bacteriocins due to being synthesized from some bacteria, and they are more effective for the treatment of microbial infections [Ambikapathy et al., 2024; Cotter et al., 2005].

In this context, animal husbandry includes aquaculture, beekeeping, sheep farming, dog breeding, horse breeding, and cattle farming. Veterinary microbiology deals with the infectious pathogenic microorganisms in animal species [McVey et al., 2022].

Poor animal husbandry practices were leading to stress and disease among animals. Hence, animal productivity decreased. Animal husbandry is typically dependent on parameters such as nutritional modes, clean drinking, safe and comfortable shelter for animals, temperature, light source, ventilation, air quality, manure management, quarantine for sick animals, and observing and reducing stress-related behaviors [Lowenstein et al., 2016].

Microbial infections pose a significant threat to animal health, food security, and human well-being. The overuse and misuse of antibiotics in animal husbandry have accelerated the emergence of antimicrobial resistance, rendering traditional treatments ineffective. This chapter explores innovative strategies to combat microbial infections in animal husbandry, focusing on natural products and alternative approaches [Rahman et al., 2022].

10.2 THE RISE OF ANTIMICROBIAL RESISTANCE

The widespread use of antibiotics in animal agriculture has led to the selection and dissemination of resistant microorganisms. This has severe consequences for human health, as resistant pathogens can be transmitted through the food chain or direct contact with infected animals [Manyi-Loh et al., 2018].

10.3 ONE HEALTH INITIATIVE

Recognizing the interconnectedness of human, animal, and environmental health, the One Health initiative aims to address the root causes of antimicrobial resistance. This global effort promotes a holistic approach to disease prevention and control (Figure 10.1) [Velazquez-Meza et al., 2022].

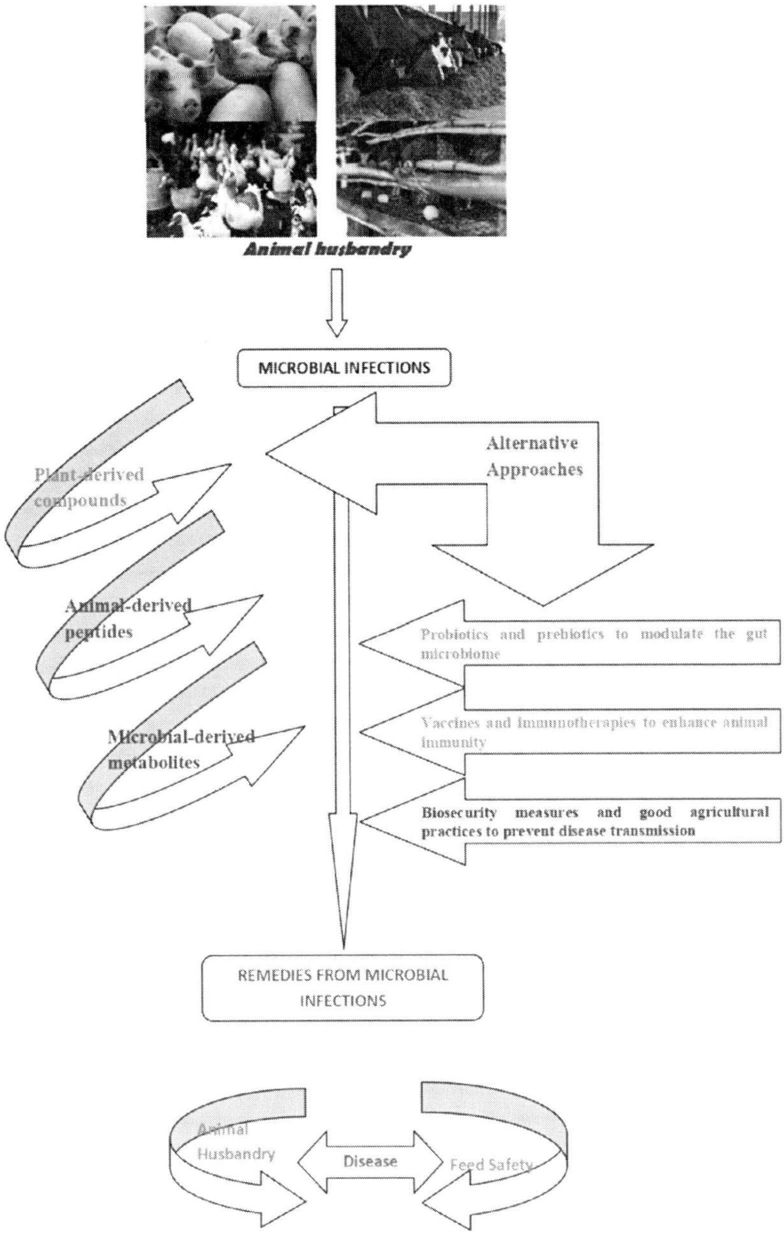

FIGURE 10.1 Methods for microbial infection remedies in animal husbandry.

10.4 NOVEL ANTIBIOTICS FROM NATURAL PRODUCTS

Nature has long been a source of inspiration for drug discovery. Recent research has identified promising natural products with antimicrobial properties [Chaachouay and Zidane, 2024], including the following:

- Plant-derived compounds (e.g., essential oils, polyphenols)
- Animal-derived peptides (e.g., antimicrobial peptides)
- Microbial-derived metabolites (e.g., bacteriocins)

10.4.1 Plant-Derived Compounds

The quest for plant-based drugs and supplements has gained momentum in recent years, driven by the search for innovative treatments for infectious diseases. Interdisciplinary teams of researchers, including ethnopharmacologists, botanists, microbiologists, and natural products chemists, are scouring the globe for phytochemicals with potential therapeutic applications. Although a significant proportion of modern pharmaceuticals are derived from plants, none are currently used as antimicrobials. However, traditional healers have long leveraged plant-based remedies to prevent and treat infections, inspiring Western medicine to explore similar approaches. Plants are a rich source of diverse secondary metabolites, such as tannins, terpenoids, alkaloids, and flavonoids, which have demonstrated antimicrobial properties in laboratory studies. This review aims to provide an overview of ongoing efforts to screen botanicals for antimicrobial activity, assess their efficacy and safety in clinical settings, and elucidate the structural and functional properties of phytochemicals. Given the increasing popularity of unregulated botanical preparations among the public, healthcare professionals must consider the implications of self-medication with these products [Cowan, 1999].

10.4.2 Animal-Derived Peptides

The escalating threat of antibiotic-resistant bacteria necessitates the discovery of innovative alternatives to conventional antibiotics. Antimicrobial peptides (AMPs) have emerged as promising candidates due to their favorable safety profile, minimal residual impact, and low propensity for resistance development. Moreover, their unique mechanisms of action offer significant potential in combating antibiotic resistance. However, the high production costs and limited potency of AMPs hinder their widespread adoption. Traditional methods for identifying and designing novel AMPs are often time consuming and labor intensive, impeding progress in this field. Fortunately, cutting-edge technologies like artificial intelligence (AI) are being leveraged to accelerate the development and design of new AMPs, presenting exciting opportunities for advancement. This chapter provides an overview of the key characteristics, antimicrobial mechanisms, advantages, and limitations of AMPs, while also exploring the application of AI in predicting and designing novel AMPs. By highlighting the pivotal role of AI in enhancing the efficiency of AMP research, this chapter aims to inform and guide the development of innovative antimicrobial therapies [Ji et al., 2024].

10.4.3 Microbial-Derived Metabolites

The search for novel antibiotic-producing microorganisms in extreme and unusual environments has yielded a plethora of innovative antimicrobials, many of which exhibit potent activity against drug-resistant pathogens like methicillin-resistant *Staphylococcus aureus* (MRSA). While some critics argue that these discoveries often rely on familiar chemical scaffolds rather than novel compound classes, the fact remains that these new antibiotics can still provide critical solutions to the growing problem of AMR. The evolutionary fine-tuning of antibiotic activities within a limited set of optimal chemical scaffolds raises questions about the necessity of discovering entirely new classes of compounds to combat AMR. However, translating these discoveries into clinically viable treatments will require significant investment from the pharmaceutical industry, which may be deterred by the risk of resistance emerging to new drugs. Nevertheless, the success of antibiotic-producing microorganisms in their natural environments offers valuable lessons. By studying these producers and exploring combinations of antibiotics or antibiotics with adjuvant, we may uncover strategies to enhance efficacy and minimize resistance. In nearly a century since Fleming's penicillin discovery revolutionized medicine, our traditional reductionist approach to antibiotic production and prescription has contributed to the AMR crisis. To break this cycle, we must adopt a more judicious approach, leveraging synergies between antibiotic compounds to create formulations that safeguard these vital medicines for future generations [Quinn and Dyson, 2024].

10.5 ALTERNATIVE APPROACHES

In addition to novel antibiotics, alternative approaches are being explored, including the following:

- Probiotics and prebiotics to modulate the gut microbiome
- Vaccines and immunotherapies to enhance animal immunity
- Biosecurity measures and good agricultural practices to prevent disease transmission

10.5.1 Probiotics and Prebiotics to Modulate the Gut Microbiome

The significance of gut microbiota in maintaining human health is a rapidly expanding area of research. Disruptions to the balance of gut microbiota have been linked to various conditions, including infections, inflammation, and obesity. The use of antibiotics is a primary driver of substantial changes in the composition and function of gut microbiota, leading to the colonization of multidrug-resistant bacteria. This review aims to examine the existing evidence on the potential of microbiota modulation and prevention of gut dysbiosis as strategies to combat antimicrobial resistance.

Given the limited discovery of new antimicrobial classes, alternative approaches are necessary to address the spread of antibiotic-resistant pathogens. Various modalities, including dietary interventions, probiotics, prebiotics, selective decontamination, and fecal microbiota transplantation, are being explored, with fecal microbiota transplantation being the most extensively studied. However, prioritizing prevention through antimicrobial stewardship programs, stringent infection control measures, and novel chelating agents can also play a vital role in reducing the colonization of multidrug-resistant organisms. Further research and development should focus on these innovative tools to combat AMR through gut microbiota modulation, which shows promise in addressing this critical health concern [Matzaras et al., 2022].

10.5.2 Vaccines and Immunotherapies to Enhance Animal Immunity

The rise of AMR poses a significant threat to global health, prompting the scientific community to explore alternative therapies beyond traditional antibiotics. Vaccines and monoclonal antibodies (mAbs) offer promising complementary approaches to combat AMR. Recent advancements have introduced innovative technologies in both fields, but a deeper understanding of their broad applicability and mechanisms of action is crucial to accelerate the development of novel interventions. To demonstrate their effectiveness, it is essential to quantify the impact of vaccination and mAb use on antibiotic consumption and AMR transmission. Further research is needed to identify optimal antigen targets, elucidate the immunologic basis of challenging infections, enhance technology adaptability, establish relevant in vitro and in vivo models, and expedite testing of novel candidates in small-scale human trials to minimize the risk of failure in larger, longer-term clinical trials [La Guidara et al., 2024].

10.5.3 Biosecurity Measures and Good Agricultural Practices to Prevent Disease Transmission

This systematic review highlights several promising interventions to mitigate infection burden, microbial loads, antibiotic use, and antibiotic resistance in animal agriculture. Future research should investigate these interventions, alone or in combination, in small-scale farming systems and low- and middle-income countries, with a focus on water sanitation, hygiene, and biosecurity. It is crucial for human health researchers to acknowledge the overlap between animal and human health, as zoonotic microorganisms can share antibiotic resistance genes, particularly in settings where humans and animals interact closely. While providing basic water, sanitation, and hygiene infrastructure is essential for good health, a comprehensive One Health approach recognizing the interconnectedness of human, animal, and environmental health is often neglected. A significant knowledge gap exists in evaluating structural interventions in agricultural communities, which future research should address. To effectively combat antimicrobial resistance among the global population engaged in animal agriculture, it is essential to build on existing evidence, such as this systematic review, which synthesizes findings from across the One Health spectrum [Chris et al., 2023].

10.6 MATERIALS AND METHODS

10.6.1 Plant-Derived Compounds

Materials

1. Plant material (leaves, stems, roots, flowers, etc.)
2. Solvents (e.g., methanol, ethanol, acetone, hexane)
3. Extraction equipment (e.g., Soxhlet apparatus, ultrasonic bath)
4. Filtration devices (e.g., filter paper, vacuum filter)
5. Chromatography equipment (e.g., column chromatography, thin-layer chromatography)
6. Analytical instruments (e.g., high-performance liquid chromatography [HPLC], gas chromatography-mass spectrometry [GC-MS], nuclear magnetic resonance [NMR])

Methods Extraction

1. Selection and preparation of plant material: Choose healthy plant material, clean and dry it.
2. Grinding: Grind the plant material into a fine powder.
3. Extraction: Use solvents to extract compounds from the plant material.

Soxhlet extraction: Use a Soxhlet apparatus for continuous extraction.
Ultrasonic-assisted extraction: Use an ultrasonic bath for efficient extraction.
Maceration: Steep plant material in solvent for several hours.

Filtration and Concentration

1. Filtration: Filter the extract using filter paper or vacuum filter.
2. Concentration: Concentrate the filtrate using rotary evaporator or vacuum evaporator.

Chromatography

1. Column chromatography: Separate compounds based on polarity and molecular weight.
2. TLC: Analyze compound separation.

Purification

1. Recrystallization: Purify compounds by recrystallization.
2. Preparative HPLC: Purify compounds using preparative HPLC.

Identification

1. HPLC: Analyze compound purity.
2. GC-MS: Identify compounds.
3. NMR: Determine compound structure.

Example Protocol Isolation of Quercetin from Onion

1. Extract onion powder with 80% methanol.
2. Filter and concentrate the extract.
3. Separate compounds using column chromatography.
4. Purify quercetin using preparative HPLC.
5. Identify quercetin using HPLC, GC-MS, and NMR.

10.6.2 Animal-Derived Peptides

Materials

1. Animal tissues (e.g., muscle, skin, blood, bones)
2. Enzymes (e.g., trypsin, pepsin, chymotrypsin)
3. Extraction buffers (e.g., phosphate-buffered saline, Tris-HCl)
4. Centrifuges and ultracentrifuges
5. Chromatography equipment (e.g., HPLC, GC)
6. MS instruments
7. Peptide purification kits
8. Cell culture media (for cell-based assays)

Methods Sample Preparation

1. Tissue homogenization
2. Protein extraction
3. Enzymatic digestion
4. Peptide separation (e.g., centrifugation, filtration)

Peptide Purification

1. Chromatography (e.g., reversed-phase, size-exclusion)
2. Electrophoresis (e.g., sodium dodecyl sulfate-polyacrylamide gel electrophoresis [SDS-PAGE])
3. Peptide affinity purification

Peptide Identification

1. MS
2. Tandem mass spectrometry (MS/MS)
3. Peptide sequencing
4. Database searching

Functional Analysis

1. Cell-based assays (e.g., bioactivity, cytotoxicity)
2. Enzyme-linked immunosorbent assay (ELISA)
3. Western blotting
4. Biochemical assays (e.g., antioxidant, antimicrobial)

Specific Techniques

1. Gelatin hydrolysis
2. Collagen peptide extraction
3. Keratin peptide extraction
4. Whey protein peptide extraction

10.6.3 Microbial-Derived Metabolites

1. Microbial cultures (e.g., bacteria, fungi)
2. Growth media (e.g., LB broth, MRS broth)
3. Incubator or shaker
4. Centrifuge
5. Extraction solvents (e.g., ethyl acetate, methanol)
6. Chromatography columns (e.g., HPLC, GC)
7. Mass spectrometer (e.g., liquid chromatography-mass spectrometry [LC-MS], GC-MS)
8. Spectrophotometer (e.g., ultraviolet and visible [UV-Vis])
9. Microplates or 96-well plates
10. Pipettes and tips
11. Biological samples (e.g., soil, water, plant tissue)

Methods Sample Preparation

1. Cultivate microorganisms in suitable growth media.
2. Harvest cells by centrifugation.
3. Extract metabolites using solvents (e.g., ethyl acetate, methanol).
4. Concentrate extracts using rotary evaporator or lyophilizer.

Metabolite Separation and Identification

1. Perform chromatography (HPLC, GC) to separate metabolites.
2. Detect metabolites using mass spectrometry (LC-MS, GC-MS).
3. Identify metabolites by comparing spectra to databases.

Quantification

1. Use spectrophotometry (UV-Vis) to quantify metabolite concentrations.
2. Utilize chromatography coupled with MS.

Biological Assays

1. Assess antimicrobial activity using microdilution assays.
2. Evaluate cytotoxicity using cell viability assays.
3. Investigate enzyme inhibition using enzyme assays.

Data Analysis

1. Analyze chromatography and MS data.
2. Use statistical software for data visualization.

Some specific protocols are as follows:

Protocol 1: Extraction of Microbial Metabolites

1. Grow microorganisms in 50 ml LB broth for 24 to 48 hours.
2. Centrifuge at 10,000 rpm for 10 minutes.
3. Extract metabolites with 20 ml ethyl acetate.
4. Concentrate extracts using rotary evaporator.

Protocol 2: HPLC-MS Analysis

1. Inject 10 μL extract into HPLC system.
2. Separate metabolites using C18 column.
3. Detect metabolites using LC-MS.

Protocol 3: Antimicrobial Assay

1. Prepare microbial suspensions (1×10^6 CFU/ml).
2. Add metabolite extracts to microplates.
3. Incubate for 24 hours.
4. Measure OD600 to determine antimicrobial activity.

10.6.4 Probiotics and Prebiotics to Modulate the Gut Microbiome

Materials Probiotics

1. Probiotic strains (e.g., *Lactobacillus acidophilus, Bifidobacterium bifidum*)
2. Probiotic supplements or fermented foods (e.g., yogurt, kefir)

Prebiotics

1. Prebiotic fibers (e.g., inulin, fructo-oligosaccharides)
2. Prebiotic-rich foods (e.g., asparagus, bananas)

Experimental Models

1. Human gut microbiome samples
2. Animal models (e.g., mice, rats)
3. In vitro gut microbiome models (e.g., gut-on-a-chip)

Analytical Tools

1. 16S rRNA gene sequencing
2. Quantitative polymerase chain reaction (qPCR)
3. Next-generation sequencing (NGS)
4. Microbiome analysis software (e.g., QIIME, Mothur)

Methods Probiotic Administration

1. Oral gavage or feeding
2. Probiotic supplementation in food or water

Prebiotic Administration

1. Dietary supplementation
2. Oral gavage or feeding

Sample Collection

1. Fecal sampling
2. Gut tissue sampling (in animal models)
3. Blood sampling (for metabolomics analysis)

Microbiome Analysis

1. DNA extraction
2. PCR amplification
3. Sequencing (16S rRNA or whole-genome)
4. Bioinformatics analysis

Statistical Analysis

1. Alpha diversity metrics (e.g., Shannon index)
2. Beta diversity metrics (e.g., Bray-Curtis dissimilarity)
3. Differential abundance analysis
4. Correlation network analysis

Functional Assays

1. Gut barrier function assays
2. Inflammation assays (e.g., cytokine measurement)
3. Metabolomics analysis
4. Immunologic assays (e.g., antibody measurement)

Study Design

1. Interventional studies (probiotic/prebiotic supplementation)
2. Observational studies (correlational analysis)
3. Randomized controlled trials (RCTs)
4. Animal studies

10.6.5 Vaccines and Immunotherapies to Enhance Animal Immunity

Materials

1. Antigens (e.g., proteins, peptides, or polysaccharides) from pathogens
2. Adjuvants (e.g., aluminum salts, squalene, or CpG motifs) to enhance immune response

3. Vaccine vectors (e.g., viral or bacterial vectors)
4. Immunomodulators (e.g., cytokines, toll-like receptor agonists)
5. Cell cultures or embryonated eggs for vaccine production
6. Animal models (e.g., mice, rabbits, or livestock) for testing

Methods Vaccine Development

1. Identification and isolation of antigens
2. Cloning and expression of antigens
3. Formulation and adjuvantation
4. Immunization and challenge studies
5. Serology and immunogenicity testing

Immunotherapy Development

1. Identification of immunomodulatory targets
2. Design and synthesis of immunomodulators
3. In vitro and in vivo testing
4. Combination therapy with vaccines or other treatments

Production and Quality Control

1. Fermentation or cell culture–based production
2. Purification and concentration
3. Sterility and safety testing
4. Potency and efficacy testing
5. Regulatory compliance (e.g., US Department of Agriculture, US Food and Drug Administration)

Animal Testing and Trials

1. Experimental design and animal selection
2. Immunization and challenge studies
3. Monitoring and analysis of immune responses
4. Efficacy and safety evaluations
5. Regulatory approval and licensing

Some key techniques used:

1. PCR
2. ELISA
3. Western blot
4. Flow cytometry
5. Gene cloning and sequencing

10.6.6 Biosecurity Measures and Good Agricultural Practices to Prevent Disease Transmission

Biosecurity Measures

1. Access control: Limit access to farms and facilities to authorized personnel.

2. Sanitation and hygiene: Regularly clean and disinfect equipment, vehicles, and facilities.
3. Personal protective equipment (PPE): Wear gloves, masks, and protective clothing.
4. Vaccination programs: Implement vaccination programs for animals.
5. Disease monitoring: Regularly monitor for disease signs and report suspicious cases.
6. Quarantine and isolation: Isolate new animals and quarantine infected ones.
7. Biosecurity signage: Display biosecurity signs and guidelines.

Good Agricultural Practices (GAPs) Farm-Level Practices

1. Crop rotation: Rotate crops to break disease cycles.
2. Soil preparation: Use proper soil preparation and fertilization.
3. Irrigation management: Manage irrigation to prevent waterborne diseases.
4. Pruning and trimming: Prune and trim plants to promote air circulation.
5. Pest management: Implement integrated pest management strategies.

Animal-Level Practices

1. Animal sourcing: Source animals from reputable suppliers.
2. Animal health checks: Conduct regular health checks.
3. Vaccination and parasite control: Implement vaccination and parasite control programs.
4. Nutrition and feeding: Provide balanced nutrition and proper feeding practices.
5. Animal welfare: Ensure proper animal welfare.

Post-Harvest Practices

1. Handling and storage: Handle and store produce properly.
2. Cleaning and disinfection: Clean and disinfect equipment and facilities.
3. Temperature control: Control temperature during transportation and storage.
4. Packaging and labeling: Use proper packaging and labeling.

Materials Needed

1. Disinfectants and sanitizers
2. PPE (gloves, masks, protective clothing)
3. Vaccines and veterinary supplies
4. Cleaning equipment (pressure washers, brooms)
5. Signage and educational materials
6. Irrigation and water management systems
7. Soil testing kits
8. Pruning and trimming tools

Methods

1. Risk assessment: Conduct regular risk assessments.
2. Training and education: Provide training for farm workers.

3. Record keeping: Maintain accurate records.
4. Regular inspections: Conduct regular inspections.
5. Disease surveillance: Monitor for disease signs.
6. Collaboration with veterinarians: Work with veterinarians for disease diagnosis and control.

10.7 EXPECTED OUTCOME

These expected outcomes highlight the potential of natural products in addressing the growing concern of antimicrobial resistance, offering innovative solutions for various industries, including pharmaceuticals, cosmetics, and food production.

10.8 CONCLUSION

The emergence of antimicrobial resistance demands innovative solutions to combat microbial infections in animal husbandry. By embracing natural products and alternative approaches, we can reduce our reliance on traditional antibiotics and promote a safer, more sustainable food-production system. Further research and collaboration are essential to realizing the full potential of these novel strategies.

REFERENCES

Ambikapathy, V., Panneerselvam, A., Babu, S., Anbukumaran, A., Shijila Rani, A.S., Jagajothi, C., Kanmani, A., Rathna, D. and Prakash, P. (2024). Dosa postbiotic food preparations. In: Dharumadurai, D. (ed) *Postbiotics: Methods and Protocols in Food Science*. Humana, New York, 403–407.

Bacanli, M. and Basaran, N. (2019). Importance of antibiotic residues in animal food. *Food Chem. Toxicol.* 125: 462–466.

Chaachouay, N. and Zidane, L. (2024). Plant-derived natural products: A source for drug discovery and development. *Drugs Drug Candidates*. 3(1): 184–207.

Chris, E.P.J., Sarai, K., Pranav, T., Oliver, C., Amy, J.P., Arshee, M. and Clare, I.R.C. (2023). Biosecurity and water, sanitation and hygiene (WASH) interventions in animal agricultural settings for reducing infection burden, antibiotic use and antibiotic resistance: A One Health systematic review. *Lancet Planet. Health*. 7(5): e418–e434.

Cotter, P.D., Hill, C. and Ross, R.P. (2005). Bacteriocins: Developing innate immunity for food. *Nat. Rev. Microbial.* 3: 777–788.

Cowan, M.M. (1999). Plant products as antimicrobial agents. *Clin. Microbiol. Rev.* 12(4): 564–582.

Cully, M. (2014). The politics of antibiotics. *Nature*. 509(7498): S16.

Ghosh, C., Sarkar, P., Issa, R. and Haldar, J. (2019). Alternatives to conventional antibiotics in the era of antimicrobial resistance. *Trends Microbiol.* 27: 323–338.

He, Y., Yuan, Q., Mathieu, J., Stadler, L., Senehi, N., Sun, R. and Alvarez, P.J.J. (2020). Antibiotic resistance genes from livestock waste: Occurrence, dissemination and treatment. *npj Clean Water.* 3(4): 1–11.

Hosain, M.Z., Kabirs, S.M.L. and Kamal, M.M. (2021). Antimicrobial uses for livestock production in developing countries. *Vet. World.* 14(1): 210–221.

Islam, A., Saifuddin, A., Al Faruq, A., Islam, S., Shano, S., Alam, M. and Hassan, M.M. (2016). Antimicrobial residues in tissues and eggs of laying hens at Chittagong, Bangladesh. *Int. J. One Health.* 2(11): 75–80.

Ji, S., An, F., Zhang, T., Lou, M., Guo, J., Liu, K., Zhu, Y., Wu, J. and Wu, R. (2024). Antimicrobial peptides: An alternative to traditional antibiotics. *Eur. J. Med. Chem.* 265: 116072.

Kirbis, A. (2007). Microbiological screening method for detection of aminoglycosides, β-lactams, macrolides, tetracyclines and quinolones in meat samples. *Slov. Vet. Res.* 44(1/2): 11–18.

La Guidara, C., Adamo, R., Sala, C. and Micoli, F. (2024). Vaccines and monoclonal antibodies as alternative strategies to antibiotics to fight antimicrobial resistance. *Int. J. Mol. Sci.* 25(10): 5487.

Lowenstein, C., Waters, W.F., Roess, A., Leibler, J.H. and Graham, J.P. (2016). Animal husbandry practices and perceptions of zoonotic infectious disease risks among livestock keepers in a rural parish of quito, Ecuador. *Am. J. Trop. Med. Hyg.* 95(6): 1450–1458.

Manyi-Loh, C., Mamphweli, S., Meyer, E. and Okoh, A. (2018). Antibiotic use in agriculture and its consequential resistance in environmental sources: Potential public health implications. *Molecules.* 23(4): 795.

Matzaras, R., Anna, N., Efthimia, P. and Eirini, C. (2022). Gut microbiota modulation and prevention of Dysbiosis as an alternative approach to antimicrobial resistance: A narrative review. *Yale J. Biol. Med.* 95(4): 479–494.

McVey, S.D., Kennedy, M. and Czuprynski, C. (2022). Microbial infections of animals. In: McVey, S.D., Kennedy, M., Chengappa, M.M. and Wilkes, R. (eds.). *Veterinary Microbiology.* Vikas Publishing House Pvt Ltd.

OECD and Food and Agriculture Organization of the United Nations. *OECD-FAO Agricultural Outlook 2021–2030. Meat.* OECD Publishing, 163.

Prescott, J.F. (2018). Hisotory and current use of antimicrobial drugs in veterinary medicine. In: Schwarz, S., Cavaco, L.M. and Shen, J. (eds.) *Antimicrobial Resistance in Bacteria from Livestock and Companion Animals.* Wiley, 1–17.

Quinn, G.A. and Dyson, P.J. (2024). Going to extremes: Progress in exploring new environments for novel antibiotics. *npj Antimicrob. Resist.* 2(8).

Rahman, M.M., Tumpa, M.A.A., Zehravi, M., Sarker, M.T., Yamin, M., Islam, M.R., Or-Rashid, M., Ahmed, M., Ramproshad, S., Mondal, B., Dey, A., Damiri, F., Berrada, M., Rahman, M.H. and Cavalu, S. (2022). An overview of antimicrobial stewardship optimization: The use of antibiotics in humans and animals to prevent resistance. *Antibiotics (Basel).* 11(5): 667.

Rushton, J. (2015). Antimicrobial use in animals: How to assess the trade-offs. *Zoonoses Public Health.* 62(Suppl 1): 10–21.

Tiseo, K., Huber, L., Gilbert, M., Robinson, T.P. and Van Boeckel, T.P. (2020). Global trends in Antimicrobial use in food animals from 2017 to 2030. *Antiobiotics.* 9: 918.

Van Boeckel, T.P., Glennon, E.E., Chen, D., Gilbert, M., Robinson, T.P., Grenfell, B.T., Levin, S.A., Bonhoeffer, S. and Laxminarayan, R. (2015). Reducing antimicrobial use in food animals. *Science.* 357(6358): 1350–1352.

Velazqueez-Meza, M.E., Lopez, M.G., Quiroz, B.C. and Aranda, C.M.A. (2022). Antimicrobial resistance: One health approach. *Vet. World.* 15(3), 743–749.

WHO. (2024). *Antimicrobial Resistance. World AMR Awareness Week 2024.*

Index

I

L

M

N

O

P

Q

R

S

T

U

V

W

Z